Geology and America's National Park Areas

Brooks B. Ellwood
University of Texas at Arlington

PRENTICE HALL, Upper Saddle River, New Jersey 07458

Library of Congress Cataloging-in-Publication Data

Ellwood, B. B. (Brooks B.)
   Geology and America's national park areas / Brooks B. Ellwood.
      p.   cm.
   Includes index.
   ISBN 0-02-332753-7
   1. Geology--United States.   2. National parks and reserves--United
States.   I. Title
QE77.E45   1996
   557.3--dc20   95-40818
                                                    CIP

Editorial/production supervision and interior design: ETP/Harrison
Copy editor: Jane Loftus
Manufacturing manager: Trudy Pisciotti
Executive editor: Robert A. McConnin
Marketing manager: Leslie Cavaliere
Marketing assistant: Amy Reed
Cover director: Jayne Conte
Cover designer: Bruce Kenselaar
Cover image: Brooks B. Ellwood

© 1996 by Prentice-Hall, Inc.
Simon & Schuster/A Viacom Company
Upper Saddle River, NJ 07458

Printed in the United States of America

10   9   8   7   6   5   4   3   2   1

ISBN 0-02-332753-7

Prentice-Hall International (UK) Limited, *London*
Prentice-Hall of Australia Pty. Limited, *Sydney*
Prentice-Hall Canada Inc., *Toronto*
Prentice-Hall Hispanoamericana, S.A., *Mexico*
Prentice-Hall of India Private Limited, *New Delhi*
Prentice-Hall of Japan, Inc., *Tokyo*
Simon & Schuster Asia Pte. Ltd., *Singapore*
Editora Prentice-Hall do Brasil, Ltda., *Rio de Janeiro*

Contents

CHAPTER 15   STRATIGRAPHIC CONCEPTS AND FOSSILS

CHAPTER 16   CULTURAL AREAS

Preface

While teaching at the University of Georgia, I was asked to prepare a course titled "The Geology of the National Parks." At that time there was only one textbook on the subject, and it was not sufficiently comprehensive to allow the coverage of those national park areas that I desired to cover. Prior to this time I had been active in research in some national park areas, so I decided to visit more park areas with the express purpose of supplementing, with photographs and observations, the materials I already had available to me.

Later I decided to write a book that would bring the field of geology and the park system together in a way that would provide broad, introductory coverage of both geology and many of the park system areas, including all of the areas known exclusively as national parks. Therefore this book is intended as a general, introductory geology text with examples taken almost exclusively from areas supervised by the National Park Service. It does not assume that students will have a background in geology, but does assume that students do have some introductory science knowledge. This book is not intended to be a comprehensive geological text, but rather it is designed to present basic geologic concepts, primarily physical geology and historical geology, and to give some examples that appear in park areas. Also given is the general background and geologic history of all the areas known specifically as national parks (there were 54 at the time this book was written), a large number of national monuments (there were 73 at the time of writing), and some other geologically significant park areas.

The book is designed to be used at colleges and universities in teaching a course on national park areas to students with no prerequisites in geology, a variation on the popular "rocks for jocks" course. It also can be used as an introductory course in geology at colleges and universities with no geology program, such as at two- and four-year liberal arts colleges. Course titles can be any variation on the title of the book, including Geology of National Parks and Monuments. Furthermore, the book was designed to be used as a general overview reference text on the park system, providing a general geologic background for many park areas and for the United States as a whole.

General references at the back of each chapter are designed to give additional reading of a broad nature and provide useful material for those actually visiting regions in the United States that contain park areas.

ACKNOWLEDGMENTS

I would like to thank the many people in the National Park Service who have helped me in my research efforts through the years and who have graciously provided information that has been used in this book. Also greatly appreciated are the many constructive critical comments by reviewers of early drafts of the manuscript, including: Jack C. Allen, Bucknell University; Jay D. Bass, University of Illinois; Philip R. Bjork, South Dakota School of Mines & Technology; David R. Hickey, Lansing Community College; Richard A. Hoppin, University of Iowa; Walter L. Manger, University of Arkansas; and Bruce H. Wilkinsen, University of Michigan. The Media Center at the University of Texas at Arlington has been very helpful in providing computer-assisted aids in preparation of this book, especially Anna L. Busboom and Robert L. Crosby. Special thanks are extended to Chuck Pratt, of UTA Television and his folks, park rangers at Grand Tetons National Park, for their help and hospitality. Finally, without the help and support of my loving family, who provided scale in many park pictures, wife Sue (field assistant and computer graphics expert), and children Amber, Robin, and Richard, this work would not have been possible.

Brooks B. Ellwood
University of Texas at Arlington

1

Introduction

THE NATIONAL PARK SYSTEM

During the 1800s, trappers and mountain men wandering the West came back with stories of a fantastic place of unusual springs and spouting geysers. Such far-fetched stories were hard to believe, but eventually an expedition was sent to check out the area. Its leader, General Henry Washburn, a surveyor from the Montana Territory, was determined to verify these incredible stories. Much to his delight he discovered that they were true! In response, Judge Cornelius Hedges, also from Montana, suggested that rather than using the Yellowstone area for mining or other activities, it should be designated as a national park so that it could be preserved. There was strong support for this idea in Montana, and efforts by members of the expedition to the Yellowstone area finally convinced Congress to set aside this land.

In 1872, before the battle of the Little Big Horn, where Custer and many of the seventh U.S. Cavalry were killed by the Sioux, the Congress of the United States designated Yellowstone as the first national park in the world (Figure 1.1). Their intent was to preserve a geologically unusual area for future generations of Americans. But this act brought Yellowstone into the public eye, and visitation increased with corresponding park vandalism. To safeguard Yellowstone and future parks, it was decided that the Army, already responsible for federal protection of frontier areas, would also police the parks. Because this was only a minor responsibility for the Army, no effort went in to improving the Park System, and protection was only minimal. There were real concerns by conservationist-industrialists in the United States that the Army was not doing its job in protecting the parks and they demanded that a new agency be set up for park protection purposes. Therefore, in 1916, during the administration of President Woodrow Wilson, Congress established the National Park Service so that the growing Park System could be adequately protected and facilities built and maintained. The new service adopted a modified version of the U.S. Army uniform of the time, and has kept similar attire ever since. Thus the "Smoky the Bear Hat" (Figure 1.2) that is worn by many park rangers today is a leftover of the old army days.

FIGURE 1.1 Old Faithful in Yellowstone National Park. (Photo by author.)

FIGURE 1.2 Grand Tetons National Park, Wyoming, with a park ranger in the foreground. Note the uniform and hat. (Photo by author.)

The National Park Service was officially designated as part of the Department of the Interior, with the Secretary of the Interior as the responsible Cabinet officer (see Table 1.1). The first Director of the National Park Service was Stephen Mather, who set the course for the service. Mr. Mather, from Chicago, was one of those unique breed of industrialists who was an outspoken proponent of conservation. It is interesting to note that he had made a fortune mining borax in Death Valley, but in the process realized how fragile were our natural resources and fought to protect them for most of his life.

Today each park has a staff consisting of a superintendent, the park director, rangers, who generally run the park and include the police force necessary to deal with the large number of people who annually visit the parks, and naturalists, who are specialists on the unique characteristics of each park. In 1916, when the National Park Service was born, there were 16 national parks and 21 national monuments. Currently there are 54 parks with the designation of National Park within the Park System. These are listed in Table 1.2 in alphabetical order.

There are more than 360 areas that are supervised by the National Park Service, including national memorials, historic and military parks, and other natural and special areas. Even the White House, where the president of the United States resides, is supervised by the National Park Service.

It is the job of the park service personnel to handle the problems, questions, and welfare of the large number of people from all over the world who visit park areas each year. Traffic problems are severe, and the rangers are responsible for regulating its flow. Visitor centers in all park areas are set up to handle literature sales, education about the park, general services, and to answer any questions that people might have. The following is a list of questions reportedly asked of park rangers by visitors to various parks recently published by Debra Shore in the May 1995 issue of *Outside* magazine.

Grand Canyon National Park in Arizona
>Was this man-made?
>Do you light it up at night?
>I bought tickets for the elevator to the bottom—where is the elevator?
>Is the mule train air-conditioned?
>So where are the faces of the presidents?

Everglades National Park in Florida
>Are the alligators real?
>Are the baby alligators for sale?
>Where are all the rides?

Denali National Park in Alaska
>What time do you feed the bears?
>How often do you mow the tundra?
>How much does Mt. McKinley weigh?

Carlsbad Caverns National Park in New Mexico
>How much of the cave is underground?
>So what's in the unexplored part of the cave?
>Does it ever rain in here?
>So what is this—just a hole in the ground?

Mesa Verde National Park in Colorado
>Did people build this or did the Indians?
>Why did they build the ruins so close to the road?
>Do you know of any undiscovered ruins?
>What did they worship in the kivas—their own made-up religion?

Yosemite National Park in California

What time of year do you turn on Yosemite falls?

What happened to the other half of Half Dome?

Where are the cages for the animals?

Yellowstone National Park in Wyoming, Montana, and Idaho

Does Old Faithful erupt at night?

How do you turn it on?

We had no trouble finding the park entrance—but where are the exits?

Park rangers certainly have their work cut out for them, but humor is a benefit of their job.

Most of the national parks are located within the contiguous continental United States (lower 48), but two are located in U.S. Territories, two are in Hawaii, and eight are in Alaska. For various reasons, Congress can change the status of a park. For example, Platt National Park in Oklahoma was removed from park status and renamed. Apparently Congress decided that Platt National Park, named for an ex-senator from Oklahoma, represented an unjustified political move that created a park that did not have the beauty and majesty characterizing most parks. Therefore, when Senator Platt's influence in Congress waned, Platt National Park was renamed and its park land was included in what is now Chickasaw National Recreation Area.

It takes an act of Congress to establish a national park, but an important piece of legislation made it possible for presidents to preserve regions of natural beauty or historical interest. In 1906 Congress passed the Antiquities Act to stop destruction of southwestern prehistoric cultural sites. Using the new antiquities law, President Theodore Roosevelt decided that there were many other areas of natural beauty that

TABLE 1.1 DEPARTMENT OF INTERIOR CHAIN OF COMMAND

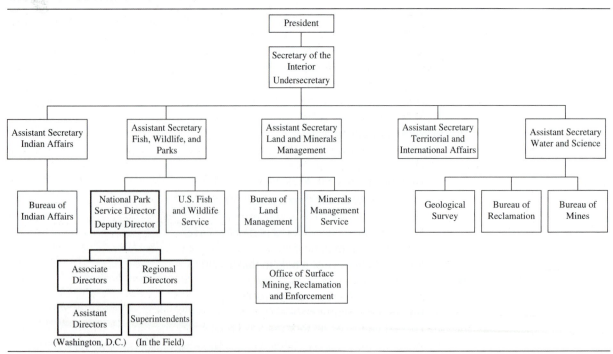

TABLE 1.2 NATIONAL PARKS

1. Acadia National Park	21. Great Basin National Park	39. Olympic National Park
2. Arches National Park	22. Great Smoky Mountains National Park	40. Petrified Forest National Park
3. Badlands National Park		41. Redwood National Park
4. Big Bend National Park	23. Guadalupe Mountains National Park	42. Rocky Mountain National Park
5. Biscayne National Park		43. Saguaro National Park
6. Bryce Canyon National Park	24. Haleakala National Park	44. Sequoia National Park
7. Canyonlands National Park	25. Hawaii Volcanoes National Park	45. Shenandoah National Park
8. Capital Reef National Park	26. Hot Springs National Park	46. Theodore Roosevelt National Park
9. Carlsbad Caverns National Park	27. Isle Royal National Park	
10. Channel Islands National Park	28. Joshua Tree National Park	47. Voyageurs National Park
11. Crater Lake National Park	29. Katma National Park and Preserve	48. Wind Cave National Park
12. Death Valley National Park	30. Kenai Fjords National Park	49. Wrangell–St. Elias National Park and Preserve
13. Dry Tortugas National Park	31. Kings Canyon National Park	
14. Denali National Park and Preserve	32. Kobuk Valley National Park	50. Yellowstone National Park
15. Everglades National Park	33. Lake Clark National Park and Preserve	51. Yosemite National Park
16. Gates of the Arctic National Park and Preserve		52. Zion National Park
	34. Lassen Volcanic National Park	**Parks Outside the United States**
17. Glacer National Park	35. Mammoth Cave National Park	
18. Glacier Bay National Park	36. Mesa Verde National Park	53. Virgin Islands National Park
19. Grand Canyon National Park	37. Mt. Rainier National Park	54. National Park of American Samoa
20. Grand Teton National Park	38. North Cascades National Park	

should be protected. Therefore in 1906 he established the precedent for setting aside and preserving such areas through presidential proclamation. He did this by proclaiming Devils Tower in Wyoming (Figure 1.3), as a National Monument (NM). Monuments today are usually designated by acts of Congress. Presently there are 73 national monuments. These are listed in Table 1.3 by the date each monument was designated.

In 1980, for the first time Congress created a new national park category, "national park and preserve." By establishing these national park and preserves it was the intention of Congress to allow special activities, such as hunting, not normally acceptable in other national parks. All the new national park and preserve areas, for example Wrangell–St. Elias National Park and Preserve, are located in Alaska. Within these national parks, hunting is legal, although many environmentalists argue it is not acceptable. In the lower 48 states, a few stand-alone preserves have been established, including the recently created Mojave National Preserve in the southern part of California, Big Thicket National Preserve in southern Texas, and Big Cypress National Preserve in southern Florida.

By 1988 the number of people employed by the National Park Service had risen to more than 10,000, with an additional 4,000 that were hired during the summer. Add to this approximately 50,000 volunteers, and the number of people doing all the park service work was still less than 70,000 to handle almost 300 million visits a year. The total budget in 1988 was almost $800 million to cover all the park areas, including over 450 campgrounds. Since that time the budget has risen only slightly, and when corrected for inflation, the real spending power of the National Park Service has dropped. In the meantime, visitation has steadily increased, placing a great strain on the Park System. Because of poor funding levels, the National Park Service is not able to pay park rangers and park naturalists a competitive salary. In 1990 the average

TABLE 1.3 NATIONAL MONUMENTS

1. 1906	Devils Tower NM, Wyoming	
2.	Montezuma Castle NM, Arizona	
3.	El Morro NM, New Mexico	
4. 1907	Tonto NM, Arizona	
5. 1908	Muir Woods NM, California	
6.	Pinnacles NM, California	
7.	Natural Bridges NM, Utah	
8. 1909	Najavo NM, Arizona	
9.	Salinas Pueblo Missions NM, New Mexico	
10. 1910	Rainbow Bridge NM, Utah	
11. 1911	Devils Postpile NM, California	
12.	Colorado NM, Colorado	
13. 1913	Cabrillo NM, California	
14. 1915	Walnut Canyon NM, Arizona	
15.	Dinosaur NM, Colorado	
16. 1916	Capulin Volcano NM, New Mexico	
17. 1918	Casa Grande Ruins NM, Arizona	
18. 1919	Yucca House NM, Colorado	
19.	Scotts Bluff NM, Nebraska	
20. 1922	Timpanogos Cave NM, Utah	
21. 1923	Pipe Spring NM, Arizona	
22.	Hovenweep NM, Colorado and Utah	
23.	Aztec Ruins NM, New Mexico	
24. 1924	Wupatki NM, Arizona	
25.	Castillo de San Marcos NM, Florida	
26.	Fort Matanzas NM, Florida	
27.	Fort Pulaski NM, Georgia	
28.	Craters of the Moon NM, Idaho	
29.	Statue of Liberty NM, New York and New Jersey	
30. 1925	Lava Beds NM, California	
31. 1930	Sunset Crater Volcano NM, Arizona	
32.	George Washington Birthplace NM, Virginia	
33. 1931	Canyon de Chelly NM, Arizona	
34. 1932	Great Sand Dunes NM, Colorado	
35.	Bandelier NM, New Mexico	
36. 1933	Chiricahua NM, Arizona	
37.	Black Canyon of the Gunnison NM, Colorado	
38.	Fort McHenry NM and Hist. Shrine, Maryland	
39.	Gila Cliff Dwellings NM, New Mexico	
40.	White Sands NM, New Mexico	
41.	Oregon Caves NM, Oregon	
42.	Jewel Cave NM, South Dakota	
43.	Cedar Breaks NM, Utah	
44. 1934	Ocmulgee NM, Georgia	
45. 1935	Fort Stanwix NM, New York	
46. 1936	Fort Frederica NM, Georgia	
47.	Homestead NM of America, Nebraska	
48. 1937	Organ Pipe Cactus NM, Arizona	
49.	Pipestone NM, Minnesota	
50. 1939	Tuzigoot NM, Arizona	
51. 1940	Little Bighorn Battlefield NM, Montana	
52. 1943	George Washington Carver NM, Missouri	
53. 1946	Castle Clinton NM, New York	
54. 1948	Fort Sumter NM, South Carolina	
55. 1949	Effigy Mounds NM, Iowa	
56. 1951	Grand Portage NM, Minnesota	
57. 1954	Fort Union NM, New Mexico	
58. 1956	Booker T. Washington NM, Virginia	
59. 1961	Russell Cave NM, Alabama	
60.	Buck Island Reef NM, Virgin Islands	
61. 1965	Agate Fossil Beds NM, Nebraska	
62.	Alibates Flint Quarries NM, Texas	
63. 1969	Florissant Fossil Beds NM, Colorado	
64. 1972	Hohokam Pima NM, Arizona	
65.	Fossil Butte NM, Wyoming	
66. 1974	John Day Fossil Beds NM, Oregon	
67. 1976	Congaree Swamp NM, South Carolina	
68. 1978	Aniakchak NM and Pres., Alaska	
69.	Cape Krusenstern NM, Alaska	
70. 1987	El Malpais NM, New Mexico	
71. 1988	Hagerman Fossil Beds NM, Idaho	
72.	Poverty Point NM, Louisiana	
73. 1990	Petroglyph NM, New Mexico	

NM = National Monument

salary for U.S. park rangers was approximately $20,000 per year, while firefighters, teachers, and other important government employees were earning an average of over $30,000 per year (Figure 1.4). The people who continue as park rangers, in spite of the low salaries, usually do so through personal dedication to individual parks. Other excellent rangers go where salaries are higher, thus creating a continuing drain on a critical National Park Service resource, its people. To supplement this loss in personnel, the Park Service has been increasingly turning to volunteers to help run the parks. The number of volunteers exceeded 70,000 people per year in the 1990s.

Besides the loss of good people, poor funding, and deterioration due to excessive visitations, many other problems confront the parks. Included in these are air and scenic pollution. In many areas, commercial zones, factories, homes, even power plants are built right up to the park boundaries (Figure 1.5). Air pollution by sulfates and other atmospheric particles has significantly reduced visibility in the parks (Figure 1.6). Visi-

FIGURE 1.3 Devils Tower National Monument in Wyoming. (Photo courtesy of the National Park Service.)

tors may even be bothered by eye and throat irritation. Also of concern is the effect visitors have on the animals and plants in the Park System. There are many serious problems that arise, including gorings by elk and buffalo (Figure 1.7).

The National Park Service now supervises over 360 units within the Park System, from the smallest, Thaddeus Kosciuszko National Memorial in Pennsylvania,

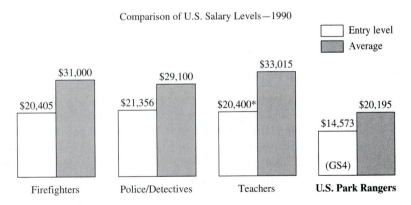

FIGURE 1.4 Salaries of rangers versus others.

Comparison of U.S. Salary Levels—1990

Source: ICMA, Municipal Yearbook; National Education Association; U.S. Dept. of Labor, Bureau of Labor Statistics; National Park Service.

*Recent college graduate with B.A. degree

FIGURE 1.5 Indiana Dunes National Lakeshore, Indiana, with a coal power plant cooling tower in the background. (Photo by author.)

with 0.02 acres, to the largest, Wrangell–St. Elias National Park and Preserve in Alaska, with about 13,200,000 acres. These units include:

a. International Historic Site   [Saint Croix Island, Maine]
b. National Battlefields   [e.g., Petersburg, Virginia]
c. National Battlefield Parks   [e.g., Richmond, Virginia; Manassas (Bull Run), Virginia]
d. National Battlefield Site   [Brices Cross Roads, Mississippi]
e. National Capital Park   [Parks in Washington, D.C.]
f. National Historic Sites   [e.g., Fort Davis, Texas; Ford's Theater, Wash., D.C.]
g. National Historical Parks   [e.g., Appomattox Court House, Virginia; Valley Forge, Pennsylvania]
h. National Lakeshores   [e.g., Indiana Dunes, Indiana (Figure 1.5)]
i. National Mall   [Washington Monument, Capitol Building]
j. National Memorials   [e.g., Lincoln Memorial, Washington, D.C.; Mount Rushmore, South Dakota (Figure 1.8); Wright Brothers, North Carolina (Figure 1.9); U.S.S. Arizona, Hawaii]
k. National Military Parks   [e.g., Gettysburg, Pennsylvania]
l. National Parkways   [e.g., Blue Ridge Parkway, Appalachian Mountains]
m. National Preserves   [e.g., Big Thicket, Texas]

Air Quality in National Parks

Ozone stings eyes, nasal passages and throat of healthy people and can worsen health problems of the elderly, asthma sufferers, and children. In 1988, national parks in the East exceeded the Federal Clean Air Act standard of 125 parts per million more often than parks in the West.

Obscured Views

Because of air pollution from many sources, including industrial sulfates, the average visual range in the United States is about 9 to 15 miles. The maximum possible, according to the Park Service, is about 250 miles. Measurements were taken in the summer of 1988.

HIGHEST OZONE LEVEL RECORDED
(parts per million)

Days in excess of U.S. standard.

Source: *National Park Service, Air Quality Division*

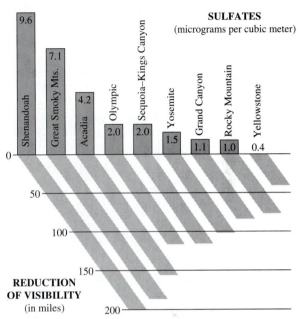

SULFATES
(micrograms per cubic meter)

REDUCTION OF VISIBILITY
(in miles)

FIGURE 1.6 Air quality in the national park areas.

n. National Recreation Areas [e.g., Lake Meredith, Texas]
o. National Reserve [City of Rocks, Idaho]
p. National Rivers [e.g., New River Gorge, West Virginia]
q. National Scenic Trails [e.g., Appalachian Trail]
r. National Seashores [e.g., Padre Island, Texas]
s. National Wild and Scenic River and Rivers [e.g., Rio Grande, Texas]
t. Parks (other) [e.g., Catoctin Mountain Park, Maryland]
u. White House [Washington, D.C.]

The park service supervises a diverse range of sites, including the Statue of Liberty, the Capitol Mall, Ford's Theater, and many other famous areas. While the national parks are generally impressive because of their scenic beauty, many other areas are dedicated to special national preservation efforts, often because of their historical interest. Military battlefields from the Revolutionary and Civil wars fall into this category. Many of the historic sites are located in the eastern part of the United States where dramatic events were occurring during our country's early years. Scenic beauty is often considered to be more dramatic in the west because of the broad vistas, striking mountain ranges, and unusual geologic forms found there. In the east these forms and mountains have been eroded away, rounded, and covered by vegetation. As a result, most of the national parks and monuments are located in the western half of the United States, while most of the historic sites are located east of the Mississippi River. Individual states were another factor in the distribution of federal lands. Many historic sites (such

FIGURE 1.7 Yellowstone National Park warning flyer showing buffalo goring a visitor. (Courtesy of the National Park Service.)

FIGURE 1.8 Mount Rushmore National Memorial, South Dakota. (Photo courtesy of the National Park Service.)

FIGURE 1.9 Wright Brothers National Memorial, North Carolina. (Photo by author.)

as the Alamo in Texas) and scenic areas were set aside by state legislatures as state parks. States that were active in preservation efforts, such as Texas, have fewer federal lands under Park Service control. In other states, such as Alaska, the federal government now controls a large portion of the state. Also, extensive private and state owned land holdings across the country, but especially in the east, block some areas from park designation.

NATIONAL PARK REGIONS OVERVIEW

Areas within the Park System have been divided into nine regions for the purpose of park governance (Figure 1.10). These regions were selected by the National Park Service, in part, due to their unique geologic character, but mainly due to their geographic location. As a result, national park regions often include distinctively different geologic terranes. This chapter discusses the geographic distribution of each park region, while a discussion of the coherent geological pattern is reserved for Chapter 8, following the introduction of many relevant geologic concepts that will make the chapter more comprehensible.

The Alaska Region (Figure 1.11) includes only the state of Alaska. The region is supervised from a headquarters in Anchorage, Alaska, and contains the largest national

FIGURE 1.10 National park regions set up for supervision of all park areas. Also given is the city in which the region headquarters is located.

park, Wrangell–St. Elias National Park and Preserve. The Aleutian Range makes up the western extension of the Alaska peninsula from mountains lying on the North American continent to an arc of islands extending out into the northern-most Pacific Ocean and separating the Pacific from the Bering Sea. This island arc, the Aleutian *archipelago*, was produced by the eruption of volcanoes forming the Aleutian Island chain. Katmai National Park (discussed in Chapter 4) is an extension of the Aleutians onto our continent. This area is part of the Pacific ring of fire, so named because of the volcanoes that form a ring rimming the Pacific Ocean basin. Volcanic eruptions in this ring are generally explosive, forming large, steep-sided volcanoes; many regions bordering the Pacific are still actively experiencing explosive volcanism. Mt. St. Helens (see Chapter 5) is an example of such a volcano, and there are many others in the National Park System.

The Alaska Range, as opposed to the Aleutians, is part of what is called the Western Cordillera that extends all the way from Alaska through Canada and into Mexico and beyond. The Alaska Range runs to the northeast from the Aleutians and then eastward across central Alaska, where it bends southeastward into Canada's Yukon Territory. *Glaciers* are large and numerous in the range. The most striking park in this area is Denali National Park and Preserve (Chapter 7), which contains Mt. McKinley, the highest mountain (20,320 ft, 6194 m) on the North American continent.

The Pacific Northwest Region (Figure 1.12) has its headquarters in Seattle, Washington. It includes a series of coastal and inland mountain ranges that are tectonically unstable, a region of many earthquakes. Excellent examples of park areas within this region include Olympic National Park (Figure 1.13) in Washington (Chapter 14), and North Cascades National Park, located in the *Cascade Range* to the east, the northern part of a 1,000-mile (1,600 km) long mountain range that is composed mainly of volcanic rocks. The Cascade Mountains are a dissected volcanic plateau with lofty volcanic cones extending to very high elevations; they include Washing-

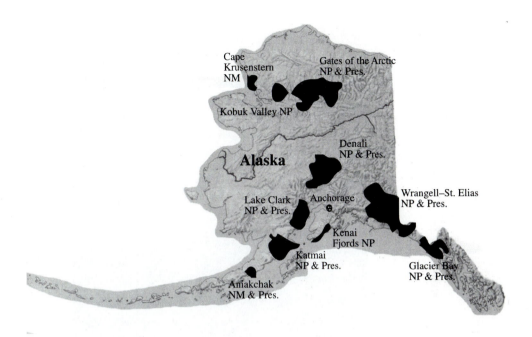

FIGURE 1.11 Alaskan National Park Locations within the Alaskan Region. (NM = National Monument; NP = National Park; Pres. = Preserve.)

ton's Mt. St. Helens, which erupted explosively in 1980. Other well-known large volcanoes, such as Mt. Hood and Mt. Baker, exist but are not controlled by the National Park Service. Also in the region are two spectacular national parks, Mt. Rainier National Park in Washington (Chapter 5), and Crater Lake National Park in Oregon (also discussed further in Chapter 5).

PACIFIC NORTHWEST REGION

FIGURE 1.12 Park locations within the Pacific Northwest Region (NM = National Monument; NP = National Park.)

FIGURE 1.13 Pacific coast in Olympic National Park, Washington. (Photo by author.)

In the eastern portion of the Pacific Northwest Region (Figure 1.12) lies the Columbia Plateau, an area of extensive basaltic lava flows erupted during the Cenozoic, which covers much of eastern Washington, eastern Oregon, and western Idaho (discussed in greater detail in Chapter 8). Craters of the Moon National Monument in Idaho (see Chapter 5) is an excellent example of the volcanic rocks that are extensive in the region.

The Western Region of the National Park System includes all of the U.S. territories lying in the Pacific Ocean basin, the states of Hawaii, California, Nevada and part of Arizona (Figure 1.14). Headquarters for the region is located in San Francisco, California. This includes the National Park of American Samoa (discussed in Chapter 14) in the south Pacific and territories such as Guam in the North Pacific. At the southern end of the Cascade Range, lying mainly in the Pacific Northwest Region, is the Lassen volcanic zone, which includes Lassen Volcanic National Park in California (Chapter 5), where eruptions occurred between 1914 and 1921. Lassen Volcanic National Park divides the Cascades to the north from the Sierra Nevadas lying further to the south. The Sierras are the exposed roots of ancient volcanic mountains, with Yosemite (Figure 1.15) and Sequoia National Parks in California (Chapter 5) being the best examples. The topography throughout this area is controlled primarily by glacial erosion, discussed in detail in Chapter 13. The Sierra Nevadas represent a large, 400-mile (650 km)-long, north-northwest trending fault block mountain range, bounded on the eastern edge by the Sierra Nevada fault system and tilted up to the west. At the southern end the block is cut off by the Garlock fault zone.

Western Region

FIGURE 1.14 Park locations within the Western Region. (NM = National Monument; NP = National Park; Pres. = Preserve; NRA = National Recreation Area.)

Included in the Western Region is part of the *Basin and Range geologic province* (discussed in greater detail in Chapter 8), a region of many mountain ranges separated by basins between, that extends through Nevada, southern California, and Arizona. It is a very arid region that includes approximately 150 small mountain ranges where the higher mountains exhibit glacial erosion features and the basins are being filled by accumulating sediments from erosion of the ranges. Large-scale tectonic activity is actively pulling the basins apart. At one time large lakes existed in the basins as a result of glacial runoff following the last glaciation. Since that time these lakes have disappeared as the result of high evaporation rates in these desert regions. A prime example of this part of the region is Death Valley National Park (Figure 1.16), lying mainly in California with a small portion in Nevada (Death Valley is discussed in greater detail in Chapter 12). To the south and east of Death Valley in California and Arizona is a large desert landscape that includes the newly designated Joshua Tree National Park in California and Saguaro National Park in Arizona (see Chapter 12 for more details).

The Colorado Plateau is an important geologic province that lies in southeastern Utah, southwestern Colorado, northeastern Arizona, and northwestern New Mexico (see Chapter 8 for a discussion). It is an area of relatively high elevation that is dry and essentially treeless, except within canyons and at higher elevations, and is drained by the Colorado River and its tributaries. It contains eight national parks, but the National Park Service has included the Colorado Plateau within three different park regions. The

FIGURE 1.15 Yosemite Valley in the center of Yosemite National Park, California. (Photo by author.)

FIGURE 1.16 Playa lake in the southern end of Death Valley National Park. (Photo by author.)

best known of the parks, Grand Canyon National Park (see Chapter 15), lies within the Western Region in Arizona.

The Southwest Region (Figure 1.17) of the National Park Service includes part of Arizona and the states of New Mexico, Texas, Oklahoma, Arkansas, and Louisiana, and is headquartered in Santa Fe, New Mexico. The region includes the extension of the Basin and Range geologic province into the southwest, where active groundwater has produced several cave systems. This includes caves in Great Basin National Park in eastern Nevada (see Chapter 7), in Carlsbad Caverns National Park in southeastern New Mexico (see Chapter 11), in Guadalupe Mountains National Park (Chapter 10) and Big Bend National Park (Chapter 7) in Texas. To the east, in the Ouachita Mountains, lies Hot Springs National Park (Chapter 11), Arkansas. These mountains are much older than those located in the western part of North America, and they are genetically related to the *Appalachian* Mountains further to the east.

Also included in the Southwest Region is the southeastern portion of the Colorado Plateau in Arizona and New Mexico. This is an area that at one time was extensively occupied by ancient peoples, and many of the national monuments in the region were established to preserve the remains of their civilization. One of the best examples of this is Chaco Culture National Historic Park (Chapter 16), where extensive ruins are preserved.

East of the Pacific Northwest and Western Regions lies the Rocky Mountain Region (Figure 1.18), with its headquarters located in Denver, Colorado. This is a region

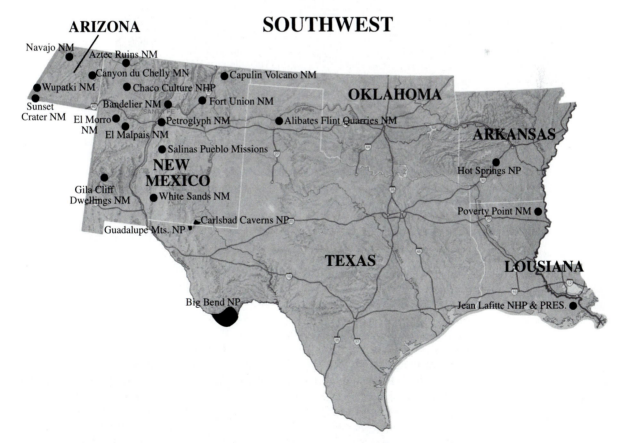

FIGURE 1.17 Park locations in the Southwest Region. (NM = National Monument; NP = National Park; NHP = National Historic Park; Pres. = Preserve.)

ROCKY MOUNTAIN

FIGURE 1.18 Park locations in the Rocky Mountain Region. (NM = National Monument; NP = National Park; Pres. = Preserve; NRA = National Recreation Area.)

encompassing the Rocky Mountains, as well as areas to the east and west. The Rocky Mountains, including Rocky Mountain National Park in Colorado (Chapter 7), are an extensive range of mountains running from Alaska to New Mexico. In the northern part of Alaska, the Arctic Rockies begin the chain. In the lower 48 states, the Northern and Middle Rocky Mountains extend southeastward from western Montana, where Glacier National Park is located (Chapter 13), into Wyoming and Colorado. The Rockies are made up of several fault block mountain ranges similar to the Sierra Nevadas, and include the Teton Mountains where Grand Teton National Park, Wyoming, is located (see Chapter 7). Two major plateaus are located in the region, the Beartooth Plateau and Yellowstone volcanic plateau, formed by recent volcanism, on which Yellowstone National Park, located in Idaho, Montana and Wyoming (discussed in Chapter 11), can be found. The Southern Rocky Mountains include the highest peaks in all the Rockies with 52 mountains having summits over 14,000 ft (4,000 m) high. These include mountains in the Front Range (Pikes Peak, just to the west of Colorado Springs, Colorado, at 14,110 ft (4,301 m), with its extension further to the south including the Sangre de Cristo Mountains, and the Sawatch Range to the west containing Mt. Elbert, the highest peak in the Rockies at 14,431 ft (4,399 m). Slightly to the west and south of the Sangre de Cristo Mountains, the Southern Rockies also include the San Juan Mountains in southwestern Colorado, and the Jemez Mountains in northern New Mexico.

Also in the Rocky Mountain Region is the northern part of the Colorado Plateau. In this geologic province are a number of the nation's most outstanding parks and monuments, including Bryce Canyon National Park (Figure 1.19) discussed in Chapter 9.

In addition to other national parks in the plateau area is Mesa Verde National Park in southwestern Colorado (Chapter 16). While known mainly for its Anasazi cultural ruins, it is also geologically quite beautiful.

Just to the east of the Rocky Mountains, the *Great Plains* begin. The states of North and South Dakota are included within the Rocky Mountain Region of the Park System, and, along with eastern Montana and Wyoming, are mainly broad, flat plains. Stream erosion has left some interesting land forms that are well exposed in Theodore Roosevelt National Park in North Dakota (Chapter 9), where there is also a very large buffalo herd (Figure 1.20); and in Badlands National Park in South Dakota (also discussed in Chapter 9). Some remnants of volcanoes exist in the region (for example Devils Tower National Monument, Wyoming (Figure 1.3). In the *Black Hills of South Dakota* are located some spectacular caves (including Wind Cave National Park, see Chapter 11). The Great Plains were an obstacle that had to be crossed by early pioneers on their way westward in covered wagons, and their trails or associated forts, such as Fort Laramie National Historic Site (Chapter 16) are now national historic sites in the region, but are not continuous.

To the west of the Rocky Mountain Region lies the Midwest Region (Figure 1.21). The National Park Service headquarters for the region lies in Omaha, Nebraska. This Region includes a small segment of the north-central continental interior of the United States, where the central, very old core of the North American continent, called the *Precambrian Shield,* is exposed. Glaciation has significantly altered the region, leaving behind the Great Lakes as dramatic proof of the giant ice sheets that once covered the land. In Lake Superior, Isle Royal National Park, Michigan (Chapter 5), is a volcanic remnant resistant to the grinding erosion of passing glaciers. Further to the west the core of the continent is exposed in Voyagers National Park, Minnesota (see Chapter 6). Part of the area is covered by glacial debris many feet thick that is

FIGURE 1.19 Unusual erosional features of Fairyland Canyon in Bryce Canyon National Park, Utah. (Photo by author.)

FIGURE 1.20   Buffalo herd in the southern section of Theodore Roosevelt National Park, North Dakota. (Photo by author.)

FIGURE 1.21   Park locations in the Midwest Region. (NM = National Memorial; NP = National Park.)

excellent for plant growth and provides the soil for the breadbasket of the U.S., important in states like Iowa, Illinois, Indiana, Ohio, and others. No other national parks lie in the region, but several national monuments are present.

To the south of the Midwest Region lies the Southeast Region (Figure 1.22), with headquarters in Atlanta, Georgia. The region includes most of the Appalachian plateau that borders on the Appalachian Mountains further to the east. It is composed of flat sedimentary rocks in which many caves are developed, including Mammoth Cave National Park, Kentucky (Chapter 11), the longest cave in the world, with over 300 miles of connected passage.

Paralleling the eastern seaboard of North America are the Appalachian Mountains, a linear chain of mountains extending from Alabama to Canada. Lying within the Southeast Region of the National Park System, and sitting along the summit of the Appalachians bordering North Carolina and Tennessee, is Great Smoky Mountains National Park (see Chapter 7). Bordering the Atlantic coast and extending around Florida and along the Gulf coast to Texas is the Coastal Plain, composed of seaward

SOUTHEAST

FIGURE 1.22 Park locations in the Southeast Region. (NHS = National Historic Site; NM = National Monument; N MEM = National Memorial; NMP = National Military Park; NP = National Park; NS = National Seashore.)

sloping lowlands. Sea-level fluctuations are very important in the region, and an extensive system of barrier islands has developed due to these fluctuations. Several national parks exemplify the Coastal Plain, but the best known is Everglades National Park, located at the southern tip of Florida (discussed in Chapter 14). Also included within the Southeast Region are National Park Service areas lying in U.S. Territories in the Caribbean Sea to the south of Florida. Virgin Islands National Park (Chapter 14), in the Virgin Islands of the Western Antilles island chain, is one of two national parks in U.S. Territories, the other being the National Park of American Samoa. Virgin Islands National Park lies in the Southeastern Region, while the National Park of American Samoa is supervised by the Western Region (Figure 1.14).

The Mid-Atlantic Region of the Park System (Figure 1.23) has its headquarters in Philadelphia, Pennsylvania. Only one national park, Shenandoah National Park (Chapter 7), lies in the region. Located to the west of Washington, D.C., in northern Virginia, Shenandoah National Park is another example of the ancient backbone of the Appalachian Mountains that run through the center of the region. While containing only one national park, the Region does have a large number of National Park Service supervised areas, including Valley Forge National Historic Park (Chapter 16).

The North Atlantic Region (Figure 1.24) includes all the New England states, as well as New York and New Jersey. The regional office is located in Boston, Massachusetts. As with the Mid-Atlantic Region, there are many areas supervised by the National Parks System, but Acadia National Park (Chapter 5) is the only national park

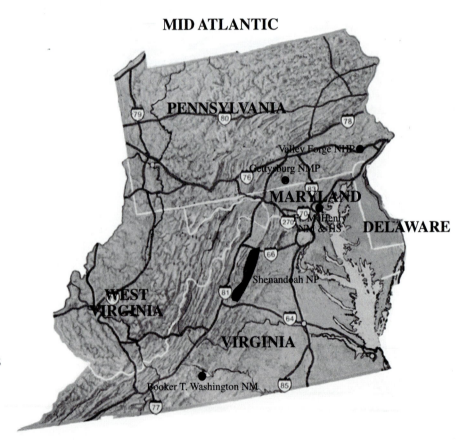

FIGURE 1.23 Park locations in the Mid-Atlantic Region. (NHP = National Historic Park; HS = National Historic Site; NMP = National Military Park; NM = National Memorial; NP = National Park.)

NORTH ATLANTIC

FIGURE 1.24 Park locations in the North Atlantic Region. (NP = National Park; NM = National Monument.)

in the region. The Appalachian Mountains continue on through the region and into Canada.

REFERENCE

U.S. Department of the Interior, Office of Public Affairs and Division of Publications National Park Service. 1995. *The National Parks: Index*. Washington, D.C.

2

Introduction to Geology

In the following chapters, some important geological processes and rock types will be discussed. Examples from the National Park System will be used to characterize individual geological concepts. It is the purpose of this book to provide a general geological background to the parks. Many park units exhibit a wide range of unusual and spectacular geological settings, but our discussion will center on particular characteristics common to several national parts.

CONCEPTS AND TERMS

First some of the concepts and terms used by geologists must be defined. Therefore the initial few chapters of this book were designed to provide the information necessary to understand the geological history of the parks. We start by loosely defining science as a means and method for arriving at explanations for natural phenomena. Geology has been considered a natural science since the late 18th century, when James Hutton in 1788 published a book titled *Theory of the Earth*. Until this time the sciences were chemistry, physics, astronomy, and biology. But in order to understand Earth processes, geologists had to first understand all the "classic" sciences and apply this understanding to an interpretation of how and why Earth formed and how it changed. As a result of these efforts, many subdisciplines have emerged under the broad topic area of physical geology, including geochemistry and geophysics. A second broad topic area, historical geology, deals with how the physical processes have operated throughout time, in part by studying fossils, the remains or evidence of ancient plants and animals, and by hypothesizing what Earth was like when these organisms were living. When both physical and historical knowledge are combined, we can best interpret past geological events, such as the assembly and fragmentation of supercontinents.

The main tools of geology are observation and interpretation, an examination of the interrelationships of rocks, *minerals*, and fossils. One of the most striking results of geological interpretation is the observation that Earth is very old, created approximately 4.6 billion years ago, and has a very complex history. A primary emphasis in geology is to understand the physical, chemical, and biological processes operating now, to apply that understanding to what happened in the past, and to predict what we may find in other geological settings. Ultimately, we may be able to predict what will happen in the future.

Geologists, as well as all other scientists, must employ care and maintain a strong ethical code in their pursuit of truth. Their observations and methodology, when applied to any research problem, must follow a reasoned scientific approach, termed the scientific method. This method entails five elements or steps of problem solving.

Formulation of the Problem.   The first step, probably the most important and most intellectually challenging, is to identify those important problems whose solutions will help us better understand the processes that govern how Earth is formed. This is not a simple task. Geologists are basically curious people, and the questions they ask are driven by their own curiosity. Identifying the important problems and asking the right questions requires understanding a great deal about geology and being able to recognize a problem's solvability. This takes into account resources, knowledge, equipment, and funding availability. An example of a very simple but important and interesting question that early geologists asked is "Were the African and South American continents connected at one time?" They asked this question because the outlines of these continents were observed to be very similar (Figure 2.1). Scientists have long worked on supporting the theory of continental drift, and in just the last 30 years the existence of continental drift has been confirmed. For over 100 years geologists worked to answer that question. Much of the data they collected argued for such a connection, but it took modern geophysics to convince many skeptics. More about this will be presented in Chapter 4 when we discuss plate tectonics.

Observation and/or Experimentation.   The second step in applying the scientific method is an effort to acquire information or data relevant to solving the problem. In their early efforts to answer the question concerning the similarity between the continental margins of Africa and South America, geologists collected data on both continents and tried to confirm similarities between those observations. One of the biggest problems they faced was in demonstrating the possibility that continents could have drifted to their present locations. Great skepticism remained in most geologists' minds concerning the possibility of continental drift until the early 1960s, primarily because of the lack of a reasonable mechanism that could explain continental motion, as well as the lack of information showing that the continents actually had physically moved. Geophysical data acquired in the late 1950s and through the 1960s provided indirect evidence of this movement.

Development of an Hypothesis.   The third step involves analyzing the available data and forming concepts. Accumulated geological data led to the continental drift hypothesis in the early 1900s, but it still took 50 years or so, and the addition of large amounts of data from seismic, magnetic, and heat flow studies to finally convince geologists that continents were actually moving across the face of Earth. Even today there are skeptics, but these become fewer with time and the accumulation of more data.

Hypothesis Testing.   During the last 30 years, the fourth step, hypothesis testing, involved performing more experiments. Those tests resulted in the development of new methods, techniques, and the acquisition of new data sets, including paleomagnetic, marine magnetic anomaly, and high resolution seismic data, which finally convinced most skeptics. For example, marine magnetic anomaly data, acquired by towing

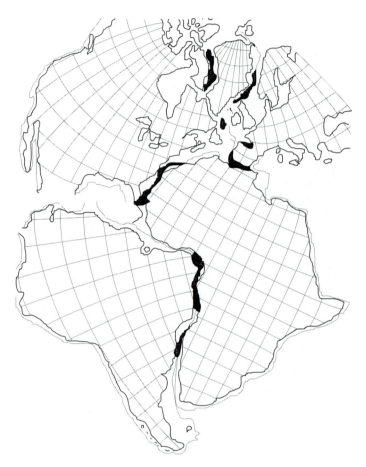

FIGURE 2.1 Fit of Atlantic continents shown by Bullard in 1955.

magnetometers behind ships, are interpreted as recordings of the tracks of continents as they moved through the ocean basins. This will be discussed further in Chapter 4.

Theory Formulation. These new observations have brought us to the fifth step, hypothesis validation, modification or rejection, or theory formulation. In the study of continental drift, we have now extensively tested, and continue to test, our hypothesis of drifting land masses; the result is that the name of the theory has changed to plate tectonics, and our understanding of the specifics of the original model has been modified. But as to the basic question, "Were the African and South American continents at one time connected?", the answer still is YES!

Our understanding of the physical and biological worlds will continue to change as the result of applying the scientific method. In geology we are constantly modifying our understanding of Earth and its processes, but the tests we perform are especially difficult, because we can't go back in time to make our observations. Furthermore, we can't construct continent-sized experimental models, and many important processes are so slow that it takes more that a single lifetime to make the observations we would like to make. The question of continental drift has taken over 100 years to answer and we are still working on the problem! We must make our tests by inference and test our models based on certain assumptions, which in turn must be tested. We assume, for example, that present physical processes are also operating in much the same way as they did in the past. This concept that the present is the key to the past is an important assumption that is constantly being tested in geology. And so it goes!

EARTH HISTORY

Due to the development of radiogenic age dating techniques over the last 50 years, we now know that the age of Earth is 4.6 ± 0.1 billion years old. However, before radiogenic dating, scientists thought Earth to be much younger. Those early age estimates were based on geological observations and the relationships between rocks and fossils. Called *relative ages*, those estimates were based on observation and intuition, with no real knowledge of the time involved in forming Earth. Today, radiogenic age dating produces *absolute ages* in years for many geological materials. Combined with relative ages, absolute ages are now leading to a much better understanding of geologic processes.

The history of Earth, as interpreted from rocks, has resulted in another important conclusion, that Earth is dynamic. The evidence is all around us and is especially obvious when we observe the result of geologic events and when geologic events are constantly being reported in the papers and on television. Volcanic or molten rock periodically pours forth from many places on Earth, often causing death and destruction. Streams erode deep into Earth leaving impressive examples of the power of rivers, such as the Grand Canyon (Figure 2.2). Major earthquakes destroy portions of cities, taking lives in the process, and geologists predict that we can expect more of these destructive earthquakes to occur. And great mountain ranges such as the Grand Tetons (Figure 2.3) are actively being built and eroded away.

Earth's dynamic processes are constantly creating new rocks and destroying others. The rock cycle (Figure 2.4) is often used to characterize the dynamic aspects of rock stability on and within Earth. Rocks are made of *minerals*, the building blocks of Earth. Minerals are stable when formed, but geological processes cause changes in mineral stability, thus altering the rocks involved. Rock burial results in elevated pressures and temperatures, restructuring the rocks involved and producing a new, *metamorphic rock*. Further heating can cause melting, thus forming a liquid, called *magma*. Later, when geological processes bring these materials toward or onto the surface of

FIGURE 2.2 View from the south rim in Grand Canyon National Park, Arizona. (Photo by author.)

FIGURE 2.3 Central portion of Grand Teton National Park in Wyoming. (Photo by author.)

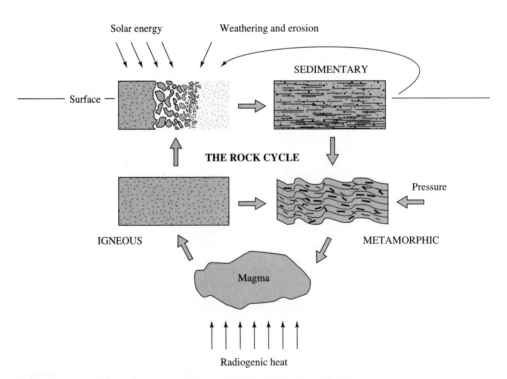

FIGURE 2.4 Very simple version of the rock cycle illustrating how one rock type can be converted into another.

Earth, the pressure/temperature decreases and the magma solidifies, forming *igneous rocks*. Physical and chemical processes at or near the surface of Earth will decompose and disintegrate exposed rocks, thus *weathering* these materials. The broken constituents of these rocks can then be picked up and *eroded* away. Later the particles will be deposited and can be reformed into *sedimentary rock*. If these rocks are buried in turn, the cycle continues.

MINERALS

Before it is possible to understand the processes active on Earth, it is necessary to understand the building blocks of Earth. Minerals, composed of elemental constituents that are bonded together, are the building blocks that make up all rocks. There are thousands of known minerals, but only about 20 of these make up most rocks.

All minerals have the same four unique characteristics. They are all (1) naturally occurring, (2) inorganic, (3) solid, as opposed to a gas or a liquid, and (4) crystalline, that is they have a definite chemical formula and internal structure. Water is not a mineral because it is a liquid. But if the temperature drops below freezing, *ice* [H_2O], a solid, forms naturally. Because it is inorganic and has a crystalline structure, ice is a mineral. Clearly, changing temperature or pressure can result in major changes in Earth's building blocks.

On occasion there can be some controversy concerning mineral classification. Minerals are defined as being inorganic, but what about those materials that are identical to minerals in every way, but have an organic source? For example, there is a marine animal, a chitin, that abrades limestone rocks with its teeth that are capped by the mineral magnetite [Fe_3O_4]. Even though this magnetite is precipitated by an animal, as calcite shells are precipitated by other organisms, the magnetite (and calcite) is still considered to be a mineral.

It is often possible to identify minerals based on their physical properties, and geologists use these physical properties for mineral identification. For example, some minerals are much more resistant to abrasion than others. This resistance is called hardness, and it was quantified by Moh, a German who lived from 1773–1839. Moh's hardness scale (Table 2.1) is a relative scale that aids in mineral identification.

Another useful property in identifying minerals is *cleavage* or *fracture*. In crystalline substances, there may be unique, parallel zones of weakness resulting from its

TABLE 2.1   MOH'S HARDNESS SCALE

Hardness Scale	Mineral	Scratched by
1	talc	fingernail
2	gypsum	fingernail
3	calcite	penny
4	fluorite	knife blade
5	apatite	knife blade
6	K-feldspar	steel file
7	quartz	steel file
8	topaz	corundum
9	corundum	diamond
10	diamond	nothing is harder

crystal structure. *Calcite* [CaCO₃] and *halite,* or common table salt [NaCl], are examples of minerals with cleavage. When a mineral exhibits cleavage it breaks uniformly, forming flat surfaces or *cleavage planes*. If the mineral does not display cleavage, or no easy direction of breakage, it will fracture unevenly, somewhat like glass, exhibiting *conchoidal* or wavy fracture patterns. A common example of a conchoidal pattern occurs in the mineral *quartz* [SiO₂].

A third mineral property is *crystal habit*, the external expression (Figure 2.5) of an internal crystal structure. Crystal habit can be identified visually, while *crystal structure* is determined by x-ray diffraction and is related to the mineral's chemical composition and internal arrangement of atoms. When crystals are grown in unrestricted environments (for example in a cavity or in a fluid), they will develop a specific geometry that is characteristic of the mineral. But if the growing crystals are constrained during growth, they may take any shape; however, this does not affect the orderly internal arrangement of the crystal. For example, it is clear that in granites crystals show growth distortion because they grow from a melt and are forced to compete for available atoms and space (Figure 2.6).

Twinning, an intergrowth of two or more single crystals, can also be exhibited by some minerals (Figure 2.7) and can be somewhat diagnostic. Some crystal faces may exhibit parallel lines or bands called *striations* that reflect twinning from multiple crystal intergrowths. Different minerals are heavier than others due to a physical property called *specific gravity*. Specific gravity depends on how much heavier a rock or mineral is than water and is given by the ratio (weight of an amount of mineral)/(weight of an equal amount of water). Specific gravity is similar to *density*, defined as mass per unit volume, with units usually given in gm/cm³. Density is difficult to measure precisely because it is difficult to measure the volume of an irregularly shaped substance. Consequently, specific gravity has been used by geologists for many years to help identify minerals. For example, in the common sedimentary rock, sandstone, sand grains may be cemented together by different minerals, but often that cement is the relatively light calcite. In many instances the cement is the much heavier iron carbonate mineral, *siderite* [FeCO₃]. It is heavier because the iron ion (Fe) contained within the siderite crystal lattice is heavier than the calcium ion (Ca) contained within the calcite lattice.

FIGURE 2.5 Crystals of the mineral halite, composed of chlorine and sodium [NaCl]. (Photo by author.)

FIGURE 2.6 A granite sample, composed mainly of K-feldspar, quartz (light minerals), and ferromagnesian minerals, including the mica, biotite (dark mineral). (Photo by author.)

When the rock is lifted, its unusual weight tells the geologist that something unusual is contained within the rock, leading to a search for the explanation.

One property that isn't always diagnostic in mineral identification is color because many minerals have many different color varieties. For example, while emeralds and aquamarines are both gems of different color, they are both crystalline varieties of the same mineral, beryl. Rubies and sapphires are another example, since both gems are corundum, mineral 9 on Moh's hardness scale. However, streak, the color of a mineral in powdered form, is more diagnostic than using just the color of a hand specimen and can be useful in mineral identification. Geologists have a small white porcelain plate, called a *streak plate*, that is used to scratch or powder minerals for identification. Because most of the color is determined by trace impurities of only a few elements in

FIGURE 2.7 Twinned calcite crystal [$CaCO_3$]. (Photo by author.)

a mineral, the streak allows the human eye to better distinguish the primary, more diagnostic, color in the sample.

Other properties include *luster*, reflection of light from a mineral's surface, magnetism, *fluorescence*, solubility, and taste. Simple tests based on these properties can be performed, even in the field, to help geologists identify minerals. For example, a magnet will pick up small fragments or grains of magnetic minerals, such as magnetite. And dilute hydrochloric acid [HCl] will rapidly dissolve calcite, causing it to effervesce. Taken together, these physical properties make it possible to identify many minerals in the field simply by careful observation.

An interesting characteristic of some minerals is that they have the same chemical composition, but exhibit a different crystalline structure. This property is called *polymorphism,* and one of the best examples is that of diamond [C] and graphite [C]. Both minerals have the same chemical composition, only carbon atoms are present, but clearly have a very different crystalline structure. It is this structural difference that makes diamonds valuable as a gem mineral and graphite valuable as a lubricant.

In those instances where it is very difficult to identify minerals in the field, they must be brought into the laboratory for identification. The principal mineral identification methods are based on optical properties, using microscopes, or by x-ray diffraction, where x-ray diffractometers characterize the unique crystallinity of minerals in powdered form.

The Building Blocks of Minerals

The building blocks of minerals are *elements* that are composed of *atoms* with a *nucleus* at the center, containing *protons* (positively charged), and, in all but the element hydrogen, contain *neutrons* (no charge). Surrounding the nucleus, like planets around the sun, are orbiting *electrons* (negatively charged). We place elements in order on the periodic table of the elements (Table 2.2) based on their *atomic number*, the number of protons contained within the nucleus. The *atomic weight* of an element is based on the number of protons and neutrons within the nucleus of the atom. Oxygen, for example, is element number 8 on the periodic table because it has eight protons in the nucleus. To denote the atomic number, we would write it as O_8. Usually oxygen also has eight neutrons in the nucleus giving it an atomic weight of 16 (formally yielding $^{16}O_8$), but the 8 is generally dropped because all oxygen ions have eight protons. In some instances oxygen has nine and at other times ten neutrons in its nucleus, increasing its atomic weight to 17 and 18, respectively. These heavier atoms of oxygen are called *isotopes,* written ^{17}O and ^{18}O, respectively. The addition of a neutron or two doesn't change the atomic number of the element, therefore its place on the periodic table remains the same. But the atomic weight does increase. However, if a proton were added, the element number (atomic number), as well as the mass (atomic weight), would change, creating a new element. Isotopes can be very useful in studying Earth processes. For example, when ^{18}O is contrasted with the normal isotope of oxygen, ^{16}O, it becomes very important in many of areas of geology and geochemistry, including studies of ancient climates. Other isotopes are important in isotopic dating, and some of these will be discussed in Chapter 3.

If the number of electrons orbiting an atomic nucleus is the same as the number of protons in the nucleus, then the negative and positive charges associated with these particles are balanced, yielding no net charge. However, if the element has an unbalanced number of electrons or protons, it will have a charge and is called an *ion*. The mass of the element remains the same even if electrons are added, because electrons are so small that their effective mass, relative to the mass of the nucleus, is zero. Therefore, even though an element has a charge, its atomic weight or atomic number doesn't

TABLE 2.2 PERIODIC TABLE OF ELEMENTS

Key:
- Atomic number
- Symbol of element
- Atomic weight
- Name of element

Example:
- 1 / H / 1.008 / Hydrogen

Legend:
- Inert gas
- Gas
- Liquid
- Solid–all others

Light Metals · Transitional Elements · Heavy Metals · Nonmetals

	IA	IIA	IIIB	IVB	VB	VIB	VIIB	VIIIB	VIIIB	VIIIB	IB	IIB	IIIA	IVA	VA	VIA	VIIA	VIIIA
1	1 H 1.008 Hydrogen																	2 He 4.003 Helium
2	3 Li 6.941 Lithium	4 Be 9.012 Beryllium											5 B 10.81 Boron	6 C 12.01 Carbon	7 N 14.01 Nitrogen	8 O 16.00 Oxygen	9 F 19.00 Fluorine	10 Ne 20.18 Neon
3	11 Na 22.99 Sodium	12 Mg 24.31 Magnesium											13 Al 26.98 Aluminum	14 Si 28.09 Silicon	15 P 30.97 Phosphorus	16 S 32.06 Sulfur	17 Cl 35.45 Chlorine	18 Ar 39.95 Argon
4	19 K 39.10 Potassium	20 Ca 40.08 Calcium	21 Sc 44.96 Scandium	22 Ti 47.90 Titanium	23 V 50.94 Vanadium	24 Cr 52.00 Chromium	25 Mn 54.94 Manganese	26 Fe 55.85 Iron	27 Co 58.93 Cobalt	28 Ni 58.71 Nickel	29 Cu 63.55 Copper	30 Zn 65.38 Zinc	31 Ga 69.72 Gallium	32 Ge 72.59 Germanium	33 As 74.92 Arsenic	34 Se 78.96 Selenium	35 Br 79.90 Bromine	36 Kr 83.80 Krypton
5	37 Rb 85.47 Rubidium	38 Sr 87.62 Strontium	39 Y 88.91 Yttrium	40 Zr 91.22 Zirconium	41 Nb 92.91 Niobium	42 Mo 95.94 Molybdenum	43 Tc 98.91 Technetium	44 Ru 101.07 Ruthenium	45 Rh 102.91 Rhodium	46 Pd 106.4 Palladium	47 Ag 107.87 Silver	48 Cd 112.4 Cadmium	49 In 114.82 Indium	50 Sn 118.69 Tin	51 Sb 121.75 Antimony	52 Te 127.60 Tellurium	53 I 126.90 Iodine	54 Xe 131.30 Xenon
6	55 Cs 132.91 Cesium	56 Ba 137.33 Barium	57 to 71	72 Hf 178.49 Hafnium	73 Ta 180.95 Tantalum	74 W 183.85 Tungsten	75 Re 186.2 Rhenium	76 Os 190.2 Osmium	77 Ir 192.22 Iridium	78 Pt 195.09 Platinum	79 Au 196.97 Gold	80 Hg 200.59 Mercury	81 Tl 204.37 Thallium	82 Pb 207.2 Lead	83 Bi 208.98 Bismuth	84 Po (210) Polonium	85 At (210) Astatine	86 Rn (222) Radon
7	87 Fr (223) Francium	88 Ra 226.03 Radium	89 to 103															

Lanthanide series

57 La 138.91 Lanthanum	58 Ce 140.12 Cerium	59 Pr 140.91 Praseodymium	60 Nd 144.24 Neodymium	61 Pm (145) Promethium	62 Sm 150.4 Samarium	63 Eu 151.96 Europium	64 Gd 157.25 Gadolinium	65 Tb 158.93 Terbium	66 Dy 162.50 Dysprosium	67 Ho 164.93 Holmium	68 Er 167.26 Erbium	69 Tm 168.93 Thulium	70 Yb 173.04 Ytterbium	71 Lu 174.97 Lutetium

Actinide series

89 Ac (227) Actinium	90 Th 232.04 Thorium	91 Pa 231.04 Protactinium	92 U 238.03 Uranium	93 Np 237.05 Neptunium	94 Pu (244) Plutonium	95 Am (243) Americium	96 Cm (247) Curium	97 Bk (247) Berkelium	98 Cf (251) Californium	99 Es (254) Einsteinium	100 Fm (257) Fermium	101 Md (258) Mendelevium	102 No (259) Nobelium	103 Lw (260) Lawrencium

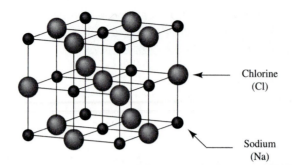

FIGURE 2.8 Structure of the mineral halite [NaCl], illustrating small sodium cations and larger chlorine anions.

Chlorine (Cl)

Sodium (Na)

change, nor does it become a new element or even an isotope, unless a proton or neutron is lost or gained by the nucleus, respectively. If the ion has a negative charge (extra electrons) it is called an *anion*. Conversely, positively charged ions (fewer electrons than protons contained within the nucleus) are known as *cations*.

Minerals are formed by combining anions and cations that balance each other's charges. For example, the element sodium, Na on the periodic table, is a cation with one missing electron. Chlorine, Cl on the periodic table, is an anion with an extra electron orbiting the nucleus. These two ions can balance each other's charge and *bond* together to form *sodium chloride*, the mineral halite (common table salt) (Figure 2.8) by the chemical reaction,

$$Na^+ + Cl^- \rightarrow NaCl.$$

This is the mineral shown in Figure 2.5. It requires the proper concentrations of ions and must occur within proper temperature and pressure ranges for specific minerals to crystallize. For example, if the temperature is very hot, the mineral ice will not crystallize. The size of the combining ions is also important. For example, in the case of the mineral halite [NaCl], the small sodium cation fits nicely between the much larger chlorine anions.

Minerals can crystallize from a liquid after evaporation has concentrated the ions in solution, which accounts for the development of cave formations and the crystallization of the mineral calcite. They can also form from melts or magmas as a result of cooling. Newly developing crystals within a liquid, growing at the same time as other crystals, compete for available space, thus forcing growth-shape modifications on individual crystals.

Crystals are destroyed by removing atoms or ions from crystal faces. This can result from dissolution by a solvent such as water, by melting at high temperatures, or by changing the pressure/temperature stability in some other way.

ROCKS

Rocks are composed of constituent minerals and named based on mineral composition/texture, or how the constituents were put together. For example, a *granite* has as its main mineral constituents quartz and *feldspar* with minor amounts of other minerals, such as mica. Such igneous rocks are the result of crystallization from a hot liquid or magma, have intergrown crystals, and are usually medium to coarse grained. Most rocks contain only a few of the 20 most common minerals, and some are composed of just one mineral. For example, limestones, very common all over the world, are composed mainly of the mineral calcite.

The most abundant minerals are the silicate minerals that form 95% of Earth's crust. All silicate minerals contain some form of the $[SiO_4]^{-4}$ complex ion (Figure 2.9), called the silicon tetrahedron, as their anion. The *silicon tetrahedron,* also explains the great abundance of silicon and oxygen in Earth's crust. An interesting example of a silicate mineral is given in the following chemical equation, where magnesium or iron atoms as cations combine with the silicon tetrahedron to produce the mineral *olivine.*

$$(Mg, Fe)_2^{(+4)} + SiO_4^{(-4)} \rightarrow (Mg, Fe)_2SiO_4.$$

Ferromagnesian minerals, such as olivine, that contain iron and/or magnesium tend to darken the color of rocks in which they are constituents.

Silicate minerals are grouped based on how the tetrahedrons are arranged, and the crystalline structure exhibited is due mainly to the efficient packing produced by the small silicon atom being enclosed by four big oxygen atoms (Figure 2.9). The result is that some silicate mineral groups, such as the pyroxenes, form single tetrahedron chains, others, including the amphiboles, form double chains, and still others form sheets (Figure 2.10). The sheet silicate structure of micas and clays explains why these minerals easily part or cleave into plates. On the other hand, some of the silicate minerals form a 3-D tetrahedron framework. Quartz, which falls into this category, does not exhibit cleavage. However, some 3-D framework silicates, such as feldspars, major constituents in granites, do exhibit good cleavage.

Minerals are classified by their anions. Other important mineral groups, besides the silicates, include the *oxides*, with magnetite $[Fe_3O_4]$, the *sulfides*, such as *pyrite* $[FeS_2]$, *carbonates*, with calcite being most abundant, and *sulfates*, with minerals such as *gypsum* $[CaSO_4 \cdot 2H_2O]$ formed from the complex sulfate $[SO_4]^{-2}$ ion. In addition to being a sulfate mineral, gypsum is also known as a *hydrated* mineral because it contains water as part of its crystal structure. Gypsum is one example of a hydrated mineral; it is important to the construction industry in the United States because it is used in the manufacture of sheet rock or wall boards. If gypsum is heated, it will lose its water and a new sulfate mineral, *anhydrite* $[CaSO_4]$ results.

All rocks may be divided into three groups, (1) igneous rocks that were crystallized from a magma, (2) sedimentary rocks composed mainly of the redeposited fragments of other rocks and cementing mineral constituents, and (3) metamorphic rocks, formed by alteration of other rocks at elevated temperature and pressure. The main constituent minerals in igneous rocks are quartz, feldspars, micas, and ferromagnesians, minerals containing iron and/or magnesium. In sedimentary rocks, the mineral constituents mainly include quartz, carbonates, clays, halite, gypsum, and feldspars. And in metamorphic rocks quartz, feldspars, micas, ferromagnesian minerals, garnet, and chlorites are the characteristic minerals. Igneous, sedimentary, and metamorphic rock types will be discussed in detail in later chapters.

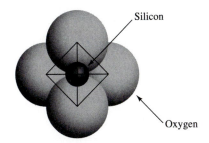

FIGURE 2.9 Silica tetrahedron, illustrating four large oxygen and one small silicon atoms.

	Arrangement of silica tetrahedra		Common minerals
Isolated tetrahedra		 SiO_4	Olivines
Isolated group example			Tourmaline
Continuous chains			Pyroxenes
			Amphiboles
Continuous sheets			Micas

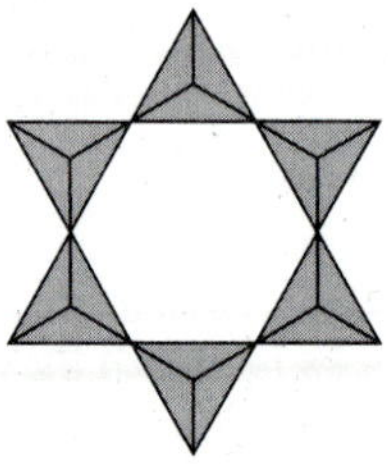

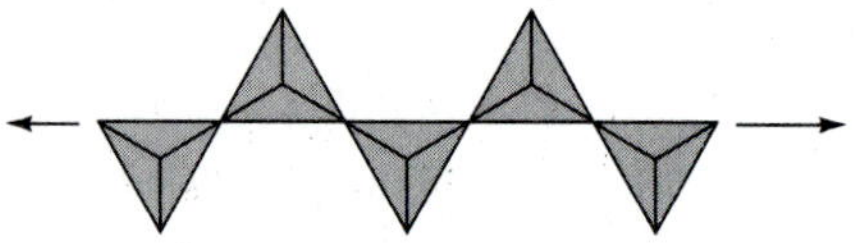

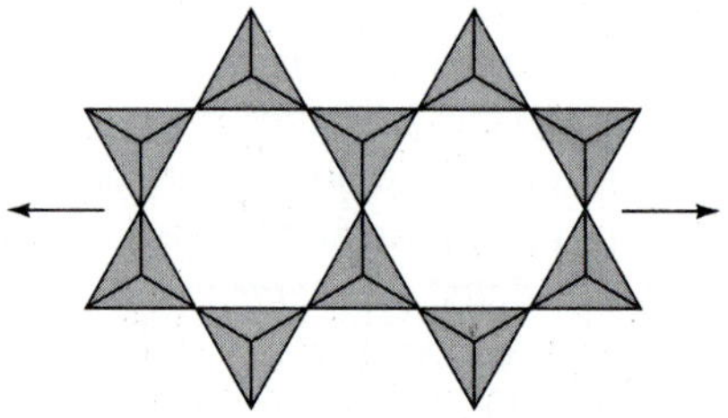

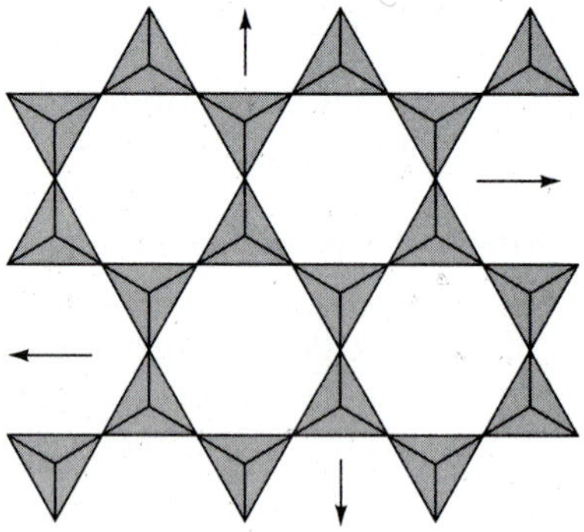

FIGURE 2.10 Structure of the silicon tetrahedrons in various silicate mineral families.

REFERENCE

DEER, W. A., HOWIE, R. A., and ZUSSMAN, J. 1983. *An introduction to the rock forming minerals*. Hong Kong: Commonwealth Printing Press Ltd.

3

Dating Earth

The science of geology deals with the interpretation of how and when Earth formed, and how it continues to change. That means that geologists are concerned not only with modern processes, but also with how those processes are recorded in the geologic record and ultimately the age of rocks on Earth. Today, and for the last 50 years or so, we have been able to directly date many types of rocks, deriving absolute ages for Earth materials. But early geologists did not have these modern techniques, and instead had to deal with how Earth materials were related to each other; from these relationships, geologists were able to derive relative ages for geologic materials.

THE CONCEPT OF TIME

There are many direct measures of time. For example, the human heart beats at a relatively constant, measurable rate. Swings of a pendulum are measures of time that have been used in the manufacture of clocks. But a standard for time was needed to provide the extreme accuracy required in today's highly technical world. In response to that need, the second was redefined in 1967. It is now based on the atomic cyclic vibrations of a cesium atom, with one second equaling 9,192,631,770 vibration cycles of cesium.

Other measures of time, like Earth's daily rotation, the day, are variable, lengthening by 1 second per year. We now know that 1.5 billion years B.P. the day was only 11 hours long. Based on daily growth rings in ancient fossils, such as corals, we have determined that 400 million years B.P., the day was 20 hours long. This change in Earth's rotation rate is due in part to the moon's gravitational pull slowing Earth. Other familiar measures of time involve a single rotation of the Moon around Earth, a lunar month, and one orbit of Earth around the Sun, a year. All these measures of time are indirectly seen in the rock record.

There are also much less precise measures of time. For example, because Earth is inclined on its axis, yearly climatic cycles are seasonal, yielding yearly growth cycles. The seasons produce growth rings in trees such as bristlecone pines, found in

FIGURE 3.1 Bristlecone pines in Great Basin National Park, Nevada. (Photo by author.)

Great Basin National Park and elsewhere in the southwestern part of the United States (Figure 3.1). These trees are the oldest living organisms, some having been alive for more than 5,000 years, and can be used to calibrate carbon 14 (^{14}C) dates (discussed later in this chapter). Seasonal variations are also reflected in growth rings produced by animals, including clam shells and fish scales.

Seasonal changes can result in light and dark-banded sedimentary sequences called *varves*, which result from winter/summer or high/low erosion cycles often found in association with glacial lakes or sediments deposited from glacial erosion. Varves have been shown to represent cyclic variations dating back approximately 15,000 years B.P. Growth cycles are also found in cave formations, such as those discovered in cores drilled from large standing cave formations in Carlsbad Caverns National Park (discussed in Chapter 11), some going back more than 100,000 years B.P. (Figures 3.2 and 3.3). In Figure 3.2 a scientist sits by an electric drill during sampling of a large standing column in Carlsbad Cavern. The results of some of his efforts are illustrated in Figure 3.3. In these cores a few growth bands can be clearly seen (white stripes in Figure 3.3). One important result emerging from such studies of many different types of materials is that we now know that seasonal variations, resulting in part from atmospheric changes, have been occurring for a very long time, indicating the presence of an atmosphere in early Earth history.

RELATIVE DATING TECHNIQUES

Relative dating techniques were the only way that early geologists were able to estimate the age of Earth and of geological materials. It was not possible, using those techniques, to resolve the actual age of rocks, only to determine if one set of materials was older or younger than another. While this doesn't seem like much information, if enough observations are available, if the processes responsible for the formation of rocks are understood, and if the rates of processes are known, then it is possible to make estimates of the age of Earth or date rock sequences. There are many problems with

FIGURE 3.2 Sampling a standing formation (the Georgia Giant) in Carlsbad Caverns National Park, New Mexico. (Photo by author.)

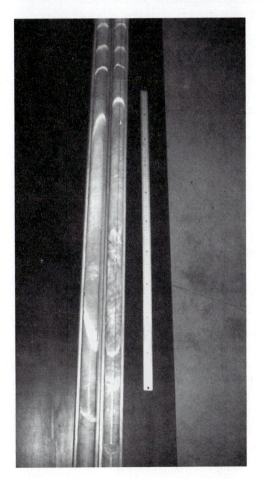

FIGURE 3.3 Cores recovered from Carlsbad Caverns National Park illustrating symmetrical growth bands. (Photo by author.)

these assumptions, such as the problem of missing beds, therefore missing comparative dates, due to erosion or nondeposition.

Relative dating is based on several general relationships. The first, the *principle of superposition*, states that in a layered sequence younger sedimentary rocks, called *strata*, are deposited over older rocks, unless these rocks have been deformed and/or overturned. It is also recognized that not all time periods may be represented by sedimentary sequences or *beds*. In *stratigraphy*, the study of sedimentary materials, the *principle of crosscutting relationships*, where an older rock layer is cut by one younger, is important (Figure 3.4 a, b). An example of crosscutting occurs when igneous magmas, called *dikes*, push their way and intrude into already existing beds. Buried surfaces of erosion or nondeposition, called *unconformities*, complicate relative dating by removing portions of the record. Many examples of this are evident in national park areas, and some of these will be discussed in following chapters.

Along *faults*, breaks in Earth's crust, motion occurs in rocks that are older than the fault, offsetting one rock unit against another. These crosscutting relations are useful in reconstructing past Earth motion by realigning offset beds that can be identified on both sides of the fault. Other important relationships include rock fragments that are older than the rocks containing them (Figure 3.4 c); order of growth, where younger

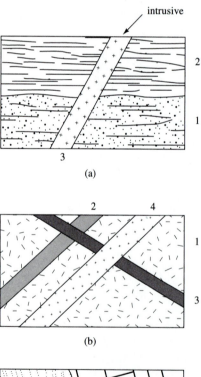

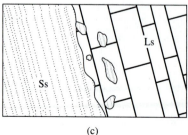

FIGURE 3.4 (a) and (b) Crosscutting relationships in rocks illustrating younger versus older relationships. The oldest units are labeled 1 with higher numbers progressively younger. (c) Sandstone (Ss) eroded and fragments incorporated in the overlying limestone (Ls), followed by tilting of the units.

materials grow on older (observed in reefs and cave formations); and landscapes, where cut and fill structures are younger than the beds that have been cut. All relative dating techniques are based on careful observations and interpretations.

An important aspect of developing an understanding of Earth history is the demonstration of a relationship between two localities, using some sort of physical correlation method. This can be done using very local or small scale indicators, such as *key* or *marker beds*, that exhibit unique characteristics in what are generally thin sedimentary beds. Individual beds can exhibit physical similarity or a sequence of beds can exhibit systematic similarities that can be traced over short distances. Such correlations assume that these similar beds are laterally continuous, but not necessarily the same age. For example, in the past, along the coastal regions of North America, sea level fluctuated widely. Approximately 22,000 years ago, the sea was as much as 100 m lower than it is today. This was due to glaciation, where vast amounts of water were tied up in ice on the continents, rather than in the oceans, thus lowering the sea level. At such times, beaches were located about 100 m lower than they are today. As the ice melted, sea level rose and the beach zone, with its accumulating sand, moved inland until it reached its present point. In the deeper water in front of the beach, finer-grained sediment was deposited. If we were to dig a trench from a modern beach off-shore to the beach that existed about 22,000 years B.P., we could trace a continuous thin bed of beach sand from today's beach to the 22,000-year-old sea-level site. Clearly this sand bed would not contain material that was all the same age. Instead it would represent a *time-transgressive bed*, a layer of similar material deposited over a long period of time in response to sea level rise or fall.

Regional correlations are also possible, using techniques such as tracing layers identified from seismic reflections, where small explosions or other causes produce vibrations that bounce off layers within Earth and are recorded in the same manner as sound waves. Patterns of these reflectors are used to delineate subsurface sedimentary layers. Seismic reflections are extensively used in oil exploration (Figure 3.5).

Key beds with unique characteristics can often be traced for long distances and in some cases can be useful in regional correlation, as can certain unique fossils known as

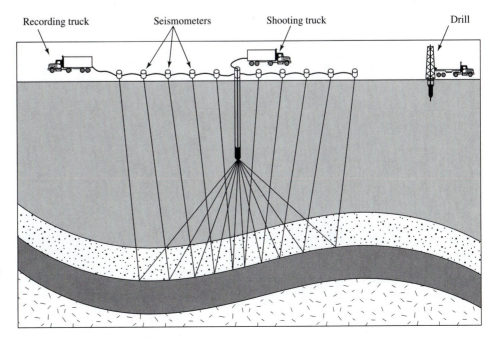

FIGURE 3.5 Oil exploration set-up used to identify beds in the subsurface. An explosion is set off in the subsurface and the seismic pulse generated bounces off subsurface layers and returns to be recorded. Because it takes longer for a wave to travel from deeper parts of the bed, the recorded returns can be used to indicate the subsurface structure.

Recording truck   Seismometers   Shooting truck   Drill

guide or *index fossils*. Guide fossils are common, easy to identify, and the organisms seen in them generally lived for only a short period of time. A group of fossil remains with overlapping ranges, called *fossil assemblages*, can also be used for regional correlations. Some of the most useful fossils are represented by successive *evolutionary stages* of organisms, such as horses in North America, where extensive fossil remains, representing detailed evolutionary changes through time, have been recovered (Figure 3.6). Horse fossils found anywhere in North American, or elsewhere in the world, can be compared to a standard or *composite sequence* made from our records of all such fossil discoveries. This comparison can be used to estimate relative age, when these animals migrated from one continent to another, and when they disappeared from a specific continent. For example, Figure 3.7 indicates that horses were first identified on the North American continent, but became extinct here during the Pleistocene. During the Pliocene, however, they found a pathway to Europe and Africa, where they continue to survive. The horse was then reintroduced to North America by the 16th century Spanish conquistadors. Our knowledge of extinct horses comes in part from Haggerman Fossil Beds National Monument, Idaho, where 150 specimens of the now extinct Haggerman horse have been identified and studied.

On a global scale, there are a few excellent correlation methods. Some guide fossils/assemblages, such as organisms that float in seawater or airborne *pollen*, can be used for global correlations. In rare cases, key beds are also useful. An example is represented in a unique, thin clay layer, rich in the element iridium found only in abundance in extraterrestrial materials. This iridium clay appears to have been deposited globally about 65 million years ago as fallout from a *bolide* (explosive comet, meteorite, or asteroid) impact. That impact event is thought by many geologists to have killed off the dinosaurs and other fossil groups, but it is the source of one of today's controversies in geology. One such bolide is believed to have struck in the northern portion of Canyonlands National Park (Chapter 11) sometime during the Cretaceous Period.

An important, global, relative dating method is based on changes in Earth's magnetic field. As it happens, this field has changed its *polarity* many times in the past

FIGURE 3.6 Examples of Tertiary horse guide fossils, which can be used to correlate beds from one locality to another. Because of the large number of species and the short length of time each species existed, they are useful in identifying the approximate age of rocks containing the fossils.

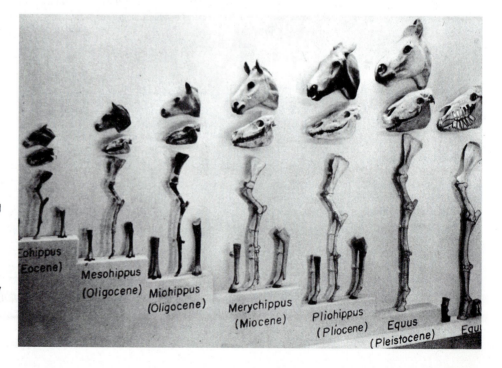

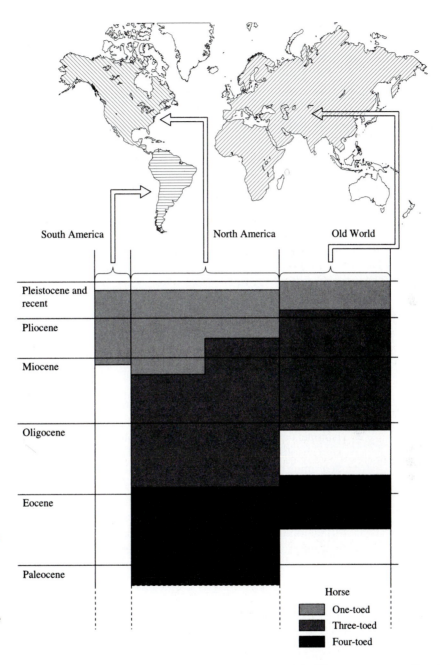

FIGURE 3.7 The global distribution of horses from the Paleocene through recent times. Charts such as this allow paleontologists to characterize when and where different organisms existed (see text).

(Figure 3.8). Today the field is *normally magnetized* and a compass needle will point toward the north magnetic pole of Earth. Such changes, or *reversals*, are characterized by a 180° shift in magnetic field direction, *declination*, and a shift in the *inclination* of Earth's magnetic field lines (Figure 3.9) from positive to negative or vice versa. For example, today, near Mammoth Cave National Park, Kentucky, Earth's magnetic field has a declination of approximately 0° and a positive inclination. That is, a compass needle will point generally toward the north geographic pole of Earth, and the needle tilts down a bit toward the pole. Today, however, in the southern hemisphere, inclination is negative and the needle will tip up a bit. Before approximately 790,000 years B.P., Earth's magnetic field had a reversed polarity, with a declination near Mammoth Cave of about 180° and a negative inclination. That is, the compass needle would have

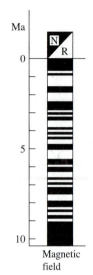

FIGURE 3.8 Polarity record of Earth's magnetic field for the last 10 million years. The black bars illustrate normal and white bars illustrate reversed polarities.

pointed toward the south geographic pole of Earth, and the tilt of the needle would change a bit.

Shifts in Earth's magnetic field have been going on for billions of years. When the magnetic field changes, the effects are seen everywhere on Earth at the same time, thus are globally synchronous. These magnetic polarity changes have been recorded in rocks, producing sequences of known order, to which other magnetic data can be compared. Thus, magnetic polarity sequences can be used for relative age estimates. Isotopic dating techniques (discussed later in this chapter) have made it possible to assign absolute ages to these sequences. The resulting *polarity time scale* can be used to determine ages for rocks that can't be dated using other methods (Figure 3.8).

Two important questions that have been asked concerning magnetic polarity reversals are, "When Earth's magnetic field reverses, do the Van Allen radiation belts (Figure 3.10), which are supported by Earth's magnetic field, collapse onto Earth?" and "If so, do mass extinctions of organisms occur as a result?" The reason these are important questions is because the Van Allen radiation belts prevent very high-energy par-

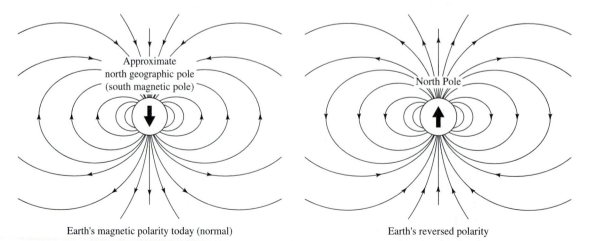

Earth's magnetic polarity today (normal)

Earth's reversed polarity

FIGURE 3.9 Earth's magnetic field during times of normal and reversed polarities, the result of periodic changes throughout Earth history.

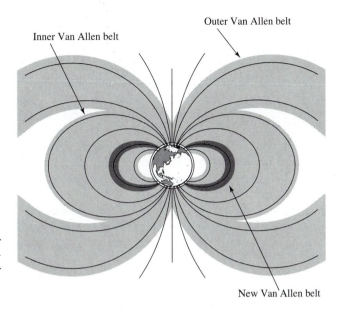

Inner Van Allen belt

Outer Van Allen belt

New Van Allen belt

FIGURE 3.10 The inner and outer Van Allen radiation belts and their relationship to Earth's magnetic field.

ticles from the Sun from hitting the surface of Earth directly. Instead, these particles, collectively called the solar wind, collect in the Van Allen radiation belts and overflow along the magnetic field lines of Earth toward the poles. The northern and southern lights are the result. Without the Van Allen radiation belts, the solar wind would strip the atmosphere from our planet and the particles would impact the surface directly, killing most life on Earth.

Geophysicists argue that the answer to both of these important questions is probably "no." When Earth's field goes through a reversal, other magnetic field elements that are independent of Earth's main field support the radiation belts. This may be a lower level of support, but it should be sufficient to protect life. Other factors appear to have been more likely causes of mass extinctions.

Another recognizable, globally synchronous effect is related to erosion caused by sea-level changes. Today, large quantities of water are tied up in ice on Antarctica and on Greenland. Slight changes in global temperature can melt this ice or tie up more water on the continents through snow accumulation, resulting in global sea-level changes. Changing shorelines, due to sea-level changes produced by glaciation and/or continental uplift, cause onshore/offshore migration of beaches and continental shelf erosion that can be recognized in the rock record. Systematic sequences representing these changes have been globally correlated.

ABSOLUTE DATING TECHNIQUES

Most rocks were formed during discrete events in time, but they do not represent continuous geologic processes that are directly datable. However, over the last 40 years isotopic dating techniques have made it possible to obtain ages in years for many kinds of rocks. In conjunction with relative dating techniques, much of the geologic record is now datable. Many rock types contain radioactive isotopes that have spontaneously decayed to other isotopes since their origin. This change results in a continuing conversion of a small amount of the original radioactive isotope, the parent, to a stable daughter isotope by emission or capture of particles (protons, neutrons, and electrons) from or by the nucleus of the parent. Isotopic dating is based on measuring the parent/daughter ratio of isotopes whose decay rates are known and requires very sophisticated instruments, primarily mass spectrometers, to make the measurements. The concept of a half-life of a radioactive isotope is important in dating and represents the time it takes for half of the parent isotope to decay to an amount equivalent to a daughter isotope (Figure 3.11). The half-life is based on the rate of decay of the parent isotope, determined from a physical constant called the decay constant, and is different for each radioactive isotope. If the half-life of an isotope is very long, such as with the isotopes of uranium [U], then it is possible to date rocks that are very old, rocks that may have been formed when Earth was very young. However, if the half-life is very short, such as in ^{14}C, then almost all of the parent isotope will have disappeared within 40,000 years of the formation of the radioactive carbon. Certain assumptions are also important in dating. The method usually requires a closed system, that is, no parent material or daughter product has left or been added to the system. For that reason, the freshest possible rocks are chosen for dating, and these are carefully analyzed to detect, if possible, whether the parent/daughter isotopes have been added, lost, or altered. Only certain minerals, such as zircons, are analyzed because it is known that these minerals contain radioactive isotopes and usually provide a closed system for dating.

Isotopic dating techniques are used for a number of different purposes. For example, minerals within very old, usually igneous rocks can be reliably dated using uranium-lead, thorium-lead, samarium-neodynium, rubidium-strontium or potassium-argon dating methods, and provide estimates of the age of Earth. This is possible because these transformations have long half-lives and the parent isotopes decay very slowly, allowing dating of very old rocks. However, ^{14}C is good for dating only the very recent geologic past because the decay rate of ^{14}C, the parent, is very fast, with a half-life of less than 6,000 years. ^{14}C is produced in the upper atmosphere from ^{14}N by cosmic radiation from the Sun (Figure 3.12). It is then taken into plants as CO_2 during pho-

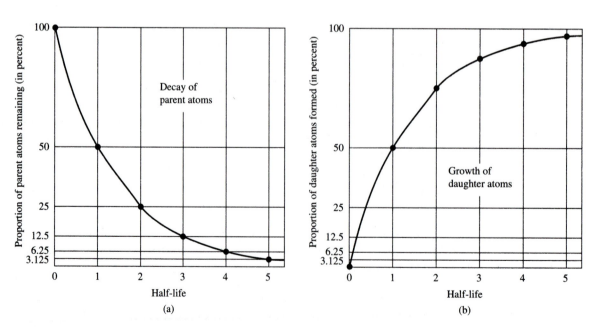

FIGURE 3.11   Buildup of a daughter product (b) after the decay of a radiogenic parent material (a).

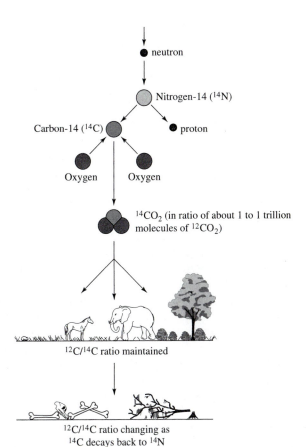

FIGURE 3.12
Production and decay of natural ^{14}C.

tosynthesis, and into animals when the plants are consumed. While the organism is alive, this ^{14}C is constantly being replaced, but after death, the ^{14}C begins to decay to ^{14}N. The radioactive disintegration rate of ^{14}C can be counted with a detector to give an estimate of the ^{14}C present, or the $^{14}C/^{12}C$ ratio can be used to give an estimate of the length of time since death of the organism. To check the accuracy of carbon 14 dating, tree rings from bristlecone pines (Figure 3.1) have been used as a check on the ages produced. There are a number of assumptions in the method, including the assumption that ^{14}C production has been more or less constant through time. In some cases these assumptions may not be valid. For example, it is assumed that the atmosphere has had the same ^{14}C concentration in the past as now, but we know that slight variations in the production of ^{14}C have occurred. Thus a number of corrections are necessary to reduce the errors in this dating method.

Table 3.1 lists common dating techniques employed in geological studies. Dating techniques are continually being tested and compared, and estimates of half-lives and decay constants are being reevaluated as better instrumentation becomes available. However, at least one important result has emerged from all the testing. When the results of different dating techniques are compared, the measurements on different rocks, from different continents, measured in different laboratories, using different isotopic dating methods yield essentially the same age for Earth. This is true even though a wide range of decay constants are used on rocks that were formed as the result of a broad range of mechanisms. These results clearly indicate that isotopic dating works!

TABLE 3.1   DATING TECHNIQUES

	Useful Age Ranges for Earth Materials
A. Stratigraphic Chronology	Relative age where sediments exist
B. Rate Processes	
1. Radiogenic isotopes	
a. ^{87}Rb-^{86}Sr	7 Ma–4,600 Ma
b. ^{14}C-^{12}C	0–100 K using AMS dates
c. ^{40}K-^{39}Ar	100 k–4,600 Ma
d. $^{235/238}$U-$^{207/206}$Pb	10–4,600 Ma
e. ^{232}Th-^{206}Pb	10–4,600 Ma
f. ^{147}Sm-^{143}Nd	10–4,600 Ma
2. Decay rate and other dating methods	
a. Fission track	70 K–1,000 Ma
b. Thermoluminescence	3 K–1,000 Ma
c. Electron spin resonance	3 K–300 Ma
3. Geologic	
a. Sedimentation	Relative age where sediments exist
4. Chemical	
a. Pedogenic soils	0–2 Ma
b. Weathering	0–65 Ma
c. Obsidian hydration	50–250 K
5. Biological	
a. Racemization of amino acids	500–10 Ma
b. Dendrochronology	0–10 K
C. Evolutionary processes	
1. Paleontology	10 K–600 Ma
2. Palynology	0–500 K
D. Phenomenological dating techniques	
1. Paleomagnetism	0–600; older in selected areas
2. Ash layer fingerprinting	0–70 Ma

Note: The abbreviation Ma means millions of years; K means thousands of years.

THE GEOLOGIC TIME SCALE

As geologists began to gather more and more relative age information about rock and fossil sequences, a global pattern emerged. This pattern, represented by the *Geologic Time Scale* (Table 3.2), has been systematically labeled and was developed over hundreds of years, completely independent of isotopic dates. Most of the early work was done in Europe as an answer to the most difficult question confronting geologists, "How do we divide geologic time?" Only recently have we been able to assign ages in years to the Geologic Time Scale. In the past geologists have recognized discrete, short, but distinctive rock sequences and trends in geologic time and grouped these within longer time sequences in Earth history. These sequences are recognized on a global scale. Now, with the addition of absolute ages, there exists a global standard that can be used in getting accurate rock age estimates. However, there are still gaps in the fossil/rock record that geologists are working to fill. This work is directed toward improving the time scale. For example, new information from rocks that are

TABLE 3.2 GEOLOGIC TIME SCALE—EARTH 4.6 ± 0.1 BILLION YEARS OLD

Eons	Eras	Periods	Epochs	Events
Phanerozoic	Cenozoic (Recent life)	Quaternary	Holocene Pleistocene ~1.6 Ma	
		Tertiary [Neogene]	Pliocene Miocene ~23 Ma	First hominid ~5 Ma
		[Paleogene]	Oligocene Eocene Paleocene ~66 Ma	First horses
	Mesozoic (middle life)	Cretaceous Jurassic Triassic	245 Ma	Last dinosaurs and ammonoids First mammals, birds, and dinosaurs
	Paleozoic (ancient life)	Permian Carboniferous [Pennsylvanian] [Mississippian] Devonian Silurian Ordovician Cambrian	570 Ma	Last trilobites First reptiles First amphibians First land plants and animals First trilobites and fish
Proterozoic	Late	Vendian [Ediacarian; Eocambrian]	800 Ma	Multicellular organisms
	Middle		1,600 Ma	Single cell organisms with a nucleus
	Early		2,500 Ma	Significant O_2 in atmosphere
Archean	Late		3,000 Ma	Aerobic photosynthesis
	Middle		3,400 Ma	
	Early		~4,000 Ma	Earliest life
Priscoan	Hadean EARTH FORMED		~4,600 Ma	Pregeologic history

Note: The abbreviation Ma means millions of years.

Proterozoic in age has resulted in efforts to define a new time period, the Vendian Period, a time when multicellular organisms were first abundant on Earth. The Vendian Period has been proposed to represent the time before the existence of trilobites (Figure 3.13), whose first appearance identifies the beginning of the Phanerozoic Eon, corresponding to the base of the *Cambrian Period*, the oldest period in the Paleozoic Era. The addition of the Vendian Period represents a major change in our understanding of rock sequences over the last century, because in the middle 1800s, all time before the Cambrian Period was known as the *Azoic*, a time of no life!

When studying geology, it is very important to know the Geologic Time Scale (Table 3.2), because an understanding of relative ages is critical to understanding the

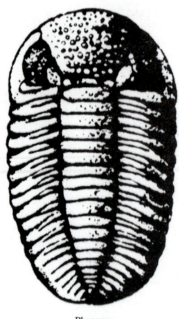

FIGURE 3.13 Trilobite, *Phacops* sp., that lived from the Silurian through the Devonian. This is one of the varieties that is often sold in stores.

Phacops
_____
(Silurian–Devonian)

sequence of geological processes responsible for forming rocks. The Geologic Time Scale is the framework to which all geologic history is tied. While the Geologic Time Scale was developed without any knowledge of absolute ages, we have now dated the critical time boundaries.

REFERENCES

COHEE, G. V., GLAESSNER, M. F., HEDBERG, H. D., eds. 1978. *Contributions to the Geologic Time Scale*, p. 388. Tulsa, Oklahoma: The American Association of Petroleum Geologists.

HARLAND, W. B., ARMSTRONG, R. L., COX, A. V., CRAIG, L. E., SMITH, A. G., and SMITH, D. G. 1989. *A geologic time scale,* p. 263. Cambridge, England: Cambridge University Press.

ROBISON, R. A., and TEICHERT, C., eds. 1979. *Treatise on invertebrate paleontology.* Part A. *Introduction*, p. 569. Boulder, Colorado: The Geological Society of America, and Lawrence, Kansas: The University of Kansas.

4

Earthquakes and Plate Tectonics

EARTHQUAKES

Earthquakes are the sudden release of energy that occurs in response to rupture of Earth along a fault. These Earth ruptures produce vibrations or elastic disturbances called *seismic waves* that originate at the point of release of energy within Earth, called the *focus* (Figure 4.1), and propagate throughout Earth. Most foci are shallow, usually less than 15 km, but earthquakes can occur to depths of nearly 700 km (deep focus earthquakes). The point on the surface above the focus is known as the *epicenter*. Recordings of seismic waves, called seismograms (Figure 4.2), are obtained using instruments called seismometers, the detector component of seismographs (Figure 4.3). We can determine the focus and epicenter of an earthquake by triangulation of seismographs

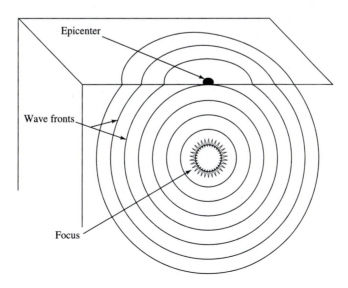

FIGURE 4.1 Diagram illustrating the location of earthquake epicenter and focus.

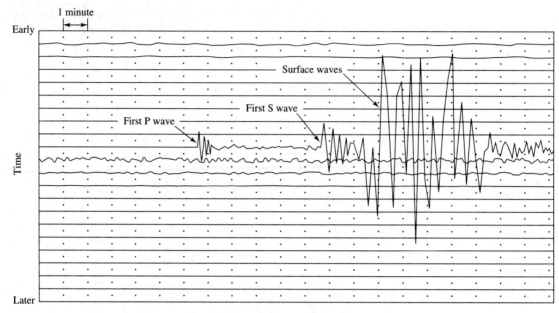

FIGURE 4.2 Seismogram, showing P, S, and surface wave first arrivals.

situated at different locations around Earth (Figure 4.4). Records of wave arrival times at seismographs tell how far they have traveled from the earthquake epicenter, but not the location of the epicenter. The location could lie anywhere along a circle with a radius from the seismometer determined by the seismograph travel time. Three records are needed to locate the epicenter, determined by the intersection of the three circles (see Figure 4.4).

Seismic waves can travel through the solid Earth (*body waves*) or along surfaces of beds (*surface waves*), as well as the surface of Earth. There are two types of body

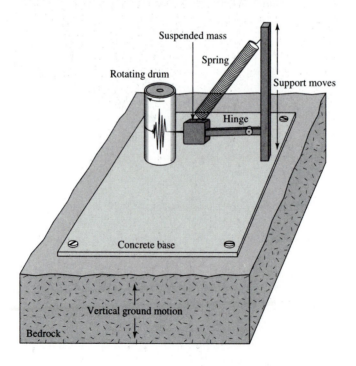

FIGURE 4.3
Seismograph used to record earthquakes.

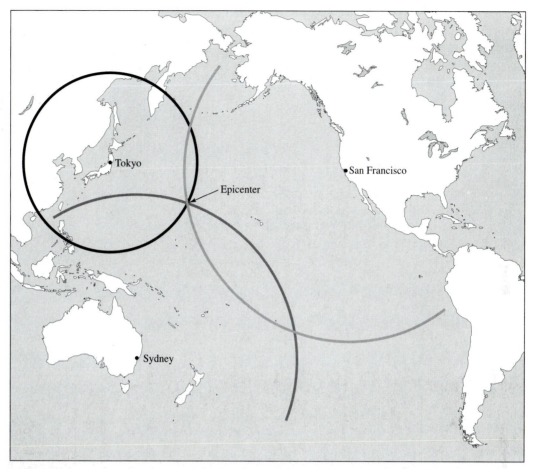

FIGURE 4.4 Location diagram, illustrating how epicenters can be identified. Seismographs at three localities can identify how far the earthquake waves have traveled, but not from which direction. By looking for intersecting circles, the epicenter can be determined.

waves. The first travels as a push-pull wave moving through Earth and is called a *P wave*, the P standing for primus or primary (Figure 4.5). These waves move by alternating between compression and dilation, with a push-pull type motion. The second, *shear* or *S waves* (secundus or secondary), travel by up and down, or side-to-side shearing motion through Earth. The properties of S waves differ from those of P waves; S waves travel slower and cannot pass through liquids. Differences in properties between these two types of body waves provide evidence of the internal structure of Earth. Also useful is the fact that seismic waves are *refracted* (bent) and *reflected* at boundaries within Earth due to density differences across these boundaries. The reflected characteristics of seismic waves, generated artificially, are the primary means used in oil exploration (see Figure 3.5).

Other types of seismic waves travel only along surface interfaces, such as the boundary between sedimentary beds or along the surface of Earth. These surface waves produce surface oscillations of Earth and can be very destructive. One type of surface wave is a *Rayleigh wave (R wave)*, a result of spherical ground-particle motion—much like ocean wave oscillations, which oscillate in the direction of motion of the wave. A second type of surface wave is a *Love wave (L wave)*, with ground motion at right angles to the direction of wave travel. Surface waves cause most of the damage produced

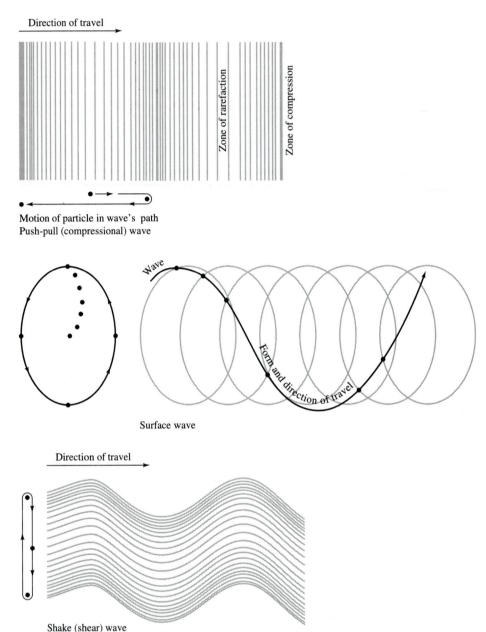

FIGURE 4.5
Earthquake waves, showing particle motion of the various waves.

during earthquakes, including cracking, breakage, and the resulting fires caused by gas main breaks.

Because of the potential for severe damage and loss of life, much of the funding in earthquake research today is directed toward predicting earthquakes. Along these lines maps of potential earthquake damage have been constructed so that communities will be aware of their seismic risk (Figure 4.6).

On shore, ground motion often causes destructive mud- and landslides, while at sea such vertical ground motion can produce *tsunamis*, seismic sea waves, often incorrectly called tidal waves. These very dangerous waves can move at speeds up to 470 km/hour and have caused a great deal of damage to communities on the Pacific Ocean shoreline.

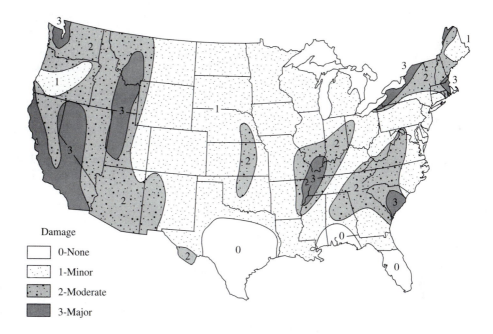

FIGURE 4.6 Seismic risk map for the United States. Least dangerous areas are 0, most dangerous are 3.

Damage

- [ ] 0-None
- 1-Minor
- 2-Moderate
- 3-Major

When earthquakes occur there is movement along a fault that obviously disturbs Earth, but this motion is poorly understood and difficult to study. Ideally, because there can be so much damage and loss of life, we would like to be able to predict when and where an earthquake will occur. But to do this it is necessary to understand the mechanism within Earth that controls fault movement and actually produces an earthquake. It has been argued that slip occurs along a fault after stress builds up to high levels, forcing the two sides of the fault to move, thus overcoming the frictional forces holding them in place. Once the stress is large enough, movement occurs along the whole fault zone at the same time. If this is true then we can estimate when stresses are high enough to predict an earthquake. However, earthquakes may result from motion in small segments along a fault, like a mole crawling very fast through Earth, the ground opening before it and closing behind it. This mechanism has recently been called *seismic fling* and the motion is technically being called a displacement pulse. With such a mechanism, very large stresses are not necessary to cause earthquakes, but rather subtle changes, like increased moisture in the subsurface, might trigger a displacement pulse. If this mechanism is actually responsible for many or all earthquakes, it may also produce greater damage to buildings than otherwise expected. Furthermore, as recent earthquakes in southern California and Kobe, Japan, have now shown, we don't adequately understand earthquake mechanisms. This makes earthquake prediction much harder.

Earthquake strength has been quantified using a number of methods. In earlier days visual phenomena were used to estimate damage, and based on the magnitude of destruction these observations were assigned numbers that then could be related to earthquake strength. A modified version of the resulting *Mercalli intensity scale* is given in Table 4.1. In 1935, *Prof. Carl F. Richter* of Cal Tech employed a more quantitative approach based on measurements taken from seismograms. Known as the Richter magnitude scale, Richter calculated M from the formula

$$M = \log_{10} A,$$

where A is the maximum seismogram trace amplitude, 100 km from the epicenter. For each magnitude change the trace amplitude is 10 times greater, but the energy released is more than 30 times greater. Modern calculations are more complex than the simple approach used by Richter, and the Japanese have their own slightly different version (see Table 4.1).

TABLE 4.1 CLASSIFICATION OF EARTHQUAKE MAGNITUDE

Modified Mercalli (MMI)		Acceleration (gals) (cm/sec^2)	Japanese (JMA)	M
I	detected only by instruments	0–0.8	0	2
II–III	felt indoors	0.8–2.5	1	3
IV	felt by most	2.5–8	2	4
V	slight damage	8.0–25	3	4.5
VI	people frightened run outdoors	25–50	4	5
VII–VIII	moderate damage	80–250	5	6
IX–X	major damage	250–400	6	7
XI–XII	severe damage	400–?	7	8+

EARTH'S INTERIOR

Early in our studies of Earth, scientists were able to determine that the interior composition of the planet must be very different from its surface materials. For example, they knew from size and rotational characteristics of Earth that the average density was approximately 5.5 gm/cm^3, much heavier than expected from measurement of surface rock densities. Now, due primarily to earthquake information, we have a good understanding of Earth's interior structure. It is divided into four major layers, crust, mantle, and outer and inner cores (Figure 4.7). The *crust*, surface and near-surface rocks with a density from 2.6 to 3.1 gm/cm^3, ranges in thickness and composition from 5–10 km in basaltic ocean basins up to 70 km or so below some granitic continental areas. At its base lies the Mohorovicic (*Moho*) Discontinuity, a distinct boundary identified by an abrupt increase in seismic wave velocity, which separates the oceanic and continental crust from the mantle below.

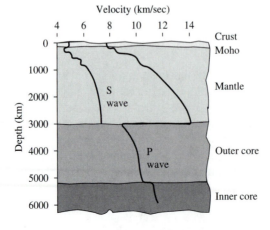

FIGURE 4.7 Earth cross section and velocity of P and S waves through Earth. Note that S waves do not pass through the outer core. Decrease in velocity below 100 km represents the low velocity layer, the asthenosphere.

Below the crust, the *mantle*, with densities ranging from 3.3 to 5.5 gm/cm^3, extends to a depth of approximately 2,900 km below Earth's surface. The upper part of Earth, from the surface to a depth of 100 km, including continental and oceanic crustal materials, as well as the upper part of the mantle, makes up the brittle exterior portion of Earth called the *lithosphere*. It is broken into regions called plates, which are associated with areas of increased earthquake activity. These plates will be discussed later in this chapter. A second layer, the *asthenosphere*, is totally within the mantle, and lies at the base of the lithosphere from a depth 100 to, in places, 650 km below the surface. The asthenosphere is believed to be a plastic zone on which the lithospheric plates move. Seismic speed slows through the asthenosphere, due in part to high temperatures resulting from convective upwelling, and these data are interpreted to indicate partial melting. The rest of the mantle, below 650 km, is called the *mesosphere*.

Below the mantle is the *outer (liquid) core*, a convecting (circulating), very hot iron-rich fluid with a density of 10 gm/cm^3. The determination that the outer core is a liquid was based on interpretation of seismic data indicating that S waves do not pass through the outer core (Figure 4.7). Studies of Earth's magnetic field indicate that it originates in the outer core, and the interaction between the convecting, outer-core fluid and the solid, "lumpy" base of the mantle is responsible for small, slight, long-term magnetic field changes (not reversals) seen at Earth's surface. These are responsible for the magnetic declination compass corrections identified on most topographic maps. Inside the outer core is the *inner (solid) core*. Unlike the outer core it is solid due to the extremely high pressures at the center of Earth, and has a density of approximately 11 gm/cm^3.

PLATE TECTONICS

A drifting continent hypothesis was first proposed in 1596 by the Dutch cartographer Abraham Ortelius. He stated in his book, *Thesaurus Geographicus*, that earthquakes and floods forced the early continents to separate. Later, based on geological evidence from rocks formed during the late Paleozoic and early mid-Mesozoic eras (Permo-Triassic; see Geologic Time Scale in Chapter 3), an early 20th century meteorologist and geophysicist, *Alfred Wegener*, argued that at one time all the continents were part of a *supercontinent* that he called *Pangaea*, meaning all lands (see Figure 2.1). Pangaea, he suggested, could be reconstructed by fitting together continental outlines, especially South America and Africa. Then, in the Mesozoic (approximately 200 Ma), this supercontinent broke into two smaller parts, *Laurasia*, named by the South African geologist Alexander Du Toit to describe a combination of North America, Europe, and Asia, and *Gondwana*, a term first used by the Austrian geologist Edward Suess and later applied by Du Toit.

A large amount of varied geological evidence supporting Wegener's hypothesis was available at that time, and more data continues to be acquired. Extensive fossil evidence, including many similar animals and plants, indicates that these strikingly similar organisms were living on what are now separated continental masses. Biologists recognize that such clear similarities do not exist unless the overall population of plants and animals is continually intermixing and interbreeding. For this to have been possible, the continents on which these similar populations exist must have been connected in some way. In addition to the fossil evidence, studies of the pre–mid-Mesozoic geologic record across the presumed supercontinental boundaries show many similarities, including structural trends such as folds, faults, and even mountain ranges, sedimentary deposits, and regions of excessive volcanism. Climatic zones reflect the

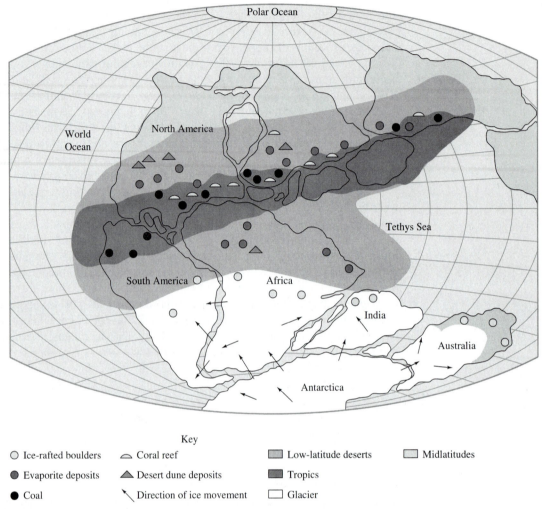

FIGURE 4.8 Similar climate indicators across continental boundaries, here represented as a Permian continental fit. These were developed when the continents were together.

pre–mid-Mesozoic conditions expected when the continents are projected back together. This evidence includes the distribution of *coal* deposits, glacial deposits, erosion indicators, dunes, and deserts (Figure 4.8).

The theory of *seafloor spreading*, the unifying explanation of global tectonic processes, was proposed in the late 1950s and early 1960s by pioneer workers such as Harry Hess and Robert Dietz. They proposed that new seafloor is created at *midocean ridges* and eventually destroyed at ocean *trenches*. This process acts like a giant conveyor belt and moves by *convection*, a concept first suggested by Arthur Holmes in 1931, carrying oceanic islands and *seamounts* (submerged mountains) along with it as it moves the continents. As new seafloor is created at midocean ridges, it acquires a magnetization in the direction of Earth's magnetic field at the time it cools. Thus, the seafloor could be characterized as a giant magnetic tape recorder that preserves changes in Earth's magnetic field (Figure 4.9). The magnetic properties of this record can then be measured from ships towing magnetometers (see discussion later in this chapter).

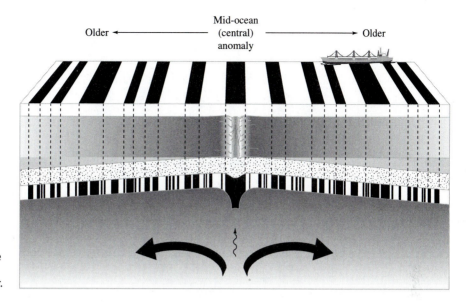

FIGURE 4.9 Diagram representing a ship towing a magnetometer. The oceanic crust is acting as a magnetic tape recorder.

Even though good geological data existed in support of spreading, it was not until modern geophysical evidence was available that most geologists began to accept the theory, today known generally as plate tectonics. The primary evidence came from seismic and magnetic studies in conjunction with absolute age determinations, through which came the development of the magnetic reversal time scale. Heat flow and gravity data also provided important supporting information. Plate tectonics differs from continental drift in that the continents move as parts of larger lithospheric plates rather than as discrete continental elements. That is, continental drift occurs as a consequence of plate tectonics.

Seismic Evidence

Earthquake data, in addition to other geological data, have provided very strong evidence in support of the motion of continents on Earth's surface. During the late 1950s and early 1960s a new network of seismic stations was built to monitor Soviet nuclear tests. Data from this *World Wide Standardized Seismograph Network (WWSSN)* became available in the 1960s, and have provided striking support for seafloor spreading. The first important information came from recognition of *deep focus earthquakes* occurring behind and below ocean trenches. These earthquakes define the *subduction zones*, where ocean crust is being reabsorbed back into the mantle (Figure 4.10). Also known as *Benioff zones*, these were named after Hugo Benioff, the Cal Tech geophysicist who first discovered Earthquake characteristics associated with subduction zones. Destruction of ocean crust at about the same rate as it is produced at midocean ridges is necessary because Earth is not growing rapidly. That Earth is approximately stable in size is supported by independent paleomagnetic evidence indicating that Earth has remained essentially the same size for at least the last few billion years.

In 1965, Professor *J. Tuzo Wilson* proposed a new type of strike/slip fault, called a *transform fault*, which allowed offset between midocean ridge segments (Figure 4.11). (Faults will be discussed further in Chapter 7.) Along midocean ridges, where new ocean crust is being created, there are areas where magma is being produced faster than at other areas. The result is that one ridge segment may be spreading faster than another. Transform faults allow this differential spreading to occur. Furthermore, ridge spreading and

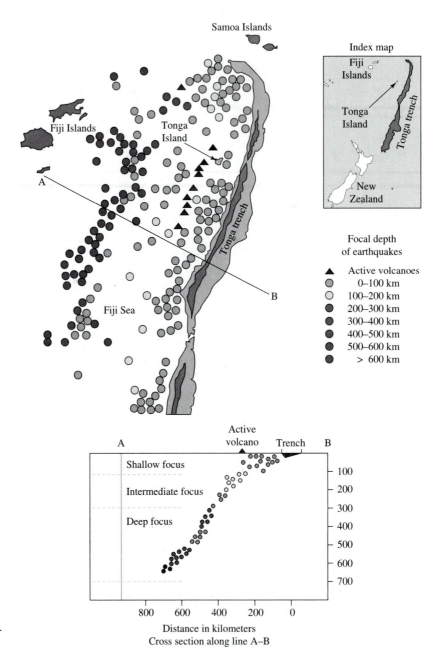

FIGURE 4.10
Subduction into the Tonga trench in the Pacific Ocean north of New Zealand. Earthquakes are shallow near the trench, but steepen as the oceanic slab penetrates into the mantle.

subduction at trenches cannot by themselves allow movement of a rigid plate over the curved surface of Earth. A special type of strike/slip fault is necessary to account for this motion. Wilson argued that earthquakes could only occur along these limited ocean floor fault segments, even though the fractures in Earth's crust appeared to extend much farther beyond the midocean ridges. In 1967 the WWSSN provided proof, confirming Wilson's hypothesis (Figure 4.12). Then, in 1969, Barazangi and Dorman published the first WWSSN global earthquake data set (Figure 4.13). (Actually the diagram was published in 1968 by other scientists, who borrowed the data with permission of Barazangi and Dorman, creating great excitement. When the 1969 paper was

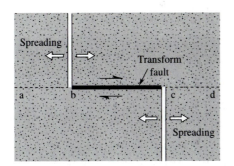

FIGURE 4.11
Illustration of fault motion in a transform fault, proposed by Wilson in 1965.

finally published, it was old news!) The striking thing about the worldwide data set was that earthquakes were found to occur mainly along very narrow zones that represent breaks in Earth's lithosphere. Regions outlined by these tectonically active earthquake zones were termed plates (Figure 4.14), by *Jason Morgan* in 1972, thus the term plate tectonics came to characterize all the related processes and effects. It is now clear that most earthquake activity occurs at predictable locations, along transform faults, midocean ridges, and in subduction zones on the opposite side of trenches, away from midocean ridges.

One of the important results of ocean crust subduction is the production of magmas. Changes in temperature and pressure associated with the slab subducting into the asthenosphere produce melting. Rising magmas result in volcanic eruptions at Earth's surface. This produces large, active volcanoes behind trenches (away from midocean ridges). On land these volcanoes are scattered in linear trends aligned parallel to the subduction zone, but at sea these volcanic eruptions produce *island arcs*. The Aleutian Islands (see Figure 1.11) are an island arc chain formed from Pacific Plate

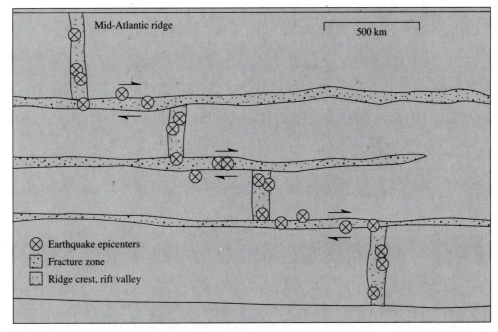

FIGURE 4.12 Successful test of Wilson's hypothesis in 1967, using WWSSN data. Arrows indicate motion direction and the X's represent identified earthquake epicenters.

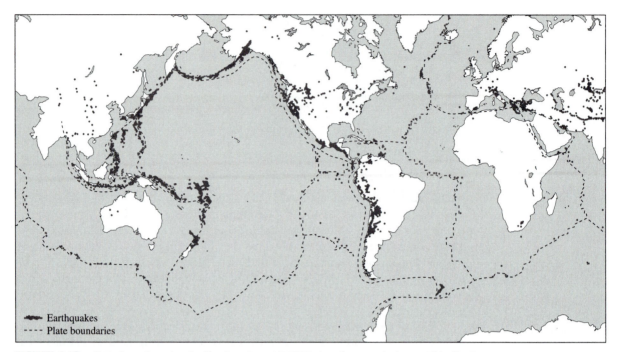

FIGURE 4.13 Global earthquake distributions from WWSSN data for a typical year. This distribution defines the plate boundaries (see Figure 4.14).

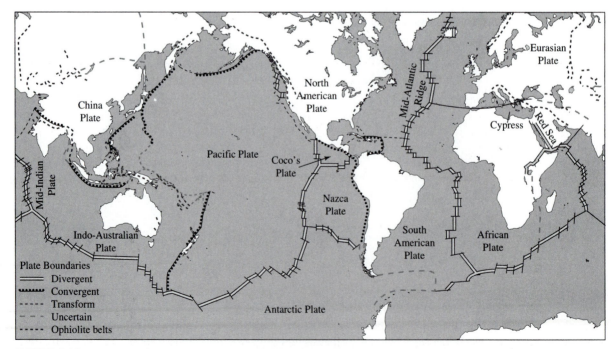

FIGURE 4.14 Plate boundaries identified from WWSSN data.

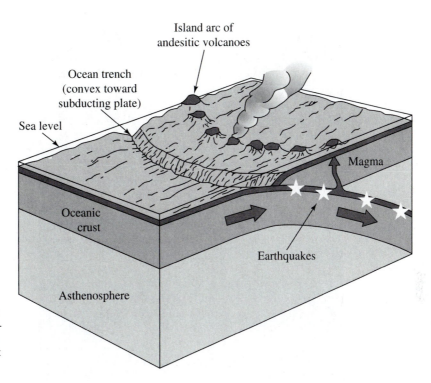

FIGURE 4.15 Andesitic volcanoes behind subduction zones, producing an island arc. Stars represent earthquakes.

subduction under Alaska (Figure 4.15). Katmai National Park (see Park Insert) is part of the continuation of the Aleutian Island arc chain onto continental crust. In turn, volcanic eruptions, such as those occurring throughout the Cascade Mountains, including Mt. St. Helens in Washington, are due to subduction of oceanic crust under the North American continent.

 KATMAI NATIONAL PARK AND PRESERVE

Katmai National Park and Preserve in Alaska (see Figures 1.11 and P5.15) was established in 1918 as a national monument and upgraded to national park status in 1980 to preserve more of the surrounding area (Figure P4.1). It is a U.S. deep-drilling site for geothermal testing. The park is only accessible by boat, plane, or helicopter.

Geology. A striking feature of the park is Katmai caldera in Mt. Katmai (Figure P4.2), formed in 1912 as a result of magma draining from beneath the volcano due to the eruption of nearby Novarupta volcano. It later filled with water, forming a lake. Still later, andesitic magmas (see Chapter 5) formed a *resurgent dome* that is now an island in the lake. Novarupta volcano erupted explosively in 1912, producing a *nuée ardente*, a cloud of incandescent ash that roared down the mountain slope and buried the valley below to a depth of over 200 m. These eruptions result from partial melting of the subducting Pacific Plate below the Aleutian trench. Currently there are 15 active volcanoes in the park that have produced 10 major eruptions during the last 7,000 years.

The rocks exposed in the park are mainly Tertiary andesitic volcanics (see Chapter 5), with some Jurassic rocks also well exposed. Severe stream erosion has cut deep canyons, developing a badlands-type topography (discussed in Chapter 9). Gravitational processes, including mass wasting (discussed in Chapter 9) and glaciers (discussed in Chapter 13), have had a major effect on the topographic relief and landscape in the park.

FIGURE P4.1 Katmai National Park, Alaska. (Photo courtesy of the National Park Service.)

FIGURE P4.2 Katmai Caldera, a resurgent dome in the flooded caldera, in Katmai National Park, Alaska. (Photo courtesy of the National Park Service.)

Besides occurring at plate boundaries, earthquakes also occur within continental areas. In the North American continent many of these earthquakes can be attributed to plate movements. Today the North American continent is moving westward, and it has overridden other oceanic plates, part of the Pacific Plate, and the seismic zones associated with these plates. The large number of earthquakes that occur in the western part of the United States are mainly associated with these overrun seismic zones. The result is a series of earthquakes associated with the San Andreas transform fault complex in California, and with extension in the Basin and Range province and elsewhere in the western United States.

Magnetic Evidence

During World War II, portable magnetometers were developed to detect German U-boats. While making measurements at sea, scientists were successful in finding submarines, but they also observed that the oceanic crust exhibited distinctive magnetic patterns, called anomalies. At that time it was not clear what these *seafloor magnetic anomaly patterns* meant. Since World War II, and into the early 1960s, geophysicists such as Cox, Doell, and Grommé, and geochemists such as Dalrymple, Hay, and McDougall were measuring the magnetic properties of rocks on land, particularly stacked volcanic piles such as the Hawaiian Islands. Isotopic dating of polarity events led to development of Earth's magnetic field reversal sequence (Figure 4.16). In 1963 Vine and Matthews brought these earlier results together in a paper, stating that the seafloor

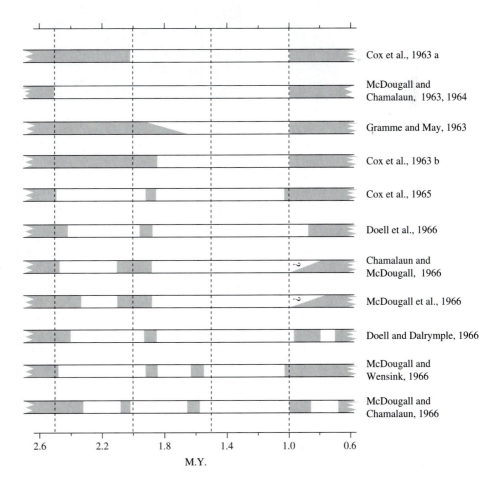

FIGURE 4.16
Development of the polarity time scale, from 1963 to 1966, for the last 2.5 million years. Shaded areas indicate zones of normal polarity and white indicates reversed polarity. (From Norman D. Watkins, 1972. "Review of the Development of the Geomagnetic Polarity Time Scale and the Discussion of its Definition," pp. 551–74. Geological Society of America Bulletin. Vol. 83. Used by permission.)

FIGURE 4.17 Mid-Atlantic Ridge where it is exposed at the surface in Iceland. (Photo by N. D. Watkins.)

magnetic anomaly patterns represented Earth's magnetic field history and that this history, in turn, could be used to date the ocean floor. Magmas are intruded along the axis of midocean ridges (Figure 4.17 shows the Mid-Atlantic Ridge, where it is exposed in Iceland). Recent lavas reflect today's magnetic field, but older intrusions preserve reversals in polarity that existed when they were emplaced and cooled. Figure 4.18 illustrates the patterns developed as a result of this process. The dashed lines in Figure 4.18 illustrate the signal that would be picked up by a towed magnetometer. Such signals identified along the Mid-Atlantic Ridge south of Iceland (Figure 4.19) were interpreted by Vine and Matthews to represent a pattern of polarity reversals preserved in the oceanic crust (Figure 4.20). Furthermore, they observed that these anomalies were symmetrically centered about midocean ridges. When they first published this idea in the journal *Nature,* they were not able to make a direct comparison to the polarity time scale being developed on land, because, in 1963, that time scale was only poorly defined. However, a paper by Doell and Dalrymple in 1966 provided an additional normal polarity interval, the Jaramillo (see Figure 4.16), that then made it possible for the first time to make direct comparisons between the seafloor magnetic anomaly patterns and the polarity time scale.

We now know that the oceanic crust acts like a magnetic tape recorder with records that can be traced back to the Jurassic Period (approximately 200 Ma). All older ocean seafloor has been destroyed by subduction, but presumably seafloor has been produced in the world's oceans for billions of years. Tests of the seafloor ages predicted

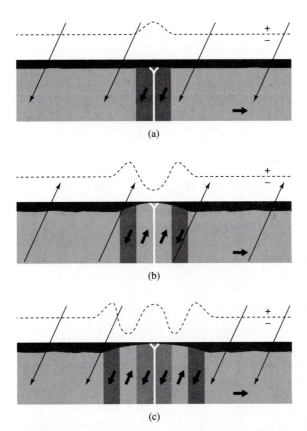

FIGURE 4.18 Diagram illustrating the acquisition of polarity reversals by intruding dikes at mid-ocean ridges. Arrows represent Earth magnetic field directions. Alternating light and dark bands represent zones of alternating normal and reversed polarity. (a) A dike intrudes into the crust and acquires a magnetic direction in Earth's magnetic field direction, producing a positive response (magnetic anomaly) identified with a magnetometer (dashed line). (b) Earth's field reverses and new dikes intrude along the ridge axis acquiring the new magnetic direction. This splits magnetic record seen in (a) with a reversed response. (c) Earth's field again reverses, producing a new positive magnetic response, and so forth. These magnetic variations are symmetrical about the midocean ridge axis.

by the spreading hypothesis have been performed by *deep-sea drilling*, where seafloor has been cored and magnetic lineations dated. The results confirm the age assignments and prove that a record of the polarity time scale is preserved by the ocean's crust. *The Ocean Drilling Program (ODP)* is continuing these tests.

Continental reconstructions and plate movements can be established when seafloor segments of the same age are removed from the ocean basins and the continental positions are moved back to reflect the removed tracks. Eventually the continental configuration shown in Figure 2.1 emerges, where the continents of South America and Africa come together (Figure 4.21). The magnetic anomaly pattern has now been extended to all the world's ocean basins (Figure 4.22), and a relatively simple pattern of spreading has emerged that can be traced away from major ridge axes. Only a few areas of seafloor have been found where the spreading pattern is very complex and still not fully understood.

Examination of Figure 4.22 shows that the polarity bands are wider in some ocean basins than in others, the wider bands indicating faster spreading. For example, the *East Pacific Rise*, the main spreading center in the Pacific basin, is producing more new

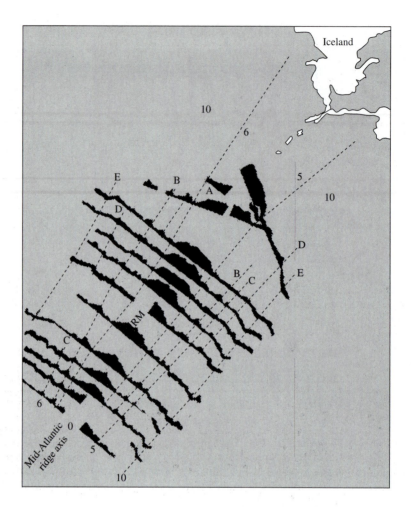

FIGURE 4.19 Marine magnetic anomaly data recovered south of Iceland, along the Reykjanes Ridge, part of the Mid-Atlantic Ridge. Vine and Matthews first published this pattern in 1963.

ocean crust, therefore the Pacific Ocean basin is spreading faster than the *Mid-Atlantic Ridge*, where new ocean floor is produced in the Atlantic. The rate of spreading can be calculated for any spreading center by comparing the distance between each center at which polarity boundaries were encountered. This is illustrated in Figure 4.23, and from these data, actual rates of spreading can be calculated. For example, the East Pacific Rise is spreading at a rate of almost 5 cm/year, while the Reykjanes Ridge, the part of the Mid-Atlantic Ridge just south of Iceland, is spreading only at a rate of approximately 1 cm/year.

Even though all ocean crust older than 200 Ma has been destroyed it is still possible to trace past continental tracks using the measurement of *paleomagnetism* in rocks on land. These reconstructions are based on the concept of *apparent polar wander (APW)*, which assumes that Earth's magnetic field is an axially centered dipole (magnetic field similar to that produced if a very large bar magnet is aligned along the rotational axis of Earth) that has remained essentially axially centered throughout time (see Figure 3.6). The magnetic properties of rocks, carefully collected anywhere on Earth, can be measured and used, with some exceptions, to calculate where the magnetic north pole was located when the rock was magnetized. The present latitude and longitude of the sample are noted. If the continent has moved, the magnetic north pole indicated from the sample will be different from its current location. The new direction of magnetic north will be recorded in many rocks that form during the history of continental

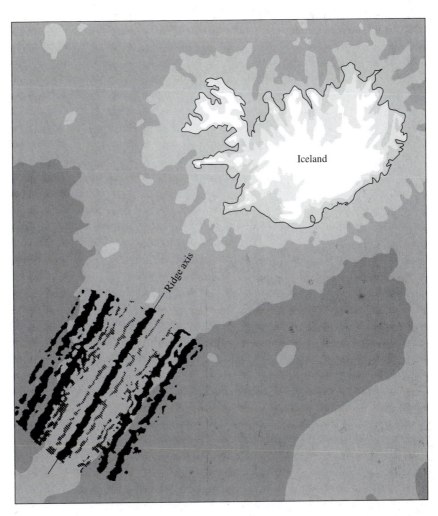

FIGURE 4.20 Magnetic anomaly stripes drawn from the data in Figure 4.19, representing increasing ages of oceanic crust away from the ridge axis. Dark bands are normal polarity.

movement. As we analyze old rocks, we find that poles calculated from their magnetic directions show an apparent shift away from today's known magnetic pole position. This shift represents continental movement. Two tests are available to evaluate these data. First, past continental positions from APW data agree with continental positions predicted by the seafloor magnetic anomaly record for rocks less than 200 Ma. And second, rocks from continents that were connected in the past give the same location for the magnetic poles of Earth. For example, Figure 4.24 shows how paleomagnetic poles are used to indicate converging continental paths for Gondwana and *Euramerica* (Europe and North America) during the lower Paleozoic (S), remaining coincident during the mid-upper Paleozoic (P) while these land masses were together, and then diverging as individual continents moved away from each other. APW data can be used to locate continental blocks even back into the Archean Eon.

Other supporting geophysical data for seafloor spreading includes *gravity* and *heat flow* measurements. Gravity data yield information concerning densities of rocks. Density contrasts determined at plate boundaries are consistent with plate models. For example, low densities are observed at heated midocean ridges, where heating from rising magma should reduce rock densities. These patterns are also consistent with heat flow models, where flow of heat out of Earth is measured. Heat flow models precisely predict ocean-floor depth, based on spreading and crustal cooling rates. Depth should increase as density increases. Combined with gravitational settling (sinking) as

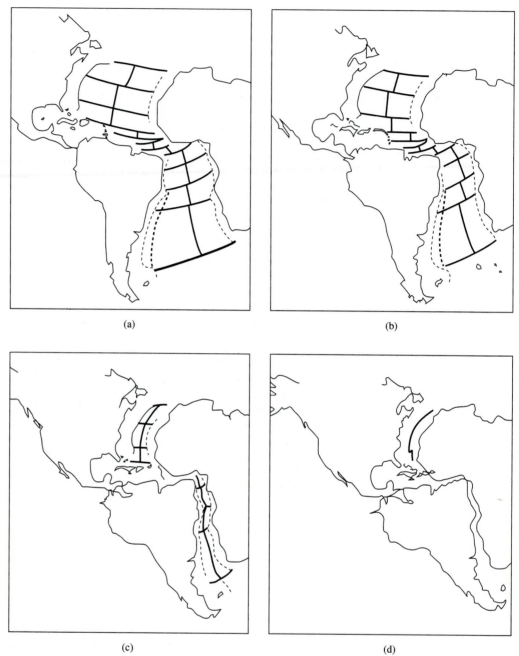

(a)

(b)

(c)

(d)

FIGURE 4.21   Atlantic reconstruction by subtracting ocean floor of equal age, using magnetic anomaly patterns to date the ocean floor. Note that after subtracting equal age segments of the ocean floor, the resulting reconstruction is that of the fit by Bullard and others, 1965 (see Figure 2.1).

the ocean crust cools, this results in depression of cooling ocean crust as it moves away from midocean ridges.

Summary of Major Plate Features

Summarizing plate tectonics, we now know that ocean crust is created at midocean ridges and destroyed at ocean trenches by subduction back into Earth (Figure 4.25).

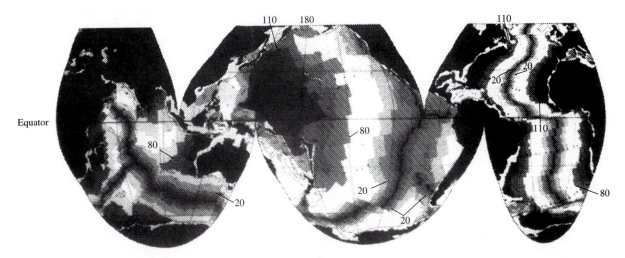

110   180   110

20
20
20

Equator

80

80

80

20

20

20

80

FIGURE 4.22   Global marine magnetic anomaly patterns like this were first published in 1981. Numbers increase away from midocean ridges and represent ages of ocean crust in millions of years.

Oceanic rocks are also lost to the ocean basins when they are pushed up into mountains by plate collisions. Such oceanic materials can be seen exposed in many areas in the Olympic Mountains within Olympic National Park, Washington (Chapter 14). Transform faults connect ridge and trench segments allowing Earth crust adjustments in response to spreading and subduction. Plate motions are directly affecting the western part of North America. For example, in 1989 Tanya Atwater showed that North America has overridden the *Farallon Plate*, now completely gone, and part of the eastern

FIGURE 4.23   Diagram illustrating the spreading rates in different ocean basins. A ship sailing from a midocean ridge and identifying magnetic anomaly boundaries will encounter those boundaries sooner if the spreading rates are slower in that ocean basin. For example, at a distance of a bit over 40 km from the East Pacific Rise the ship will find 1-million-year-old ocean crust, while in the Atlantic the ship will travel less than 10 km to find crust of the same age. Pattern at bottom represents the magnetic polarity log.

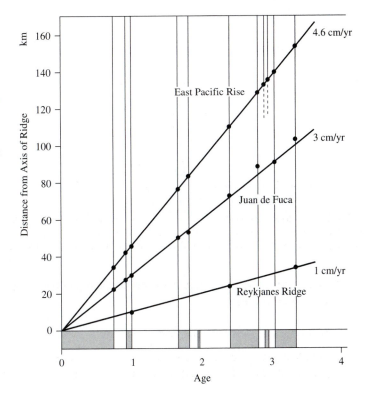

km

160

140

120

100

80

60

40

20

0

Distance from Axis of Ridge

4.6 cm/yr

East Pacific Rise

3 cm/yr

Juan de Fuca

1 cm/yr

Reykjanes Ridge

0   1   2   3   4

Age

FIGURE 4.24
Distribution of the Pangaea continents, and paleomagnetic data for these continents. The magnetic data indicate that the continents came together during the middle Paleozoic, stayed together through the rest of the Paleozoic (P), and then began to break up in the Mesozoic (M).

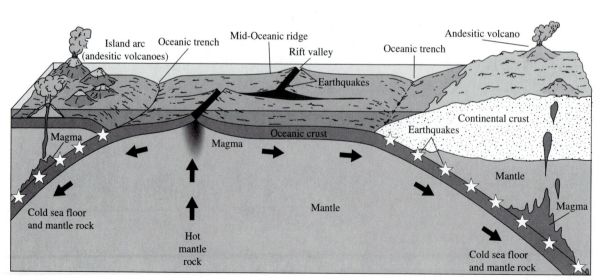

FIGURE 4.25 Summary of ridge spreading, transform faulting, and subduction processes on Earth. Earthquakes at subduction zones are represented by stars.

PINNACLES NATIONAL MONUMENT

Pinnacles National Monument (see Figure 1.14) was designated by presidential proclamation in 1908. Located in central California, it sits just to the west of the San Andreas (transform) fault, and is being progressively carried to the north as this part of the Pacific Plate moves northward along the fault. The primary features in the monument are volcanic rocks carved by erosion into *pinnacles* (Figure P4.3).

Geology.   Explosive volcanic eruptions early in the Miocene Epoch (Tertiary Period) produced a large *stratovolcano*, consisting of rhyolitic and andesitic lavas and ash flow tuffs, like those erupted from the more recently formed Mt. St. Helens. (These types of igneous rocks are further discussed in Chapter 5.) Located just to the west of the San Andreas fault and south of San Francisco, it is part of the Pacific Plate. The monument is bounded by faults on the east and west, both related to motion along the San Andreas fault zone. Identical sequences of rocks that appear to be the other half of the volcano are now offset over 300 km to the south toward Los Angeles. Erosion along fractures in the rock has produced the pinnacles.

FIGURE P4.3   The Pinnacles, in Pinnacles National Monument, California, located just to the west of the San Andreas transform fault. (Photo by author.)

Pacific Plate boundary as well. Adjustments to these complexities under western North America have caused strike/slip motion along the San Andreas transform fault within California. Pinnacles National Monument (see Park Insert) lies right on the fault.

There is still some question about the physical mechanism driving plate motion. Two primary mechanisms have been proposed; (1) convection (see Figure 4.26) in the

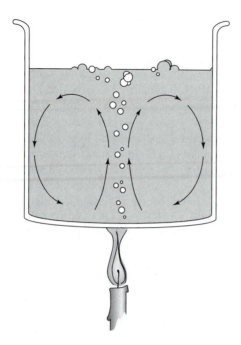

FIGURE 4.26 Simple convection illustration.

mantle, where plates are driven by frictional drag on the underside of the lithosphere and aided by *magma pulses*, pushing by hot magma emplacement; or (2) subduction/gravitational (Figure 4.27), where plates are pushed at one edge by the heated and elevated newly forming ridge axis and pulled at the other edge as the heavy, cold plates sink under lighter, hotter plates. Convection may be very complex and could involve whole mantle convection or a two-cell convection system, an upper cell in the asthenosphere, and a lower cell in the rest of the mantle below (Figure 4.28).

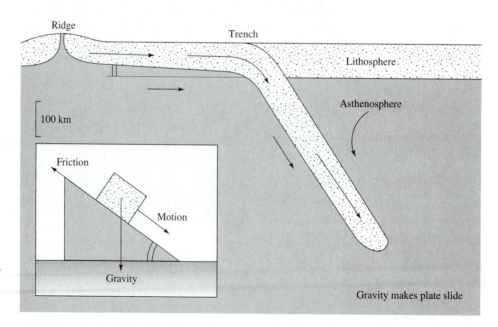

FIGURE 4.27 Gravity plate driving mechanism, analogous to a block sliding down a plate (insert).

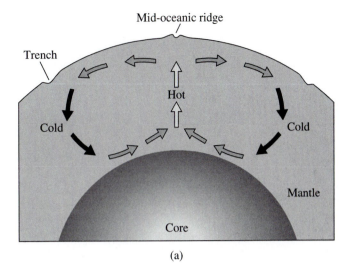

(a)

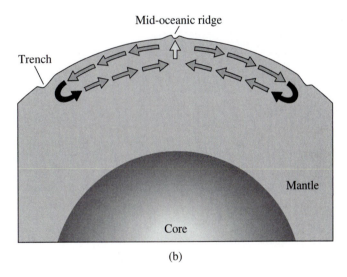

(b)

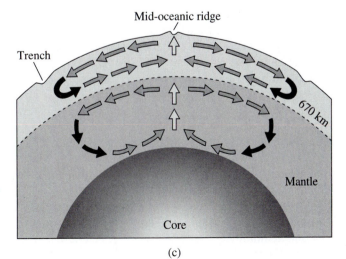

(c)

FIGURE 4.28 Three possible models for convection. (a) whole mantle convection, (b) asthenosphere convection, or (c) a model for both upper and lower mantle convection.

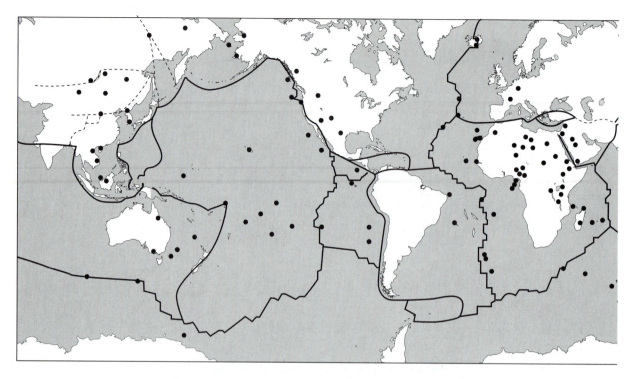

FIGURE 4.29 Global distribution of hot spots.

In some plate interiors excessive magma is generated at a fixed point deep in the mantle. These *hot spots* or *mantle plumes* (Figure 4.29) leave a magma trace as the plate moves over it (see Figure P4.4). This trace produces seamount or island chains on the plate's surface, including the Hawaiian Islands, where Hawaii Volcanoes and Haleakala National Parks are located (see Park Inserts). Hot spots under continents, as at Yellowstone National Park, mainly in Wyoming (Chapter 11), cause upward crustal warping due to heating. Cooling causes collapse resulting in breaking and tilting, and locally, chains of volcanoes develop.

In continental plate interiors, a number of mountain ranges can be found that are evidence of ancient zones of plate collision. The Appalachian Mountains in the eastern United States are a prime example, representing the collision between North America and Europe and Africa. Elsewhere, *failed rifts*, such as the Proterozoic Amargosa Aulacogen in Death Valley National Park, mainly in California (see Chapter 12), show where continents were almost torn apart.

We can now directly test to see if continents are moving by measuring the motion using lasers and satellites. Figure 4.30 gives some of these measurements in rates of motion in millimeters per year. The data are consistent with seafloor spreading rates determined in the 1960s using seafloor magnetic anomaly data.

GEODESIC RATES FROM LAGEOS DATA

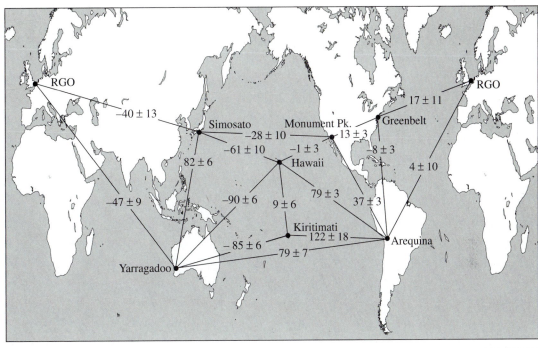

NASA/GSFC SL7.1:ANF44N 880608

FIGURE 4.30 Satellite direct measurement of plate movement rates in mm/year. This confirms rates identified from magnetic anomaly data (Figure 4.23).

 ## HAWAII VOLCANOES NATIONAL PARK

Hawaii Volcanoes National Park was established in 1916, over 40 years before Hawaii became a state. It was originally called Hawaii National Park and included Haleakala National Park on the island of Maui, but with the name change in 1961 it was limited to the "big" island, Hawaii (Figure P4.4). Because of its many unique characteristics and diverse and unique life forms, it has been designated a world heritage site and a biosphere reserve. The park includes two active volcanoes, Mauna Loa, with a caldera 3 × 1.5 miles (mi) (5 × 2.5 km) across and approximately 600 ft (200 m) deep, and Kilauea, 2.5 mi (4 km) across and approximately 450 ft (150 m) deep (Figure P4.5). The Kau desert, located on the leeward side of the island, is the result of high rainfall on the windward side caused by water-laden air moving onshore. Most moisture has been lost before the air reaches the leeward side of the island, thus the formation of the Kau desert.

Geology. Basaltic lavas (see Chapter 6) from hot spot volcanism, the result of the Pacific Plate moving over the hot spot, have formed the Hawaiian Island chain (Figure P4.4) as well as the Emperor Seamount chain further to the northwest. The lavas, some erupting as more viscous spatter cones leaving mounds of lava after they cool (Figure P4.6), have accumulated into a series of low-angle slope *shield volcanoes* (Chapter 5) that are the result of the eruption of large amounts of usually low viscosity (runny) basaltic magmas (Figure P4.7). *Lava tubes*, volcanic caves, are abundant in the park and are produced by lava draining beneath an already cooled lava exterior. While not all lava tubes run to the sea, lava tubes can be important as a major way of getting lavas to the ocean thereby building up an island (Figure P4.8).

The park is tectonically active, with earthquakes and eruptions common. Many gases are given off during these eruptions, steam (H_2O) being the most abundant (approximately 70%). CO_2 makes up about 14% of the gases and some of the remaining 16% include chlorine gas, sulfur dioxide (SO_2), and hydrogen sulfide (HS), which are very deadly. Such gases can be seen escaping from the Kilauea volcano in Figure P4.5.

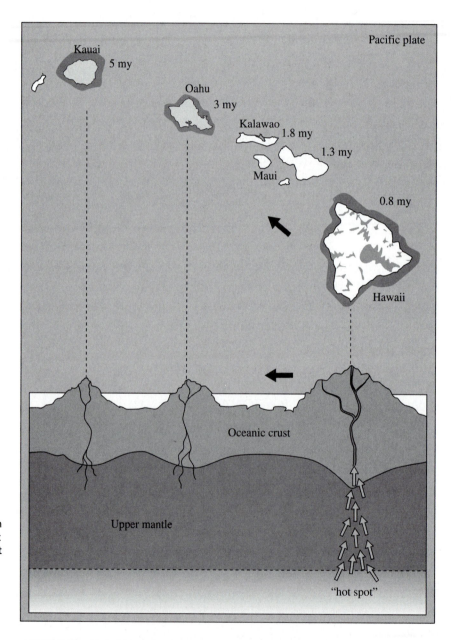

FIGURE P4.4 Hawaiian Island chain in the Pacific Plate showing movement of the islands over a Pacific Plate hot spot. Note increasing ages of the islands away from the big island of Hawaii.

FIGURE P4.5   Kilauea caldera illustrates a smaller caldera nested inside the larger caldera, whose rim is on the right in the photo, in Hawaii Volcanoes National Park, Hawaii. (Photo by author.)

FIGURE P4.6   Kilauea Iki volcano in Hawaii Volcanoes National Park, Hawaii. (Photo by author.)

FIGURE P4.7 Low viscosity rivers of magma flowing from an eruption of Kilauea. (Photo courtesy of the National Park Service.)

FIGURE P4.8 Lavas from an eruption of Kilauea that have flowed into the sea. (Photo by author.)

HALEAKALA NATIONAL PARK

Haleakala National Park, on the island of Maui in the Hawaiian Islands (Figure P4.4), was first designated as part of Hawaii National Park in 1916. It was given unique standing as a national park in 1960. An unusual yucca-type plant, the silversword (Figure P4.9), is abundant here and found only on Maui and Hawaii. Due in part to the silversword, in 1980 the park was designated a biosphere reserve.

Geology. The island of Maui is older than the island of Hawaii. Haleakala is on its eastern end and has a 10,023 ft (3,055 m) eroded volcanic caldera, 7 × 3 mi (11 × 5 km) in diameter and 1/2 mi (800 m) deep. Haleakala is no longer active and is dotted with volcanic cinder cones, scattered around the interior of the caldera. Weathering has altered the iron silicates in the lavas to soils containing secondary iron oxides that yield a range of red and yellow colors (discussed in Chapter 9). No volcanoes have erupted on the island since two lavas were erupted in 1790. Since the last major eruption erosion has significantly altered the landscape within the caldera (Figure P4.10).

FIGURE P4.9
Silverswords at Haleakala National Park, Hawaii. (Photo courtesy of M. Wilcox.)

FIGURE P4.10
Haleakala caldera in Haleakala National Park, on the island of Maui in Hawaii. (Photo by author.)

REFERENCES

BALLY, A. W., and PALMER, A. R., eds. 1989. *Decade of North American geology, DNAG, The Geology of North America*. Vol. A. *The geology of North America; An overview*, p. 619. Boulder, Colorado: The Geological Society of America.

BENNETT, R., ed. 1980. *The new America's wonderlands, our national parks*, p. 463. Washington, D. C.: National Geographic Society.

BOLT, B. A. 1988. *Earthquakes,* p. 282. New York: W. H. Freeman and Company.

BULLARD, E. C., EVERETT, J. E., and SMITH, A. G. 1965. Fit of continents around Atlantic. In *A symposium on continental drift, Royal Society of London, Philosophical Translations*, Ser. A, vol. 258, ed. P. M. S. Blackett, E. C. Bullard, and S. K. Runcorn, pp. 41–75.

CHAMALAUN, F. H., and McDOUGALL, I. 1966. Dating geomagnetic polarity epochs in Reunion. *Nature* 210:1212–14.

CHRONIC, H. 1986. *Pages of stone, geology of western national parks and monuments*. Vol. 2. *Sierra Nevada, Cascades & Pacific Coast*, p. 170. Seattle, Washington: The Mountaineers.

COX, A., DOELL, R. R., and DALRYMPLE, G. B. 1963a. Geomagnetic polarity epochs and Pleistocene geochronometry. *Nature* 198:1049–51.

COX, A., DOELL, R. R., and DALRYMPLE, G. B. 1963b. Geomagnetic polarity epoch: Sierra Nevada II. *Science* 142:382–85.

Cox, A., Doell, R. R., and Dalrymple, G. B, 1964. Reversals of the Earth's magnetic field. *Science* 144:1537–43.

Cox, A., Doell, R. R., and Dalrymple, G. B. 1965. Quaternary paleomagnetic stratigraphy. In *The Quaternary of the United States*, ed. H. E. Wright and D. G. Frey, pp. 817–30. Princeton: Princeton University Press.

Cox, A., Hopkins, D. M., and Dalrymple, G. B., 1966. Geomagnetic polarity epochs: Pribilof Islands, Alaska. *Geological Society of America Bulletin* 77:883–910.

Doell, R. R., and Dalrymple, G. B. 1966. Geomagnetic polarity epochs: A new polarity event and the age of the Brunhes-Matuyama boundary. *Science* 152:1060–61.

Doell, R. R., Dalrymple, G. B., and Cox, A. 1966. Geomagnetic polarity epochs: Sierra Nevada data no. 3. *Journal of Geophysical Research* 71:531–41.

Evernden, J. F., Savage, D. E., Curtis, G. H., and James, G. T. 1964. Potassium-argon dates and the Cenozoic mammalian chronology of North America. *American Journal of Science* 262:145–198.

Grommé, C. S., and Hay, R. L. 1963. Magnetization of basalt of bed I, Olduvai Gorge. *Nature* 200:560–61.

Harris, A. G., and Tuttle, E. 1990. *Geology of National Parks*. 4th ed., p. 652. Dubuque, Iowa: Kendall Hunt Publishing Company.

Harris, D. V., and Kiver, E. P. 1985. *The geologic story of the national parks and monuments*. 4th ed., p. 464. New York: John Wiley & Sons, Inc.

Hill, M. L., ed. 1987. *Decade of North American Geology, DNAG, Centennial field guide*. Vol. 1. *Cordilleran Section of the Geological Society of America*, p. 490. Boulder, Colorado: Geological Society of America.

McDougall, I., and Chamalaun, F. H. 1966. Geomagnetic polarity scale of time. *Nature* 212:1415–18.

McDougall, I., and Tarling, D. H. 1963. Dating of polarity zones in the Hawaiian Islands. *Nature* 200:54–56.

McDougall, I., and Tarling, D. H. 1964. Dating geomagnetic polarity zones. *Nature* 202:171–172.

McDougall, I., and Wensink, H. 1966. Paleomagnetism and geochronology of the Pliocene-Pleistocene lavas in Iceland. *Earth and Planetary Science Letters* 1:232–36.

McDougall, I., Allsop, H. L., and Chamalaun, F. H. 1966. Isotopic dating of the new volcanic series of Victoria, Australia, and geomagnetic polarity epochs. *Journal of Geophysical Research* 71:6107–18.

Sullivan, W. 1992. Continents in motion, the new Earth debate. 2d ed., p. 430. New York: American Institute of Physics.

Watkins, N. D. 1972. Review of the development of the geomagnetic polarity time scale and discussion of prospects for its finer definition. *Geological Society of America Bulletin* 83:551–74.

5

Igneous Rocks

MAGMA

Magma originates in the subsurface as the result of a series of complex effects that are mainly dependent on two factors, temperature and pressure. The logical and simplest explanation for the origin of a liquid rock is that the temperature is sufficiently high and pressure sufficiently low to cause the rock to melt. Such effects have been observed experimentally in the laboratory. But the process is much more complex than that. Temperatures build up within Earth as a by-product of the decay of radioactive minerals. Rock fracture and frictional processes associated with rock fracture also release heat. As a result of these factors, at progressively deeper levels within Earth, temperature increases in a fairly systematic way. This increase in temperature is called the *geothermal gradient*. At some localities, the temperature increase is relatively low, 10° C/km, while in other regions that are associated with hot spots, such as the Yellowstone area, the geothermal gradient may be 50°C/km or higher. On average, however, temperature increases about 30°C/km. This means that temperatures within Earth may reach 300°C or more just 10 kilometers below the ground surface! And, of course, temperature rises higher and higher with increasing depth, eventually causing the partial melting of rocks.

The next most important factor for the production of magma is pressure. Compression can stop the formation of a magma, and the release of pressure, such as that produced during motion along a fault, can cause rocks to melt or increase the rate at which melting occurs. The temperature of the system can remain steady, but rock can melt or solidify merely as the result of pressure changes (Figure 5.1). Such changes over long geologic time can be produced by erosion that unloads the crust, thus reducing overburden pressures, resulting in a corresponding reduction in pressure and the partial melting of rocks already at elevated temperatures.

Another important factor in the production of magma includes the amount of water present in the subsurface. Water, when present, will reduce the temperature at which magmas form or solidify (Figure 5.1). Also important is the primary *rock composition*. Obviously, if the primary rock is less resistant to temperature increases, then

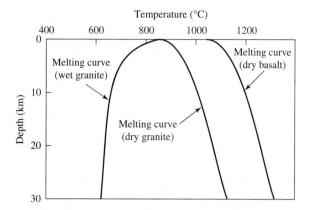

FIGURE 5.1 Melting curves for granite and basaltic magmas; pressure and temperature relationships and an illustration of the effect of water on the granite melting curve.

melting will occur at lower temperatures. Finally, as magma is produced, the amount of contamination by the intruded *country rock* will influence continued melting and the temperature at which magmas will be maintained.

General Magma Characteristics

All magmas are silica-rich liquids, due mainly to the fact that silicon (Si) and oxygen (O) are the most common elements on Earth. SiO_2 makes up 35–75% of a magma. Crystals, solids that separate from the fluid, and gases, primarily carbon dioxide [CO_2] and water vapor [H_2O], make up the rest of a magma. Some of the other gases can be quite toxic, even in the low concentrations found in magmas. They can make volcanic eruptions dangerous, as clouds containing these gases are given off when magma is extruded onto Earth's surface.

The flow behavior, or *rheology*, of magmas is controlled by the silica and water content, with high silica magmas flowing much more slowly than low silica magmas. Conversely, the addition of water decreases the viscosity of the magmas, thus allowing them to flow faster.

Minerals will begin to crystallize from a cooling magma at approximately 1200°C and continue until the liquid reaches about 600°C. The crystallization process is very systematic, with some minerals crystallizing at very high temperatures and others crystallizing at much lower temperatures. *Petrologists*, geologists who study minerals and rocks and how they form, observed that homogeneous magmas could differentiate, allowing a variety of minerals to crystallize at different temperatures. This process, known as *magmatic differentiation*, results in the formation of minerals along two compositional pathways in the same magma. One pathway, known as the *continuous reaction series*, results in new minerals forming during continued cooling, continuously changing their composition, but not their crystalline structure (Figure 5.2). The continuous series is formed by the mineral plagioclase, a feldspar, that is rich in calcium (Ca) and poor in sodium (Na) at high temperature, but poor in calcium and rich in sodium at lower temperatures. However, the crystal structure of plagioclase remains the same during these chemical changes. Along the other pathway, known as the *discontinuous reaction series*, new minerals forming in the magma during continued cooling will change both their composition and their structure (Figure 5.2). As an example, the mineral olivine, which forms at very high temperatures in the magma, will be replaced by a pyroxene mineral as the magma continues to cool. With further cooling other minerals will form, each with a different composition and crystal structure. In

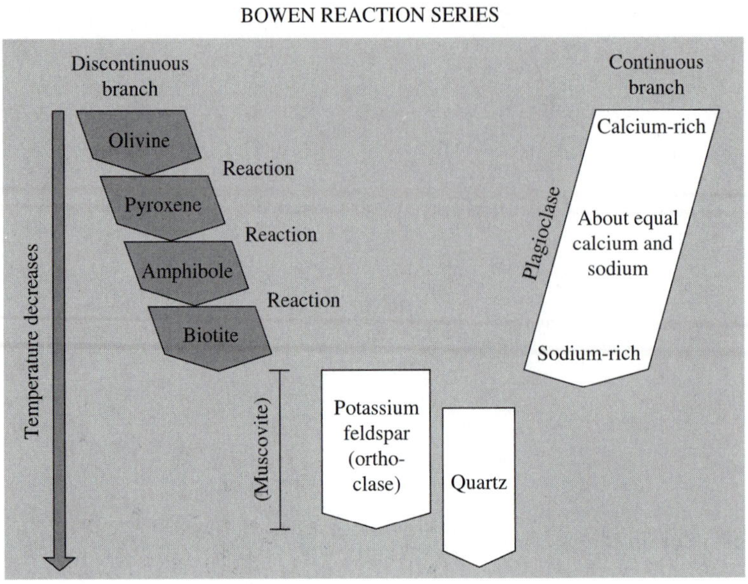

BOWEN REACTION SERIES

FIGURE 5.2 Bowen's reaction series illustrating the mineral changes during cooling of magmas. Last to crystallize is quartz.

summary, Figure 5.2 illustrates the *Bowen reaction series* and provides visual simplification of the reaction trends that occur during cooling of magmas.

The Bowen reaction series was developed as an answer to the problem of how high-temperature crystals are able to be preserved if they disappear during cooling to form other minerals. For example, how can olivine be extracted from the melt at high temperatures before it changes so that it continues to exist for us to study in rocks now at Earth's surface. The question was answered by *N. L. Bowen* just after World War I, when he expanded the concept of magmatic differentiation to include the idea of *fractional crystallization* of the magma, where crystal fractions are separated by eruption, settling, and other ways, thus no longer reacting with the liquid, therefore no longer changing composition.

Magma Types

Magma types and the rocks they produce are identified based on chemical differences, mainly on differences in silica content. Two primary magmas, originating from partial melting in the lower crust and/or upper mantle, have been identified. These are *basaltic* and *granitic* magmas. Magmas of *andesite* are intermediate in composition.

Basaltic magma cools to form the basalt-*gabbro* rock family, which is dominated by ferromagnesian minerals, such as pyroxene, amphibole, and olivine. The magma originates from partial melting (to 40%) of mantle *peridotite* (mainly composed of olivine), and has relatively low silica, approximately 50% SiO_2. The major distinguishing minerals of the basalt-gabbro family are calcium plagioclase and pyroxene, with varying amounts of olivine or amphibole. Crystallization occurs over a range of temperatures from 1,200–900° C. These are usually water poor, relatively dry magmas. Because the silica content is low and the temperature of these magmas is high, they are of low viscosity (very runny) and can travel rapidly over long distances. Basaltic magmas dominate oceanic crustal materials and are emplaced along midocean ridge spreading centers.

Granitic magmas cool to form the granite-rhyolite rock family that is dominated by silica rich minerals with aluminum, also called *sialic* minerals. Silica content is high,

60%–79% SiO_2, and temperature of crystallization in the magmas is low, less than 800°C, resulting in quartz and *potassium feldspar* (K-feldspar) as the major minerals, with minor amounts of Na-plagioclase, mica, and amphibole. Because of the high silica content, granitic magmas exhibit high viscosity and move only very slowly. Granitic magmas usually have a high water content and are generated by partial melting of oceanic crust (basaltic and sedimentary materials) in subduction zones and within continental crust due to elevated temperatures associated with metamorphism in mountain belts.

Andesitic magmas produce, upon cooling, the andesite-*diorite* rock family, intermediate in composition (there is a question as to their primary nature) between basaltic and granitic magmas. Andesites originate from partial melting of wet basalt, or as the result of complete melting of continental crust. Other clearly secondary magmas have also been identified, which result from fractional crystallization or other processes that modify primary magmas. Andesites contain little or no quartz, are extrusive, and usually porphyritic (see textures below). They are found along continental margins associated with plate boundaries.

IGNEOUS ROCK TEXTURE

The *texture* of igneous rocks is determined by the size, shape, and arrangement of constituent mineral grains in the rock and is a reflection of the rock's cooling history. Five terms are commonly used to define texture. *Glassy* texture indicates that the rock does not have distinct crystals that can be identified, even when using a microscope, and when the rock breaks, it exhibits conchoidal fracture similar to that exhibited by glass. Glassy texture develops as a result of very rapid cooling, or *quenching*, often due to extrusion into water. *Aphanitic* textures characterize rocks containing crystals that are only visible using a microscope. These rocks often contain gas pockets, called *vesicles*, that result from rapid cooling and trapping of gas as the magma's "out-gas." Aphanitic rocks are generally extrusive.

A third type of texture, *phaneritic* or granular, is observed in rocks where the crystals can be seen without a microscope and results from slow cooling. These are probably intrusive rocks that have been exposed by erosion. In cases where there are two different crystal sizes, well formed large crystals, called *phenocrysts*, surrounded by a fine-grained mass of crystals, or *groundmass*, the texture is called *porphyritic* and is indicative of two cooling stages.

A very coarse-grained igneous rock is said to have a *pegmatitic* texture. In such cases the minerals present, mainly quartz, K-feldspar (typically *orthoclase*) and micas, crystallize slowly from late stage hydrothermal (water rich, hot) solutions. Some pegmatites are rich in rare minerals, such as tourmaline and topaz, and are the source of many of the really outstanding, large gem quality stones seen in museum collections. Pegmatites can be seen well exposed in the faces of the presidents at *Mount Rushmore National Memorial* in the Black Hills of South Dakota (Figure 5.3).

Pyroclastic rocks are broken, bent, or squeezed volcanic fragments that have been blown out of volcanoes, initially as hot pieces of *ash* that have since cooled. The presence of these fragments represents clear evidence of a volcanic eruption. Usually these materials have a very fine-grained, even glassy, texture as the result of quenching in air as the particles were falling. Because these materials are often quite hot and semi-liquid, larger pieces may be bent as they twist through the air, before they cool enough to be truly solid. Most deformation, however, results from compaction after deposition.

FIGURE 5.3 Pegmatites in the face of President Lincoln in Mt. Rushmore National Memorial. (Photo courtesy of the National Park Service.)

CLASSIFICATION OF IGNEOUS ROCKS

The igneous magmas and the rocks that are produced by them are classified by mineralogy and texture. Table 5.1 (see also Figure 5.4) gives a simple classification scheme that is based on silica mineralogy and texture. Because higher silica content usually re-

TABLE 5.1   CLASSIFICATION OF IGNEOUS ROCKS

Texture	Silica Content/Color/Specific Gravity			Emplacement Mode
	High/light/light	**Intermediate**	**Low/dark/heavy**	
Phaneritic	**Granite**	**Diorite**	Peridotite/dunite	Intrusive
		Granodiorite	**Gabbro**	
Aphanitic	**Rhyolite**	**Andesite**	**Basalt**	Extrusive or intrusive
Glassy	**Obsidian**		**Basalt glass**	Extrusive and quenched
Vesicular	**Pumice**		**Scoria**	Extrusive
Pyroclastic	**Ash, Tuff, Breccia**			**Extrusive**
	Sodium rich/calcium poor	Plagioclase: Na poor/Ca rich		
	Quartz, K-feldspar, micas	**Pyroxene, olivine**		
		Amphibole		

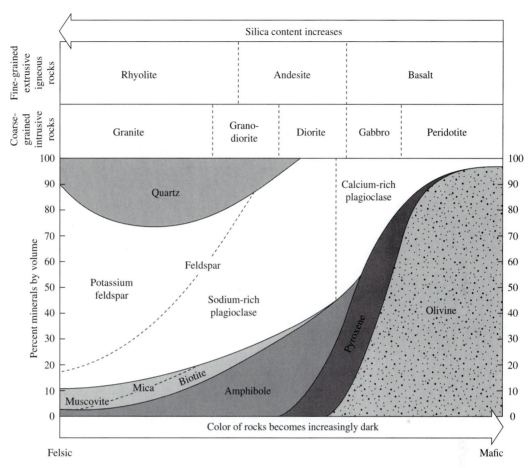

FIGURE 5.4 Igneous rock classification, illustrating silica content and mineralogy of rocks formed from igneous magmas.

sults in a rock lighter in color and specific gravity, while lower silica content rocks are usually darker and have a higher specific gravity, the igneous classification scheme in Table 5.1 also lists these characteristics. They are often helpful for diagnostic purposes. The actual rock names are given in bold letters in the columns below Silica Content / Color / Specific Gravity in the table. *Emplacement mode* is deduced from texture and indicates whether the rock reached the surface of Earth, an *extrusive* rock, or if it cooled at depth within Earth before it could reach the surface, an *intrusive* rock. Pyroclastic rocks, tuffs, ash flows, and breccia (broken rock fragments cemented together) listed in the table can be almost any composition. At the bottom of the table are given the general range of minerals found in the rock type listed above them. For example, we see in Table 5.1 that a gabbro is a low silica, intrusive rock, generally dark and dense, that is phaneritic. It contains calcium-rich and sodium-poor plagioclase, pyroxene, perhaps with minor amounts of olivine or amphibole.

EXTRUSIVE VERSUS INTRUSIVE ROCKS

Extruded volcanic materials are erupted onto the surface from volcanic *vents* usually located in a *crater*. After a significant amount of material has been erupted and removed from the subsurface, the whole area may collapse, forming a large basin or *caldera*. Crater Lake National Park in Oregon is such a collapsed volcano, with many other examples in the national parks, including Katmai, Hawaii Volcanoes, and Haleakala National Parks (all discussed in Chapter 4).

CRATER LAKE NATIONAL PARK

Crater Lake National Park (see Figures 1.12 and 8.4), located in southern Oregon, was established by Congress in 1902. It is what remains of a large stratovolcano, named Mt. Mazama. After an explosive eruption the peak collapsed into the area vacated by the magma, forming a large collapse caldera (Figure P5.1). The caldera has since been filled by rainwater, producing a lake that is 1932 ft (589 m) deep and 6 mi (9.7 km) across. The water is a spectacular blue that most photographs have not been able to capture. On the southwestern edge of the lake is a *cinder cone* that erupted and formed after the caldera. Flooding formed what is now Wizard Island (Figure P5.2).

Geology.   Development of the predominantly andesitic stratovolcano occurred during the late Cenozoic with a series of volcanic eruptions that built Mt. Mazama. Basalts are also common as volcanic materials in the park. The caldera was created after explosive volcanism had erupted large volumes of ash and pumice. Then subsidence along faults, accompanied by explosions, produced the caldera. Magma was also erupted from the sides of the volcano producing small parasitic cones on the volcano's outer slopes. Erosion has modified the earlier caldera rim (Figure P5.3), and at one time glaciation was active. However, no glaciers exist today in the park.

FIGURE P5.1   Diagram illustrating the formation of Crater Lake at Crater Lake National Park, Oregon. The series begins with the development of a stratovolcano, Mt. Mazama, then ultimate collapse of the caldera. A cinder cone then was extruded onto the floor of the caldera that was later flooded, producing Wizard Island.

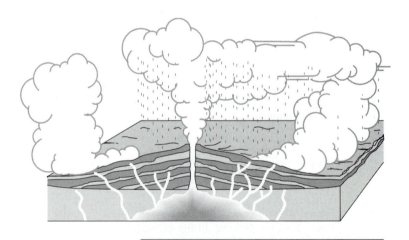

Wizard Island

FIGURE P5.2 Wizard Island, a cinder cone in Crater Lake National Park. (Photo by author.)

FIGURE P5.3 Crater Lake National Park rim, illustrating the large debris slopes developing as the rim is destroyed by weathering and erosion. (Photo by author.)

FIGURE 5.5 Fast flowing basaltic lava erupted from Kilauea Volcano in Hawaii Volcanoes National Park. (Photo by author.)

Basaltic lavas are very hot when they are first extruded, consequently they cool very quickly after extrusion. Initially they may run at speeds up to 40 km/hr (Figure 5.5), but with cooling their character changes. While they still have relatively low viscosity (runny) they may form a ropy appearing surface called *pahoehoe* (Figure 5.6), but upon further cooling of the lava crust, a blocky, broken crust form

FIGURE 5.6 Craters of the Moon National Monument, Idaho; *pahoehoe* lava flow. (Photo by author.)

FIGURE 5.7 Craters of the Moon National Monument, Idaho; *aa* lava. (Photo by author.)

called *aa* develops (Figure 5.7). (The terms *pahoehoe* and *aa* come from the Hawaiian language.) If the lavas run into water or are extruded under water or even under ice (the ice will melt), rounded or elongated forms develop called *pillow lavas*. Interesting examples (Figure 5.8) can be seen exposed in a road cut at approximately 5,000 ft (1,500 m) in Olympic National Park (Chapter 14), where pillows were encased in sediments containing marine fossils and later uplifted.

pillows

FIGURE 5.8 Pillows exposed at 5,000 ft in Olympic National Park, Washington. (Photo by author.)

FIGURE 5.9 Craters of the Moon National Monument, Idaho. Fissure eruption lavas and small cinder cones that have built up along the fissure. (Photo by author.)

When small-scale eruptions occur from single vents, small *spatter cones* or larger *cinder cones* may develop around the vent (see Capulin Volcano National Monument and Sunset Crater National Monument park inserts). With more extensive lava extrusion, the low silica, low viscosity basaltic lavas build up low, broad cones called *shield volcanoes* like Kilauea and Mauna Loa volcanoes in Hawaii Valcanoes National Park. Sometimes eruptions occur as magma spouts along segments of a fracture in Earth. Such eruptions are called *fissure eruptions* (Figure 5.9).

In some areas on Earth, for example in the Columbia River region of Washington, Oregon, and Idaho (including Craters of the Moon National Monument), and in India, South America, Africa, and elsewhere, extensive floods of low viscosity lavas have produced large basaltic plateaus. These are called *flood* or *plateau basalts*. Such extensive volcanism has produced major islands, such as Iceland on the Mid-Atlantic Ridge in the North Atlantic Ocean. There are also some very old, Precambrian flood basalts found on the North American continent, and some of these are exposed at Isle Royale National Park.

 ## CRATERS OF THE MOON NATIONAL MONUMENT

Craters of the Moon National Monument (see Figures 1.12 and 8.5) was established in southern Idaho in 1924. It lies between the Cascade Mountains to the west and the Rocky Mountains to the east.

Geology. Lavas accumulated in the area from Eocene to Pleistocene time, but Miocene lavas are the most abundant. The monument primarily contains lavas that were erupted along fractures in the Earth's crust as fissure eruptions. These lavas were very hot, with low viscosity at the time of the eruption, and some flowed for great distances. Locally recent eruptions have formed small cinder and spatter cones along an extended fracture trending NW-SE (Figure 5.9). These

eruptions are approximately 2,000 years old and extend for 55 km across southern Idaho.

There are 80 known of lava tube caves in the monument (Figure P5.4) that result from flow after the exterior has solidified. Lava often continues to flow through these tubes long after the rest of the flow has cooled, and many lava tubes show "curbs" that resulted from lavas extruded during later eruptions through the same tube (Figure P5.5). In the middle of the ocean, in places such as the Hawaiian and Azores Islands, lava tubes provide a mechanism for bringing very hot lava to the margins of eruptions, thus becoming significant contributors to the growth of the islands.

Some of the lava tubes at Craters of the Moon contain permanent ice that results from two factors. In most caves the air remains approximately at the mean annual temperature for the area; in Idaho that is cold. As wind circulates through these lava tube caves, wind-chill effects drop the temperature further, maintaining ice year round.

FIGURE P5.4 Lava tube in Craters of the Moon National Monument. (Photo by author.)

FIGURE P5.5 Lava tube showing the curbs that often develop; here in a National Park in the Azores, located in the middle of the Atlantic Ocean. (Photo by author.)

FIGURE 5.10 Cross section of a segment of a six sided column at Devils Tower National Monument. These columns are produced by thermal contraction during cooling. (Photo by author.)

As lava cools the liquid slowly becomes more and more viscous. At a critical point in the cooling process, fractures may propagate through the fluid due to thermal contraction if the viscosity becomes high. Those fractures divide the material into columns with between three and eight sides. These columns may be quite large, and may form in cooling intrusives, like those that make up Devils Tower at Devils Tower National Monument (see Figure 1.3; Figure 5.10). Columns are also often quite small, such as those exposed at Mt. Rainier National Park (Figure 5.11).

FIGURE 5.11 Columnar basalt exposed at Mt. Rainier National Park, Washington. (Photo by author.)

CAPULIN VOLCANO NATIONAL MONUMENT

Capulin Volcano National Monument, previously Capulin Mountain National Monument, in northeastern New Mexico (see Figure 1.17) was established in 1916. It is a large cinder cone, *Holocene* in age (approximately 7,000 years old) that formed as basaltic cinders and ash accumulated from an erupting vent (Figure P5.6). It has a crater at its center and the top can be visited by road.

Geology. Capulin volcano lies on the eastern edge of a series of late Cenozoic extrusives extending into north central New Mexico. It is part of a series of volcanic eruptions that have dotted the landscape with lava flows, cinder cones, and shield volcanoes. Intermittent volcanism in northeastern New Mexico occurred from the Pliocene Period through Holocene time.

FIGURE P5.6 Capulin Volcano at Capulin Volcano National Monument. (Photo by author.)

 ISLE ROYALE NATIONAL PARK

Isle Royale National Park (see Figures 1.21 and 8.14) was established by Congress in 1940. The islands composing the park (Figure P5.7) are located in Lake Superior very near Canada, but are part of Michigan. During the summer they can only be reached by boat or sea plane.

Geology. Composed of ancient flood basalts, Isle Royal is part of the exposed north limb of the *Lake Superior syncline*, a down-folded structure (see Chapter 7) that has been grooved by NE to SW-flowing *glaciers*. Only the more resistant *lavas* remain, and these are dominated by *Keweenawan* basalts of *Precambrian* age. Weathering and the growth of vegetation has obscured the rocks at the surface (Figure P5.8), but the weathering is very shallow, with fresh lavas lying just below the surface. In some areas the vesicles in the lavas are filled with native copper. Such filled vesicles are called *amygdules*, and the amygdules at Isle Royale were mined for the copper by Native Americans living in the area.

FIGURE P5.7 Isle Royale National Park in Lake Superior, Michigan. The island is a remnant of glacially carved Precambrian lavas, now above lake level in Lake Superior. (Photo courtesy of the National Park Service.)

FIGURE P5.8 Isle Royale vegetation and weathering has destroyed surface rock exposures in most areas. (Photo by author.)

Eruptions of *silicic lavas*, composed of granitic or andesitic magmas, are extremely explosive due to gas buildup. These magmas have medium to high silica content, and therefore have high viscosity. Gases trapped in the magma will blow out volcanic ash and tuff, which accumulates to form steep-sided, small-scale cinder cones. When lavas and pyroclastic debris alternate in large quantities, these eruptions form large *composite volcanoes,* also called *stratovolcanoes*. The more explosive of these volcanoes are andesitic, intermediate silica magmas. Examples include Mt. Lassen, Mt. Rainier, Mt. St. Helens, and many others, all within the Cascades mountain region of the Pacific Northwest.

Tephra is a general term for all the pyroclastic material blown out of volcanoes. More specific terms include *volcanic ash*, made up of the smallest particles, volcanic cinders or *volcanic lapilli*, gravel-size fragments of lava; *volcanic blocks*, including large angular blocks; and *volcanic bombs*, which are large rounded fragments. When first extruded, this material moves as unconsolidated volcanic gas and debris, some still molten. After it stops moving, it will often become welded and lithified as a result of final crystallization from its retained heat. The term *tuff* is used to refer to lithified, fine-grained accumulations of ash fall. Sometimes this material is erupted as pyroclastic flows that are rapidly flowing, destructive, hot-air–fragment mixtures (Figure 5.12) such as those erupted from Mt. St. Helens in 1980. The name for these ash-flows is *ignimbrite* or nuée ardente, the term first being applied by the French after this type of eruption destroyed the town of St. Pierre on the island of Martinique.

Intrusive igneous rocks are those rocks formed from intruding magma that does not reach the surface. They are classified primarily by size and their relationships with the country rocks that they are intruding. Intrusive rock classification is independent of magma composition, so that all the classification elements involve both basaltic and granitic rock types.

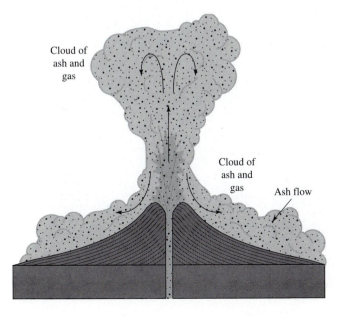

Cloud of ash and gas

Cloud of ash and gas

Ash flow

FIGURE 5.12
Illustration of an ash flow eruption.

MT. ST. HELENS NATIONAL VOLCANIC MEMORIAL

Mt. St. Helens National Volcanic Memorial is a famous stratovolcano because of its 1980 explosive eruption, typical of silicic lavas (Figure P5.9). Several years after the eruption, the effects are still quite evident (Figure P5.10). Located in southern Washington, Mt. St. Helens is not governed by the National Park Service, but rather is controlled by the U.S. Forest Service.

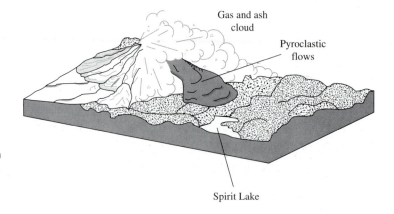

FIGURE P5.9 Diagram illustrating the eruption of Mt. St. Helens in Washington.

FIGURE P5.10 Mt. St. Helens one year after the eruption. The trees on the flanks of the mountain were blown flat. (Photo by author.)

FIGURE 5.13 Dike exposed in Acadia National Park, Maine, crosscutting the Devonian granites exposed in the park. The dike is Mesozoic in age. (Photo by author.)

The most common type of intrusions are called *dikes*. These are generally narrow, tabular bodies that are *discordant;* and they crosscut layering or *bedding* previously developed in the country rock (Figure 5.13). Dikes are very common in the national parks, often being identified from a distance as a black ribbon cutting across the face of a mountain such as Middle Teton in Grand Tetons National Park (Figure 5.14; see also Chapter 7) or are exposed in park road cuts. Groups of dikes, usually parallel to each other, are called *dike swarms*, and they make up the feeder systems to many volcanoes.

Narrow tabular intrusive bodies that are *concordant* (features parallel to bedding) are called *sills*. These are not as common as dikes and are often misidentified as dikes, when bedding is not easily distinguishable. When exposed as part of a sedimentary sequence of rocks, sills can also be misidentified as lava flows. But careful examination of the upper contact with the country rock will reveal that if a sill intruded the section, the rocks both above and below will be baked and altered (metamorphosed, see Chapter 6) by the heat associated with the intruding sill (Figure 5.15). This baking effect is also evident at both margins of intruding dikes. Lavas may flow over sedimentary rocks, baking and altering the material immediately below, but any rocks on top of the lava will have been deposited after it cooled; therefore these later sediments will not be baked or altered by the heat from the lava. Furthermore, no vesicles, formed at the surface in lavas as the result of degassing at the reduced pressures found at Earth's surface, are observed in either dikes or sills.

FIGURE 5.14 Precambrian dike exposed at Middle Teton, in Grand Tetons National Park, Wyoming, cutting the Precambrian metamorphic and granitic rocks. (Photo by author.)

FIGURE 5.15 Sill exposed in Glacier National Park, Montana. The whitened zones above and below the sill are limestones baked by contact and metamorphosed into marbles. (Photo by author.)

MT. RAINIER NATIONAL PARK

Mt. Rainier National Park in south-central Washington (see Figures 1.12 and 8.4) was established by Congress in 1899. (Figure P5.11). It is still geologically active and has the potential for a major eruption that could, in turn, generate large *lahars* (*mudflows* due to melting snow and ice generated by the heat from the eruption), which could affect population centers in Washington. The possibility of lahars presents a major problem in the area. These *autosuspensions* move very rapidly downhill and maintain their momentum over flat terrain. They can travel long distances, can be very destructive, and have been responsible for much loss of life following andesitic eruptions.

Due to the fact that it could erupt at any time and its spectacular appearance, distinctive geology, and easy accessibility, it was designated as a *decade volcano* in 1992. This designation means that it will be intensively studied during the decade ending in the year 2002. Mt. Rainier is 14,410 ft (4,392 m) high and had a major eruption 2,000 years ago. Between 1820–1854 there were a series of minor eruptions, but like Mt. St. Helens, a few miles to the south (Figure P5.12), it could erupt explosively at any time. The mountain has two summit craters with 26 glaciers and a small ice cap, thus most of the summit is covered by snow and ice year round (Figure P5.12).

Geology. Only Cenozoic rocks are exposed in this stratovolcano sitting on an eroded volcanic plateau. The oldest rocks in the area are Eocene volcanics. In the Miocene, the granodiorite Tatoosh pluton was intruded into the subsurface. Later, during the Pliocene, the Tatoosh was uplifted and exposed by erosion. It was rounded further by *mass wasting*, a process where the force of gravity causes weathered materials to tumble downhill (see Chapter 9). Carving by a large number of glaciers is responsible for the shape of the mountain that we see today.

FIGURE P5.11
Mt. Rainier National Park, Washington, a view from the flanks of the mountain of the partially obscured twin summit. (Photo by author.)

FIGURE P5.12 Mt. St. Helens from Mt. Rainier, pre-1980. (Photo courtesy of the National Park Service.)

 ## LASSEN VOLCANIC NATIONAL PARK

Lassen Volcanic National Park (see Figures 1.14 and 8.4) was initially established as a national monument in 1907 by President Theodore Roosevelt, who was responsible for establishing several monuments to protect some of America's natural wonders. The monument was upgraded to park status by Congress in 1916. Located in northern California, it is still active and has erupted several times between 1914 and 1921. One of these eruptions was an ignimbrite, an explosive cloud of gas and ash, that roared down the mountainside. There is a thermal area in the park (Figure P5.13) that includes active geysers heated by magmas in the subsurface. These lie below the exposed, very large *volcanic dome* that makes up the summit of Mt. Lassen today (Figure P5.14).

Geology. Initially, as the North American Plate moved over subducting oceanic plate segments to the west, a stratovolcano formed at the Mt. Lassen site as the result of the alternating eruption of volcanic lavas and tuffs. Then explosion and collapse formed a caldera that was later intruded by viscous magmas forming a plug in the caldera, called a *plug dome*. Mass wasting processes have cluttered the slopes of the dome with debris and hydrothermal alteration has discolored the plug, giving it a yellowish-red hue, the result of oxidation of iron minerals to red hematite [Fe_2O_3] and yellow goethite [$FeO(OH)$].

FIGURE P5.13 Thermal area in Lassen Volcanic National Park, California. (Photo courtesy of the National Park Service.)

FIGURE P5.14 Mt. Lassen in Lassen Volcanic National Park. (Photo by author.)

Very large scale, intrusive igneous rocks, derived primarily from granitic and andesitic magmas, are generally classified as *plutons.* There are a number of different types of plutons, categorized mainly by their size and geometry. The classification of these bodies is important in mining geology, but will not be discussed in depth here. Extremely large plutons are called *batholiths.* One of the best examples of an exposed batholith in the park system is the Sierra Nevada *Batholith* complex (composed of many intrusions) in California, which can be seen in Yosemite, Kings Canyon, and Sequoia National Parks.

The magma from which plutons are derived forms at great depths within Earth and moves upward by a process called *stoping,* where blocks of the country rock fall into magma, thus making room for the magma to move upward. Some of these blocks are melted (assimilated) into the magma, but others, with higher melting temperatures than contained in the intruding magma, remain as *xenoliths* (Figure 5.16), which can often be easily distinguished because of the very different appearance of these blocks in an otherwise uniform igneous granitic or andesitic rock. Figure 5.17 illustrates intruding and extruding magmas (Figure 5.17a) and their cooled and lithified equivalents. Small plutons like the one on the right in Figure 5.17b are *stocks,* identified as usually having less than 100 km^2 of exposed area.

Granite is the most common intrusive (subsurface) rock and results from deep-seated melting of crustal rocks, with very slow cooling producing phaneritic textures (see Figure 2.6). During the cooling process, the granite shrinks, causing the surrounding country rocks to crack. Into these cracks fluids are injected from the magma, resulting in the production of intrusive granitic dikes. Such dikes can travel long distances, and are seen in granitic and metamorphic rock outcrops as contrasting, usually lighter bands. In addition to pegmatite dikes, many finer grained granitic dikes can be seen in the Presidents' faces at Mount Rushmore (see Figure 5.3).

FIGURE 5.16 Xenoliths (dark elements) enclosed within a granitic (lighter material) batholith intrusion at Yosemite National Park, California. (Photo by author.)

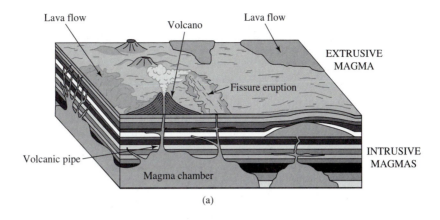

(a)

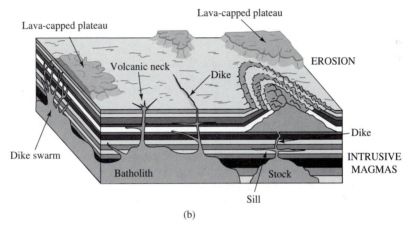

(b)

FIGURE 5.17
(a) Diagram illustrating magma intrusions and extrusive elements.
(b) Those igneous rocks after cooling and partial exposure by erosion.

 ## LAKE CLARK NATIONAL PARK AND PRESERVE

Lake Clark National Park and Preserve in southern Alaska (see Figure 1.11) was established as a national monument in 1978 (Figure P5.15). It was upgraded by congress in 1980 to park status, along with many other parks at that time. The preserve classification was added to most of the national parks in Alaska as a special designation that denotes, among other things, that hunting is allowed in the park.

Geology. The park contains two active volcanoes, Mt. Iliamna (Figure P5.16), which last erupted in 1947, and Mount Redoubt, which erupted in 1989/1990. Also serving as evidence of igneous activity in the park are some major exposures of granites emplaced mainly during the Jurassic, although some granite intrusion did

occur during the Tertiary. The oldest rocks in the park are Triassic metamorphics that include *schists* and *marbles*. Uplift in the Cretaceous was followed by deposition of the eroded clastic sediments. Explosive volcanism began as a result of subduction of the Pacific Plate under Alaska, from Middle Tertiary to the Quaternary Period. Tectonic forces have tilted the sediments exposed in the park, and the terrain has been modified by glaciation, producing unique glacially carved granite exposures. An example is the "tusk" in Figure P5.17. Many glaciers are still active, and evidence of extensive glaciation can be seen throughout the park. For example, the clear U-shaped glacial valley in Figure P5.18 now contains Kontrashibuna Lake, where a glacier once ground across the landscape.

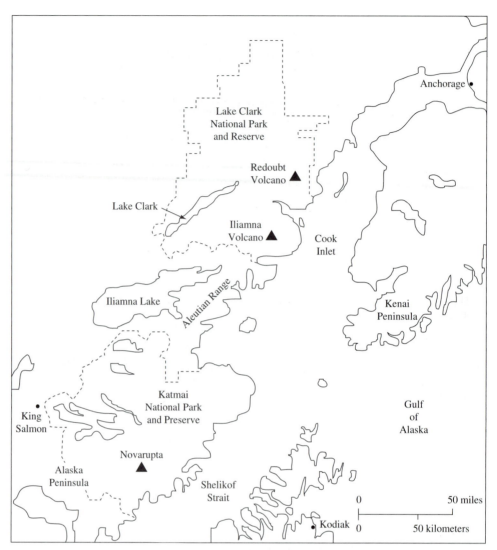

FIGURE P5.15   Map of Lake Clark National Park and Preserve and Katmai National Park and Preserve, in Alaska.

FIGURE P5.16
Mt. Iliama in Lake Clark
National Park and
Preserve, one of two
active volcanoes in the
park. The cloud at the
summit is a steam cloud.
(Photo courtesy of the
National Park Service.)

FIGURE P5.17   Granite
"tusk" exposed and
sculpted by glaciation in
Lake Clark National Park
and Preserve. (Photo
courtesy of the National
Park Service.)

FIGURE P5.18 Kontrashibuna Lake, now filling a glacially carved valley in Lake Clark National Park and Preserve. (Photo courtesy of the National Park Service.)

YOSEMITE NATIONAL PARK

Yosemite National Park (see Figures 1.14 and 8.4) was established in 1890 in the Sierra Nevadas. Located in central-eastern California, it has been designated a World Heritage Site because of its spectacular beauty. John Muir, a tireless environmentalist of the last century and protector of the area, was mainly responsible for bringing the park to the attention of Congress. Today it is a place of unique grandeur, molded by ancient glaciers that have left behind evidence of their passing. Just three small glaciers remain as a reminder of their past dominance in the area. Spectacular waterfalls fall away from cliffs undergoing stream erosion (Figure P5.19), an ongoing process continuing to mold the park.

Geology. The park is an area of exposed plutons on part of the Sierra Nevada Batholith, intruded during the Mesozoic (mainly Cretaceous), with compositions ranging from granites, quartz diorites, granodiorites, to diorites. During the Tertiary, concentric fracturing, produced by unloading due to erosion and by the weathering effects of alternating freezing and thawing, has produced very large *exfoliation domes* that are symmetrical in shape. Half Dome (Figure P5.20) is one example that has been further undercut by glaciers after which collapse occurred along vertical fractures called *joints*. The result is that only half of the exfoliation dome remains. Other domes are glacially carved, producing asymmetrical shapes. Glaciers have also cut deep U shaped valleys throughout the park. Yosemite Valley was carved by the Merced Glacier leaving in places 3,000 ft steep walls, of which El Capitan (Figure P5.21) and Half Dome are prominent examples. Glacial debris, left behind as the glaciers melted, covers the valley floor and has formed dams producing glacial lakes. Uplift during the late Tertiary was also accompanied by andesitic volcanism, and these lavas are exposed in the northern part of the park.

Sadly, a reportedly spectacular glacial valley in Yosemite National Park, Hetch Hetchy Valley, is currently flooded by a dam built to store water for the city of San Francisco (Figure P5.22). Environmentalists and park supporters have been trying to have the dam removed and the valley restored. Valley restoration would take a very long time, and the water problems in California make any effort to remove this lake very difficult. It remains to be seen when Hetch Hetchy will again emerge as a national wonder. Eventually, of course, geologic forces will do the job for us, because nothing built by human beings can withstand the forces active over geologically long periods of time. Eventually, the dam will be breached and erosion will form a new Hetch Hetchy Valley.

FIGURE P5.19 Waterfall in Yosemite National Park, California, flowing over a granite wall carved by glacial erosion of Yosemite Valley, and part of the exposed elements of the Sierra Nevada batholith. (Photo by author.)

FIGURE P5.20 Half Dome in Yosemite National Park, made up of granite carved by glacial erosion of Yosemite Valley, and part of the exposed elements of the Sierra Nevada batholith. (Photo by author.)

FIGURE P5.21 El Capitan in Yosemite National Park made up by granite and carved by glacial erosion of Yosemite Valley, and part of the exposed elements of the Sierra Nevada batholith. (Photo by author.)

FIGURE P5.22 Hetch Hetchy Valley in Yosemite National Park. (Photo taken by and courtesy of Jeannette Stewart.)

SEQUOIA AND KINGS CANYON NATIONAL PARKS

Sequoia National Park and Kings Canyon National Park (map Figures 1.14; 8.4) are two parks that are joined together and interwoven (Figure P5.23). Located in southern California, Sequoia was established in 1890. It contains most of Mt. Whitney, at 14,495 ft (4,418 m) the tallest mountain in the lower 48 states, but its flat top lies just outside the park boundary. The summit is flat, first due to erosion, then later uplifting of the flat summit. At the margins of the park are marbles, metamorphic rocks (see Chapter 6) formed during intrusion of the Sierra Nevada batholith (Figure P5.24) and heating of the limestone country rock. Caves have developed as the result of dissolution of the marble. There are several examples of caves in Sequoia and Kings Canyon. One is Crystal Cave, a commercial cave in Sequoia. A second example, Lilburn Cave in Kings Canyon, lies in an unusual blue and white banded marble.

Of course, Sequoia National Park is famous for its giant sequoia trees (*Sequoia gigantea*), including the General Sherman, which is the largest tree in the world. These trees live to be over 3,000 years old and have an unusual life history. To reproduce, they need fire to open their pine cones, so the Park Service will, on occasion, set small controlled fires to help the seeds along. One of the biggest problems with the stability of these trees is their very shallow root system. The trees have a long trunk with most of the branches restricted to their tops. During winter storms, snow and ice accumulate in these upper branches and make the trees very top heavy. With a very shallow root system, the trees are subject to falling when overloaded, and they do fall, especially during ice storms. When the trees fall they violently disintegrate when they hit the ground, taking out many smaller trees in and near their path.

Kings Canyon National Park was initially established as General Grant National Park in 1890. The name change occurred in 1940 to better reflect the major feature of the park, Kings Canyon, a river valley cutting through metamorphic and some igneous rocks

FIGURE P5.23 Map of Sequoia-Kings Canyon National Parks, California, showing the common boundary of the two parks. Note small segment of Kings Canyon National Park containing the General Grant Sequoia grove.

(Figure P5.25). The park has a number of giant sequoias, including the General Grant tree, also called the Nation's Christmas Tree (Figure P5.26), and down the hill from it, the General Lee tree.

Geology. As with Yosemite National Park, Sequoia and Kings Canyon National Parks include an area of exposed plutons on part of the Sierra Nevada Batholith, the result of intrusion and uplift during the Mesozoic (mainly Cretaceous). Compositions of these rocks range from granites, quartz diorites, granodiorites to diorites. Throughout the Tertiary, erosion sculped the mountains until in the late Miocene the only remnants were gently rolling hills, covered in areas by lush forests of sequoias. Uplift began again in the Pliocene, eventually bringing Mt. Whtiney, with its flat eroded top, to the elevations we see today. This area was then further sculpted by stream and glacial erosion, and now only pockets of the sequoias remain. The area is still tectonically active, and uplift and erosion continue in the Sierra Nevada.

FIGURE P5.24 Sequoia granites, Sequoia National Park, part of the exposed elements of the Sierra Nevada batholith. (Photo by author.)

FIGURE P5.25 Kings Canyon in Kings Canyon National Park. (Photo courtesy of the National Park Service.)

FIGURE P5.26 General Grant Tree in Kings Canyon National Park. (Photo by author.)

 ## ACADIA NATIONAL PARK

Acadia National Park, located on the central coast of Maine (see Figures 1.24 and 8.16), was initially established in 1916 as Sieur de Monts National Monument, with name changes in 1919 to Lafayette National Park and in 1929 to Acadia National Park. Most of the park is located on Mt. Desert Island, just off the coast of Maine, and on several other small islands. One of the most interesting aspects of this park is that all the land was privately donated, with John D. Rockefeller, Jr. donating about one-third of the park. No money from Federal sources was used for acquisitions. Acadia National Park has beautiful wave cut coastal features (Figure P5.27) and glaciation has carved and rounded much of the upland terrain.

During the last ice age affecting North America, around 18,000 B.P., sea level was about 100 m lower than it is today. A major continental ice sheet covered the New England region, and extensive glaciers cut deeply into what is now the coastline of Maine, gouging glacial valleys well below today's sea level. When these glaciers melted and sea level rose, the valleys were inundated by the ocean, forming *fjords*. Somes Sound is an example of a fjord in Acadia National Park.

Geology. The rocks that are exposed in Acadia National Park are mainly Devonian granites and diorites that were generated during the *Acadian Orogeny*. The Mt. Desert Island Granite, extensively exposed on Mt. Desert Island (Figure P5.28), is an example. Also, large granite dikes are abundant as are basaltic dikes (see Figure 5.13). The basaltic dikes were intruded during opening of the Atlantic Ocean in the early Mesozoic.

Contact metamorphism during intrusion of the granites has altered and metamorphosed the early Devonian siltstones and shales, now located along the southern margin of the park, producing fine-grained quartzites (Figure P5.27).

FIGURE P5.27
Lighthouse at the southern end of Acadia National Park, Maine. The rocks in the foreground are metamorphosed sedimentary rocks, altered by granite intrusions into the area. None of the granites are exposed at this location. (Photo by author.)

FIGURE P5.28
Mt. Desert Island granite in Acadia National Park. (Photo by author.)

REFERENCES

BALLY, A. W. and PALMER, A. R., eds. 1989. *Decade of North American geology, DNAG, The geology of North America.* Vol. A, *The geology of North America; An overview*, p. 619. Boulder, Colorado: The Geological Society of America.

BENNETT, R., ed. 1980. *The new America's wonderlands: our national parks,* p. 463. Washington, D.C.: National Geographic Society.

BEUS, S. S., ed. 1987. *Decade of North American geology, DNAG, centennial field guide.* Vol. 2, *Rocky Mountain Section of the Geological Society of America*, p. 475. Boulder, Colorado: Geological Society of America.

CHRONIC, H. 1986. *Pages of stone, geology of western national parks and monuments.* Vol. 2, *Sierra Nevada, Cascades & Pacific Coast*, p. 170. Seattle, Washington: The Mountaineers.

CHRONIC, H. 1986. *Pages of stone, geology of western national parks and monuments.* Vol. 3, *The Desert Southwest.* p. 168. Seattle, Washington: The Mountaineers.

HARRIS, A. G. and TUTTLE, E. 1990. *Geology of national parks, 4th ed.,* p. 652. Dubuque, Iowa: Kendall Hunt Publishing Company.

HARRIS, D. V. and KIVER, E. P. 1985. *The geologic story of the National parks and monuments.* *4th ed.,* p. 464. New York: John Wiley & Sons, Inc.

HILL, M. L., ed. 1987. *Decade of North American geology, DNAG, Centennial field guide.* Vol. 1, *Cordilleran Section of the Geological Society of America,* p. 490. Boulder, Colorado: Geological Society of America,

ROY, D. C., ed. 1987. *Decade of North American geology, DNAG, Centennial Field Guide.* Vol. 5, *Northeastern Section of the Geological Society of America*, p. 481. Boulder, Colorado: Geological Society of America.

6

Structure and Metamorphic Rocks

ROCK STRENGTH

The concept of rock strength is very important in understanding the structures that develop in rocks. Rock strength is characterized in terms of two parameters. These are *stress*, the deforming force applied to a rock, and *strain*, the magnitude of deformation (change in shape and/or volume) that a rock experiences. It takes time to deform a rock, so strain magnitude is, in part, a function of time.

Stresses are basically of three types. There are *tension* stresses that cause the rocks to be stretched, thus producing dilation. There are *compression* stresses that result in contraction due to the rocks being squeezed. These stresses involve *lithostatic pressures* that result from uniform compression deep within Earth. And there are *shear* stresses that result in changes in shape but may not include a volume change. An example would be a deck of cards with one card sliding past another.

Strain can be divided into three types. Initially, as forces are applied, rocks exhibit *elastic deformation*. When the stress is removed, the rock returns to its original shape and the deformation disappears. This is somewhat like stretching a spring. When the stress is removed, the spring returns to its original shape. In the case of elastic deformation of rocks, stress is proportional to strain. This relationship is linear, and is illustrated by the straight line in Figure 6.1. If the stress gets too large and the deformation becomes nonlinear (curved line in Figure 6.1), then the rocks will no longer regain their original shape when the stress is removed. When the rocks exhibit viscous and/or plastic behavior we call this *ductile* deformation. Molding clay exhibits ductile behavior in the hands of the potter. Last, if the rock breaks under stress, we call this behavior rupture (Figure 6.1).

When geologists first began to look carefully at rocks they discovered a number of structures that were the result of rock deformation. Some of these structures, such as folds, clearly resulted from ductile deformation. Other rocks were broken and displaced along faults, while still others were broken but no displacement was evident. It became clear that to understand the physical forces involved in rock forming processes, it was necessary to understand these various structures.

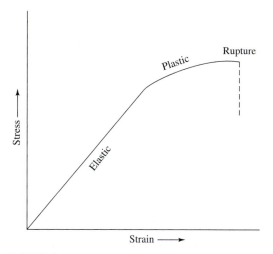

FIGURE 6.1   Stress-strain diagram, illustrating the relationship between applied stress and strain (deformation). In the elastic region, if stress is removed, the strain also disappears. Seismic waves moving through Earth do not cause permanent deformation as they pass because the stress they apply falls in the elastic region. However, in the plastic region, applied stress results in permanent deformation even after the stress is removed. At rupture, the rock breaks. This is the case at the earthquake focus.

FOLDS

Geologist's observations of bent but not broken rocks resulted in the conclusion that there are at least two physical processes that form folds, and both of these involve compressive stress or pressure. Folds form due to bending (*flexural slip*) of beds (Figure 6.2) or due to ductile flow into folds. In both cases, the characteristic features formed can be used for classification purposes (Figure 6.3). The simplest fold is a *monocline,* where flat beds are offset by a double flexure bend connection. To visualize this, place a book on a desk, place a piece of paper on the book, then slide it over until half of the paper drapes over onto the desk. The paper ends up with one end flat on the desk and the other flat on the book, with a double flexure, a monocline, between the two segments.

If that same piece of paper is placed flat on a desk, you can place your hands palms down 6 inches apart on each side of the paper, and then slowly slide your hands together. The center of the paper will begin to bend upward, forming an arch, or *anticline.*

FIGURE 6.2
Compression and folding of sedimentary layers with applied stress. (a) Undeformed sedimentary beds. (b) The same beds after deformation.

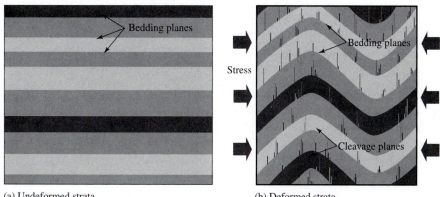

(a) Undeformed strata

(b) Deformed strata

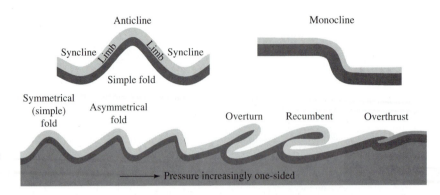

FIGURE 6.3 Diagram of fold types. Simple folds and monoclines represent relatively low stresses. Increasing stresses produce progressively deformed folds until ultimately the rocks rupture and an overthrust is the result.

Anticlines have bedding surfaces where all the beds within the folds are convex upward (Figure 6.3). To describe folds geologists often refer to the orientation of the sides or *limbs* of the fold. In the case of the paper anticline, as the space between the hands narrows, the anticline will begin to change from a *symmetrical fold* which is uniform on both sides, to something that is an *asymmetrical fold*. To characterize the deformation, geologists divide the fold with an imaginary plane that bisects the angle between the fold limbs. This imaginary plane is called the *axial plane* (Figure 6.4) because it bisects the fold along the *fold axis*, which is the surface intersection with the axial plane. The opposite of anticlines are *synclines* (Figure 6.3), where the beds are bowed downward instead of upward. Synclines, concave downward folds, are also characterized by having limbs, axes, etc. In each of the folds discussed above, the fold axis may be horizontal, or it may plunge downward into the ground.

There are some additional complex fold features that are worth mentioning, and these are illustrated in Figure 6.3. When folds begin to lean over as they become more and more asymmetrical, the axial plane also leans over. Eventually the fold is pushed so far over that beds from one limb of the fold begin to lie over beds in the other limb. Such folds are called *overturned folds*. If the process continues further and the axial plane becomes nearly horizontal, the fold is called *recumbent*. Eventually, continued compression may cause the rocks to break, and an overthrust develops (discussed below).

Generally the folds described above are relatively small features, although some might be quite large. There are also deformation features of relatively low intensity, but

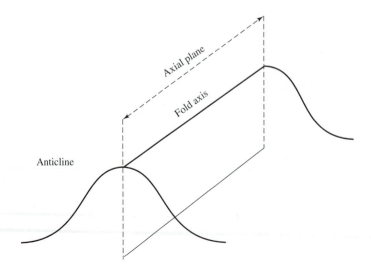

FIGURE 6.4 Diagram of an anticline illustrating the axial plane and the fold axis (see text).

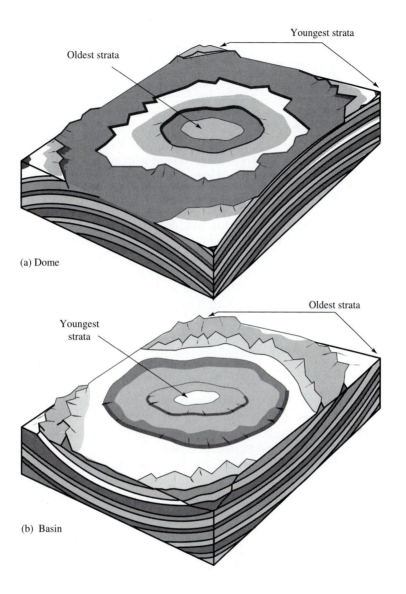

Oldest strata

Youngest strata

(a) Dome

Youngest strata

Oldest strata

(b) Basin

FIGURE 6.5 Structure of an eroded (a) dome and (b) basin.

very large scale, that develop from compressive stresses. These features are called structural *domes* and basins (Figure 6.5). Basins can also result from extension. Furthermore, high intensity of deformation produces very large-scale, complex folds, along with thrusts, called *nappes*, that include overturned and recumbent shapes. The Alps in Europe are an example. Locally, nappes have also been recognized in the Appalachian Mountains and in many other ranges.

FAULTS

When we look at areas where rocks are exposed on Earth, we find that there are many *joints* (fractures without displacement) exposed almost everywhere. These fractures are the result of a number of processes, such as cooling and fracturing due to thermal contraction of igneous bodies. Unlike joints, *faults* are fractures in the crust along which displacement has occurred and are the result of brittle deformation due to differential stress. They are identified based on the type of motion that is observed along the fault.

Because there are two blocks moving past one another during fault motion, each block is identified for descriptive purposes. One block, called the *hanging wall*, lies above the fault plane and the other block, the foot wall, lies below the fault plane. Fault types are illustrated in Figure 6.6. These are miners' terms and reflect the perspective of someone standing and facing the fault. When relative motion of the hanging wall is downward, which is what we might normally expect, the faults are called normal faults. Normal faults represent crustal stretching due to tensional stresses.

If the hanging wall moves upward these faults are called reverse faults, because the sense of motion is the reverse of what we might expect. Thrust faults are low angle reverse faults, a special case where the fault plane lies at an angle less than 45° from the horizontal. The term overthrust is used to represent a great deal of horizontal motion along the thrust fault. Reverse faults result from crustal compression such as that produced by continent to continent collisions.

A third major type of fault is a strike-slip fault (or transcurrent fault). These are characterized by very high angle faults with horizontal displacement based on the frame of reference of the observer. For example, if we happened to be standing at the edge of a strike-slip fault, and while looking across it we were to notice that things on the other side of the fault were displaced to our left, this motion identifies the fault as a left-lateral strike-slip fault (Figure 6.6). If we were to hop over to the other side of the fault, objects on the other block would also be displaced to our left. It doesn't matter which block we are standing on, the sense of displacement is the same from both sides of the fault.

When we look at rocks associated with faults we often find a zone of unusual and unique features produced by faulting. These include *fault drag*, where beds adjacent to faults have been bent by motion along the fault; *slickensides*, where the fault has abraded, striated, and polished the surfaces along which the fault motion occurred (Figure 6.7; see Pinnacles National Monument, Park Insert in Chapter 4); and *fault breccia*, where the rock has been crushed, and jagged fragments remain, often with fine-grained material filling in between blocks. A fine powdered rock "flour," called *mylonite*, may also result from this fault motion, as rocks are ground up along the fault plane.

FIGURE 6.6 Diagram of fault types. (a) An undeformed block with a fault plane (dashed line). Motion along the fault produces (b) normal fault motion, the result of tension; (c) reverse fault motion, the result of compression; (d) lateral fault motion; and (e) thrust fault motion.

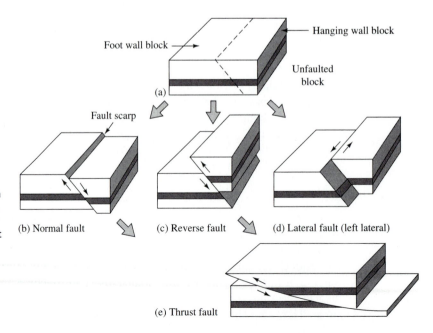

FIGURE 6.7 Slicken-sides, a scraped surface resulting from fault-block motion, from Pinnacles National Monument. (Photo by author.)

Direction of fault motion

MEASUREMENT OF STRUCTURE

Geologists use a compass to measure the structural elements they see in the field. While they are mapping, they describe the orientation of faults, beds, and other structures that they identify. From these measurements, and by identifying the types and extent of rocks, geologic maps and cross sections of the subsurface can be constructed. Two primary measurements are used. The strike of a fault or bed is defined as the trend (the angle between the line and true north) and represents an imaginary horizontal line on a dipping bed that results from the intersection of an imaginary horizontal plane with that bed (Figure 6.8). The dip, an indication of the bedding or fault plane inclination, is measured at an angle of 90° to the strike direction.

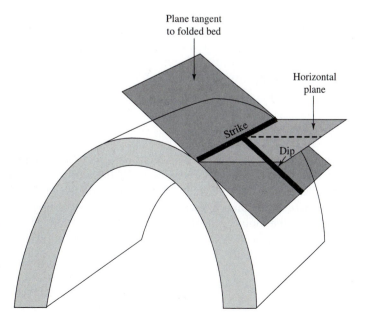

FIGURE 6.8 Illustration of the measurement of strike and dip. Geologists use strike and dip, measured using a compass, to map structure.

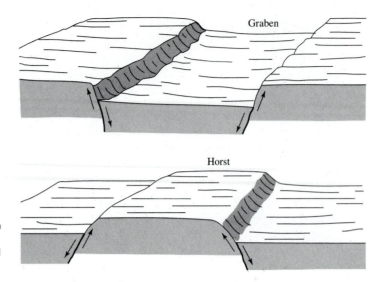

FIGURE 6.9 Illustration of a horst and a graben, both the result of normal fault motion.

CRUSTAL DEFORMATION AND STRUCTURE

Very large-scale block movements often result from regional stresses producing valleys, called *grabens,* and mountains, called *horsts* (Figure 6.9). Usually, horsts and grabens result from tensional stresses causing normal fault displacement. The Basin and Range province in the western part of the United States (see Figure 8.6) contains examples in a number of national parks.

It has been clear for a very long time that on a continental scale the crust is in motion and has always been in motion. When we look at continental *shield* rocks, representing the core of continents, we see that these rocks are very old (Precambrian), with the oldest dates found in gneissic rocks at about 3.96 billion years B.P. But there are even older greenstones that have been intruded by these gneisses, and still older undated lavas. Continents slowly grow from an initial shield, primarily by *orogenesis* and the deformation of sediments bordering the continents into mountain belts. Orogenesis also produces crustal warping, upwarps or downwarps. But with time the central continental core becomes stable because it is generally protected from orogenic effects occurring along its margins. It is not protected from erosion and isostatic effects, or from impact by extraterrestrial objects. Eventually continents break apart due to plate tectonic forces and orogenesis starts all over again.

MASS MOVEMENTS OF EARTH MATERIALS

Mass wasting is a process by which loose particles and rocks move downslope. The driving force behind mass wasting is the force of *gravity.* Of course with every Earth process, energy is expended in some way to do the geological work. The types of energy involved in the mass wasting process are interesting. For example, solar energy is stored in the moisture in clouds through evaporation. Moisture in the clouds has potential (waiting) energy until rain falls, when the raindrops take on *kinetic* (moving) energy. Falling rain dislodges particles that then move downhill. Orogenesis, and associated uplift, stores energy where uplifted rocks represent potential energy. Later, as erosion progresses, the particles moving downhill represent kinetic energy.

TABLE 6.1   CLASSIFICATION OF MASS WASTING

	Name	Particle Size	Particle Movement
highest energy	rockfall	large	downward
	debris fall	smaller	downward
	rockslide	large	downward and outward
	debris slide	smaller	downward and outward
	slump (slope failure)	all sizes	downward and outward
	slump-generated flows		
	landslides		
	rock avalanches		
	debris flows		
	mudflows		
lowest energy	*creep*		

An understanding of mass movements of particles is very important to many activities performed by humans. These activities include soil conservation in agriculture, structural design in engineering, building site location in construction, and many others. Therefore mass movements have been classified based on the magnitude of observed effects, including particle size and flow velocity, inferred from the type of fall. For example, a particle falling straight down would have a fast flow. This classification is given in Table 6.1, and is illustrated in Figure 6.10.

Slump generated flows (Figure 6.11), while descriptive of particle size, include flows that are suspended by their own turbulent motion and may flow very fast and far. These slump-type flows include the following from highest to lowest flow velocity:

- *Landslides* is a general term used to characterize very large tumbling rock flows and large discrete blocks. They include rock avalanches and, in some instances, debris flows.
- *Rock avalanches* are fast, extensive, fluid-like flows that can move at velocities up to 100 km/hr and therefore are very dangerous.
- *Debris flows* are much less dynamic, and flow like a relatively slow moving fluid, with speeds from less than 1 cm/hr to 1 km/hr. They are tongue shaped and may be very extensive or quite small, and the grain sizes carried in the flow are usually smaller than carried by rock avalanches.

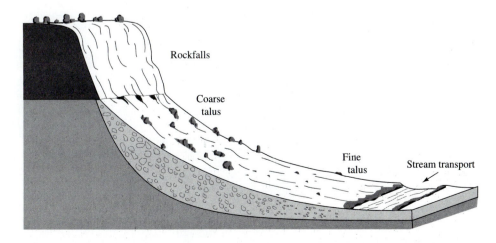

FIGURE 6.10
Movement of materials downslope by mass wasting processes.

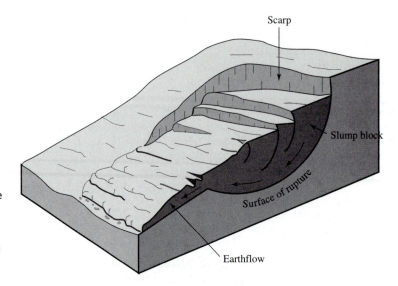

FIGURE 6.11 Structure of a slump block, representing simple ground failure that can generate earthflows and result in the development of scarps.

- *Mudflows* may or may not be slump generated and involve very fine particle flows with up to 30% water. These flows can dislodge and carry large rocks and therefore they can be quite destructive.

Creep is another interesting mass wasting process because it involves the imperceptible downslope movement of particles (Figure 6.12). Evidence of this motion does not come from direct observation because the movement is too slow. Rather it comes from observations of leaning trees and fence posts, buckled pavements and walls, bent strata, and other factors. This imperceptible flow has several causes. *Frost heaving* by freezing lifts particles in the soil as ice crystals form. Then, during thawing, the particles drop back but move slightly downslope. Volume changes in sediments, such as those resulting from wetting and drying of clays, or the heating and cooling of minerals, cause these materials to expand and contract. As particles contract the sediment moves downhill. When snow accumulates on a hillside, its weight drags the upper sur-

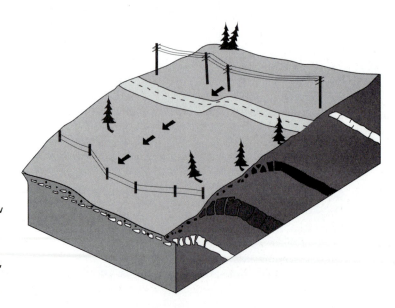

FIGURE 6.12 Surface movement associated with creep, showing how the upper surface can bend and even break. This may cause trees to grow in a deformed way, fences to bend out of line, and roads to move.

face of the rocks downslope. Dissolution in subsurface rocks creates voids that collapse and the upper surface of the material then moves downslope.

MASS WASTING DEPOSITS

Deposition of the jumbled, angular particles that result from mass wasting produces a unique deposit, *colluvium,* that can be recognized by the particles it contains and by the fact that usually no layering is observed. In contrast, alluvial fans are deposited from intermittent stream flow (see Chapter 11). A special type of colluvium is called talus. *Talus,* an apron of rock found at the base of cliffs, accumulates by rock falls and slides. The largest particles remain nearest the cliffs from which they have fallen, while further away from the cliff the accumulating particles are finer grained. Sliderocks are the rocks contained in a talus apron and they are moving very slowly downslope. This very slow movement allows weathering to affect the rocks in the slide.

In addition to colluvial material, there may also be frozen materials that are the result of mass wasting processes. In areas of permafrost (permanently frozen ground), *solifluction,* downslope motion of water-saturated debris, can become very important. Also important in some areas are *rock glaciers* that are mainly made up of rock debris bound together by interstitial ice, producing a driving mechanism that is very similar to that of a glacier.

METAMORPHISM

The word metamorphic means change, and some rocks now exposed at Earth's surface have been altered as the result of changing conditions. Temperature, pressure, and active fluids are the causes of these changes. Increases in temperature and pressure (stress) produce several effects that can be identified in outcrops, in hand samples, and under the microscope. The primary changes include *plastic deformation* of mineral grains identified by changes in their shape. Changes in pore fluid chemistry also occur due to the mobility of water and CO_2 during metamorphism. Furthermore, new ions from external sources migrate into rocks and combine with the minerals already present to produce new metamorphic minerals. This process is called *metasomatism.* Minerals also experience recrystallization, forming new minerals, structures, and textures resulting from elevated temperatures and pressures. The recrystallization process occurs in slow stages producing interlocking grains that are distinctively metamorphic.

Metamorphic Rock Types

Metamorphic rocks are divided into several types based primarily on the size of grains within the sample, and whether they are foliated. *Gneisses* are banded light/dark rocks with relatively coarse grain sizes; the lighter grains are quartz and feldspar. The dark bands are mainly ferromagnesian minerals, primarily biotite. Micas are some of the first metamorphic minerals that form, first as microscopic grains; then with higher temperature and pressure, larger mica grains appear. This change to greater metamorphism starts with slates, which have microscopic mica flakes. Phyllites have grains that are larger than in slates, containing barely macroscopic mica flakes. *Schists* have larger, visible mica flakes. These and other metamorphic terms are contrasted with the original rock type in Table 6.2.

As temperatures and pressures increase, distinctive mineral assemblages are produced from preexisting *protoliths.* These assemblages are used in classifying

TABLE 6.2 COMMON PROTOLITHS AND METAMORPHIC ROCK TYPES

Protolith (Original Material)	Metamorphic Rock
sandstone	quartzite
conglomerate	metaconglomerate
limestone	marble
shale, mudstone	slate-phyllite-schist-gneiss
granite	granite gneiss
basalt	schist, amphibolite
peat	bituminous coal→anthracite

Note: The protolith sedimentary rock types will be discussed in Chapter 10. Two gneisses are listed in Table 6.2; one originating from a sedimentary protolith (*S-type gneiss*) and the other from an igneous protolith (*I-type gneiss*).

metamorphic rocks (Figure 6.13). This classification is based on the concept of *metamorphic grade*; how extreme were the conditions that produced the characteristic (index) minerals observed in the metamorphic rock? For example, *low grade* metamorphism produces rocks that have been subjected to small temperature increases at low pressures. Low grade metamorphic rocks have small grain sizes. The minerals formed include, in order of increasing metamorphic grade, chlorite, biotite, and garnet. *High grade* metamorphic rocks form at high temperatures and pressures and exhibit large grain sizes. The minerals in these rocks include staurolite, kyanite, and sillimanite, with sillimanite being the highest-grade metamorphic mineral in this group.

Metamorphic rock texture or *fabric* is developed as minerals deform and grow under metamorphic conditions. The most easily identified effects are produced by planar mineral growth and alignment at right angles to applied stresses (Figure 6.14). The resulting planar *foliation* can be quite distinctive and is often used to differentiate some igneous rocks, such as granites, from metamorphic look-alikes, such as gneisses. Other planar features include *slaty cleavage*, developed at relatively low temperature

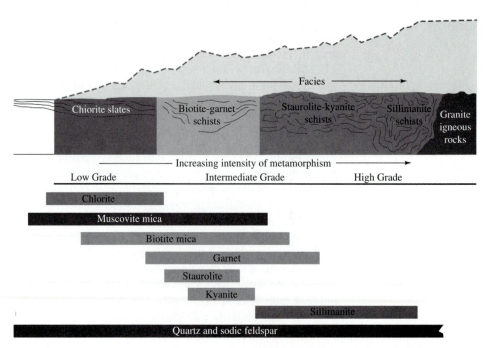

FIGURE 6.13 Diagram illustrating metamorphic grade, metamorphic facies, and the corresponding minerals.

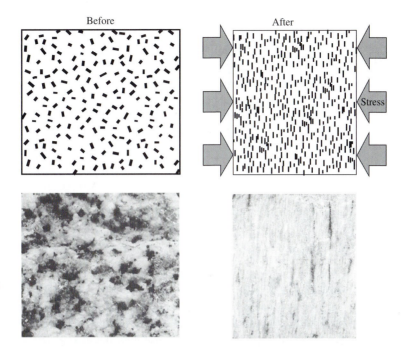

Before   After

Stress

FIGURE 6.14 Diagram illustrating how stress applied to a granite develops foliation, producing a gneiss.

and pressure. The result is planar distributions of fine grained minerals that are easily split apart. Medium to coarse, dark mica grains produced at higher temperatures and pressures form a textural character called *schistosity*. Coarse grained metamorphic rocks with alternating light and dark bands or layers exhibit what is called *gneissic layering* (Figure 6.15; see also Rocky Mountain National Park Park Insert in Chapter 7).

FIGURE 6.15 Banded gneiss in Rocky Mountain National Park. (Photo by author.)

TABLE 6.3 METAMORPHIC ROCK CLASSIFICATION

Facies	Grade	Minerals
zeolite	very low	zeolites, illite, chlorite
greenschist	low	chlorite, epidote
amphibolite	intermediate	garnet, amphibole
granulite	high	sillimanite
blueschist	high	low temperature and high pressure minerals

Note: While zeolite to granulite facies rocks represent relatively uniformly increasing temperatures and pressures, blueschist minerals are unusual in that they are developed in the very low temperature, but relatively high pressure, environments associated with subducting plate margins.

Gneissic layering may be deformed, as in Figure 6.15, or it may be relatively undeformed as in Figure 6.14.

Over broad regions, metamorphic rocks often exhibit a distinctive, unique character and appearance that is termed the *metamorphic facies*. The development of facies is controlled by temperature and pressure, but also by the availability of ions, as well as the original rock composition. The facies concept can be useful in comparative geological studies, in part because the same facies usually reflect common conditions present during formation of the rock. Facies are also useful when studying sedimentary rocks (Chapter 10). Table 6.3 gives the common metamorphic facies, the metamorphic grade, and the minerals usually found in such rocks.

The following is an example of how these facies may develop. Low temperature alteration can change rocks by the addition of a few new mineral constituents. The magnitude of this alteration may be very small, but the color of the minerals can be distinctive and diagnostic. For example, a yellow/white coloration is imparted by *zeolite* minerals and is found in *zeolite* facies metamorphic rocks (Table 6.3). With little greater alteration a green coloration results from the formation of the mineral *chlorite,* and the metamorphic facies becomes *greenschist* (Table 6.3).

Besides facies classifications, metamorphic rocks are also characterized by the style and the size of the area affected by metamorphism. For example, large-scale meta-

BLACK CANYON OF THE GUNNISON NATIONAL MONUMENT

The Black Canyon of the Gunnison National Monument (see Figure 1.18) was established in 1933. Located in west central Colorado, it is a 2,700 ft (about 800 m) deep gorge that has been cut by a Late Tertiary stream diverted here by a volcanic dam. As the result of the change in direction of the stream, now represented by the Gunnison River, a steep gradient was created, resulting in high water velocities that have rapidly cut a canyon, in places only 1,100 ft (335 m) across. The final product is a very impressive drop from a flat plateau into a very deep canyon (Figure P6.1).

Geology. Exposed here are Precambrian metamorphic rocks, mainly resistant gneisses and quartz-mica schists. At one time these rocks were sedimentary and volcanic. During late stages of metamorphism, in the later Precambrian Period, the rocks were intruded by many granitic dikes and sills leaving a complex network of light igneous rock bands cutting through the dark metamorphic country rock (Figure P6.2). The Gunnison River in the deep canyon below gives off a bright green contrasting color that is quite striking. The color is due to sunlight reflecting off the abundant particles being carried by the stream.

FIGURE P6.1 Gunnison River flowing through Black Canyon of the Gunnison National Monument. (Photo by author.)

FIGURE P6.2 Dark metamorphic rock walls (where the monument gets its name) intruded by light igneous dikes in Black Canyon of the Gunnison National Monument. (Photo by author.)

morphic effects are characterized as *regional metamorphism*, with large-scale shearing pressures found at the base or roots of mountains. When such large-scale effects occur, blocky minerals are changed to platy minerals. For example orthoclase (a feldspar) may change to muscovite (a mica). A second type, *contact metamorphism*, is formed by intense heating at low pressures. Neither metamorphic foliations or facies are associated with contact metamorphism. Instead, it may produce *hornfels* rocks, a fine-grained, hard, dense, and dark mineral assemblage found in the contact zone surrounding intrusions, called an *aureole* (Figure 6.16; see also Figure 5.15), or beneath lava flows.

Cataclastic metamorphism results from grinding during faulting, which produces coarse-grained *fault breccias* as well as distinctive fine-grained materials called *mylonites*. A fourth type of metamorphism is termed *convergent boundary metamorphism*, where blueshist facies (Table 6.3), produced in subducting marine sediments and volcanic materials, is associated with a plate boundary.

There are also some unusual metamorphic effects produced by meteor impacts. *Impact metamorphism* produces distinctive cone-in-cone shapes, like alternating inverted and upright "ice cream cones," that show the development of fine mica flakes. Other shock-related effects are also observed, including the formation of glasses

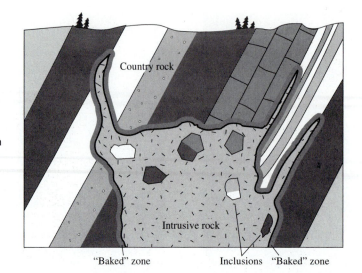

FIGURE 6.16 Diagram illustrating an intruding granite baking the intruded country rock and producing a baked or chilled zone. Inclusions, if not assimilated into the magma, become xenoliths.

(*tektites*) in some ejected material as the result of melting, flash cooling, and *breccia-tion* (breaking and lithification of the pieces) of some material. There are many examples of meteor impact features on Earth. One of the best preserved of these is the Arizona Meteor Crater. Recently, an impact feature was identified in Canyonlands National Park (Chapter 11), but such features are often very difficult to identify, thus go unnoticed or misinterpreted by geologists. For example, the ejecta from meteorite impacts (Figure 6.17) are often misidentified as *volcanic tuffs* because of the similarity in appearance between the two. There may even be lavas that fill the crater as often happened on the Moon (Figure 6.17b).

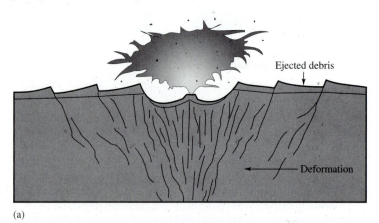

(a)

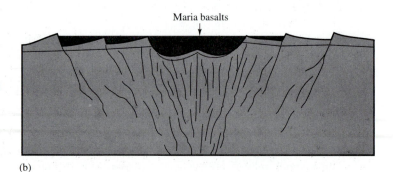

FIGURE 6.17 Meteorite impact effects, (a) where ejecta are thrown out of the crater formed by impact and collect around the area. Collapse faulting also accompanies impact, and in some cases, such as on the Moon, (b) the crater may fill with lava caused by pressure changes and corresponding melting during impact.

(b)

VOYAGEURS NATIONAL PARK

Voyageurs National Park (see Figures 1.21 and 8.14) in northern Minnesota was established by Congress in 1971. It lies on the *Precambrian shield* rocks of North America, the ancient core of our continent. Exposed within the park are ancient granites, basalts, and metamorphic rocks, but weathering and growth of vegetation have obscured most of the rocks (Figure P6.3). The North American ice sheet covered this area during the Pleistocene, and as a result, the surface has been scoured by these massive glaciers. A number of depressions gouged in the rock are now the complex of lakes that make up the park. Voyageurs was mainly designed as a boater's park and shares a common boundary with Canada. On occasion, a U.S. Customs Service boat can be seen patrolling the waterways looking for stray, unauthorized visitors into the United States.

Geology. During the Archean (early Precambrian time) deposition of marine clastic sediments occurred. This was followed by the extrusion of submarine lavas with occasional pyroclastic eruptions. Small-scale granitic intrusions were emplaced in this sequence and the pile was capped by rapid deposition of a series of clastic sediments. Then late in the Archean, the *Kenoran Orogenic* event (see Chapter 7) uplifted the area with accompanying metamorphism and the intrusion of granites. In the middle Precambrian, mafic dikes intruded the area. Since that time the area has experienced prolonged erosion, exposing the Precambrian metamorphic and igneous rocks that at one time were deep below the surface.

FIGURE P6.3 Gneiss outcrop, intruded by granitic pegmatites (light bands) in Voyageurs National Park, Minnesota. (Photo by author.)

COAL

Coal is an important source of energy that is formed from plant remains accumulating in swamps that have then undergone burial (Figure 6.18). As a result of this burial, and tectonic forces in some cases, they have been altered by increasing temperatures and pressures. Coals range from the lowest grade, *peats*, to the highest grade, *anthracites*; grade is determined by how hot the material burns and how many impurities it contains.

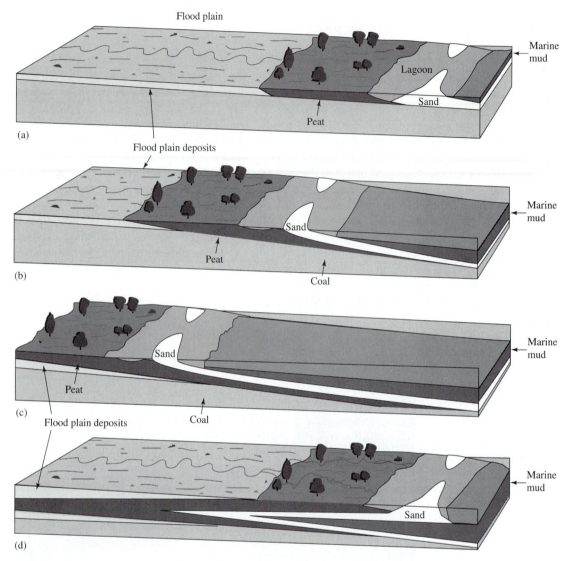

FIGURE 6.18 Diagram illustrating development of cycles (cyclothems) of plant-coal-sediment accumulation associated with fluctuating sea level and shore lines. Coal is produced from plant material (peat) as it is buried and compacted. (a) Initial sea level is low, and peat develops behind the beach or lagoonal system. (b) Deepening water causes the shoreline to move inland, burying the peat to deeper depths, increasing pressure and producing coal. (c) Maximum extent of shoreline incursion. (d) Dropping sea level and burial of peat by floodplain deposits.

Coal grades are represented in Figure 6.19. The best burning coal is anthracite, but very little anthracite exists in the U.S. (Figure 6.20). Instead, most coals are lower grade, do not burn nearly as hot as anthracites, and produce a lot of nonburnable waste products.

Coal is very important, economically launching the Industrial Revolution in Great Britain, thus geologists worked hard to locate and exploit it. In the process, they developed an understanding of the "Coal Measures" that became the basis for the Carboniferous Period of the Geologic Time Scale. Carboniferous rocks contain abundant coals in the United States and a major mid-Carboniferous break in the rock record (*unconformity*). For better resolution in dealing with these rock sequences in the United

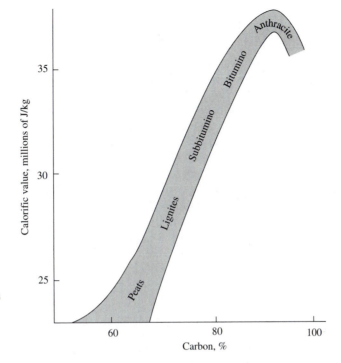

FIGURE 6.19 Coal grade is defined by heat generated during burning versus impurities produced. The lowest grade material is peat, and the highest grade material is anthracite.

States, early American geologists found it expedient to replace the Carboniferous Period with two new periods, the Mississippian and Pennsylvanian. However, because these periods are restricted to the United States, many modern American geologists have dropped these names in favor of the more globally applicable Carboniferous Period.

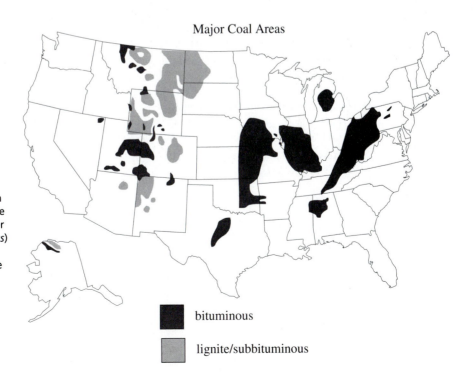

FIGURE 6.20 Location and quality of coal in the United States. The better quality coals (*bituminous*) are mainly found in the East, while poorer grade coals (lignite and sub-bituminous) are mainly found in the West. The best quality coal, anthracite, is not common in the United States.

There are many environmental problems associated with coals. One of these is that pyrite [FeS_2], found in coal as a by-product of the metamorphic process, breaks down during burning, releasing sulfur that then combines with water to produce sulfur-rich pollutants responsible for acid rain. These coals have been identified and are no longer burned in the United States, but we still mine them and send them to other countries, where they are burned. Other problems associated with coals come from the mining process. Elemental contaminants, like cadmium, are concentrated in sediments associated with coals; thus when coals are mined, these contaminants are released into the environment, becoming a serious problem.

Haze from air pollution caused by coal-fired electric generating plants and other sources is obscuring the view in important park areas such as Grand Canyon National Park (Chapter 15). A cooling tower is a prominent background feature for visitors to Indiana Dunes National Seashore (Figure 1.5). In addition, coal fires, either in coal mines or caused naturally by lightning strikes, can burn underground for decades. In Theodore Roosevelt National Park (Chapter 9) there are areas where the ground is bright red, caused by oxidation and baking of surface sedimentary rocks from burning underground coals.

REFERENCES

BALLY, A. W., and PALMER, A. R., eds. 1989. *Decade of North American geology, DNAG, Geology of North America.* Vol. A, *The geology of North America; An overview*, p. 619. Boulder, Colorado: The Geological Society of America.

BENNETT, R., ed. 1980. *The new America's wonderlands our national parks,* p. 463. Washington, D.C.: National Geographic Society.

BEUS, S. S., ed. 1987. *Decade of North American geology, DNAG, Centennial field guide.* Vol. 2, *Rocky Mountain Section of the Geological Society of America*, p. 475. Boulder, Colorado: Geological Society of America.

BIGGS, D. L., ed. 1987. *Decade of North American Geology, DNAG, Centennial field guide.* Vol. 3. *North-Central Section of the Geological Society of America*, p. 448. Boulder, Colorado: Geological Society of America.

CHRONIC, H. 1986. *Pages of stone, geology of western national parks and monuments.* Vol. 4, *Grand Canyon and the Plateau Country*, p. 158. Seattle, Washington: The Mountaineers.

HARRIS, A. G., and TUTTLE, E. 1990. *Geology of national parks.* 4th ed., p. 652. Dubuque, Iowa: Kendall Hunt Publishing Company.

HARRIS, D. V., and KIVER, E. P. 1985. *The geologic story of the national parks and monuments.* 4th ed., p. 464. New York: John Wiley & Sons.

RAMSAY, J. G., and HUBER, M. I. 1987. *The Techniques of Modern Structural Geology.* Vol. 2, *Folds and Fractures*, p. 700. London: Academic Press.

Mountain Building

Before mountain building can be discussed effectively, there are a few simple terms that need to be defined. *Orogenesis* is the term that characterizes the totality of the mountain-building processes, while *orogeny* refers to a particular mountain-building event. *Cratons* or *shields* are those relatively stable, ancient continental cores or platforms around which the rest of the continent is growing. For example, Figure 7.1 illustrates that the ancient core of the North American continent is older than 2.5 billion years, with the oldest rocks being almost 4 billion years old. The rest of the continent, in general, gets progressively younger as the continental margins are approached. This is illustrated along the A–B line in Figure 7.1, where the continental core gets progressively younger toward the southeast. These various zones are the result of orogenic events. Only in the Superior province are the ancient core rocks exposed. Instead, elsewhere they are covered by a thick layer of Phanerozoic sediments. Along the eastern margin of our continent, the Coastal Plain province is composed of Mesozoic and Cenozoic sediments, lapping up onto the old continental core.

Seaward of continental margins, eroded sediment accumulates as an accretionary wedge. When plates collide, during the resulting orogenesis these sedimentary packages may be folded, faulted, and uplifted to be exposed as new continental margins. Two continental masses may be welded or *sutured* together during the process forming a much larger continent. Or, ocean crust material, including sea-floor basalts and the overlying sedimentary cover, may be uplifted and exposed on land as *ophiolites*. *Displaced* or *accreted terranes* are fragments of continents or island arcs that have collected along a continental margin as one plate margin overrides another. The western Cordillera of North America (Figure 7.1) is composed, in part, of many accreted terranes.

ISOSTASY AND CRUSTAL THICKNESS

Why are continents above sea level? The answer to that question comes from theories by the Greek mathematician, Archimedes, who showed in the 2d century B.C. that the

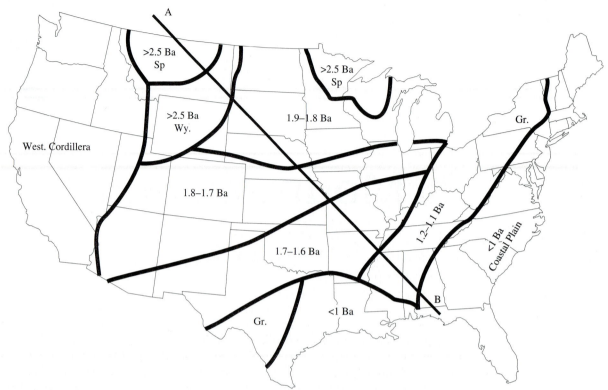

FIGURE 7.1   Ages of basement (core) rocks in the United States. Note that the ages roughly decrease from the central part of the continent, along a line from A to B or from Lake Superior to B. Everywhere but the Superior province (sp) around Lake Superior these ancient rocks are covered by Phanerozoic sediments. The western Cordillera is made up of displaced terranes, fragments of land collected as the North American continent moved westward. Ba = billion years; Gr = Grenville terrane; Wy = Wyoming terrane.

supporting force of objects placed in a fluid is equal to the weight of the displaced fluid. Objects will float if the supporting force is greater than the force of gravity pushing down on the object. This is how icebergs float in water; the force of the water is displaced by the ice and supports it. If you try to push a small beach ball underwater, the deeper you push it, the greater the opposing force. However, after it is completely submerged, no additional force is needed to keep it under water. We know that Earth materials behave as highly viscous fluids. For example, glass window panes in 17th and 18th century U.S. homes are thick at the bottom and often have a gap at the top, because the glass, a very high viscosity fluid, has been very slowly flowing downward under the influence of gravity.

Archimedes' principle applies to Earth, resulting in flotation balance among segments of the lithosphere (Figure 7.2). Average continental rocks are less dense than oceanic and mantle rocks; therefore lighter continental blocks "float" in the more dense material in which they are sitting. As a result of the fact that Earth materials flow, elevated, low-density crustal areas, such as mountains floating in the mantle, behave much like icebergs floating in water. The mountains are supported by the displaced mantle rocks. As mountains are eroded away, they behave like melting icebergs; the entire mass is pushed upward by the displacing fluid because they are no longer as heavy as they were and they don't exert as much downward force against the opposing mantle forces. These flotation effects applied to Earth materials are known as *isostasy*. When mountains float up after their mass is reduced by erosion (Figure 7.3), the effect is

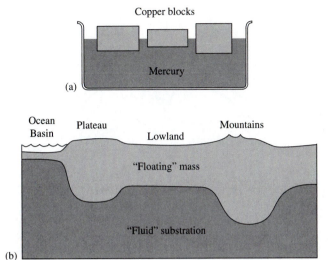

FIGURE 7.2 (a) Copper blocks floating in mercury are analogs to (b) continental materials floating in the mantle, both illustrating Archimedes' principle.

called isostatic rebound. An interesting case of observable isostatic rebound is occurring today as the result of the rebound of a good part of eastern Canada after the great North American continental ice sheets melted. Today, the Hudson Bay region of Canada is rebounding at the rate greater than 10 mm/yr (Figure 7.4), due to the fact that the weight of the continental ice sheets is now gone, and, like a cork, the North American shield region in Canada is rising.

The early observations by two British surveyors in India, Pratt and Airy in 1855, led to our realization that mountains are somewhat like icebergs floating "in" the Earth. They discovered that when they were near the Himalaya Mountains, a mass hanging from a string did not hang straight down, but rather was deflected toward the mountains. The higher the mountain, the greater the deflection. Airy argued that this meant that mountains have roots, and that they are floating in the mantle, with higher mountains having deeper roots. In contrast, Pratt postulated a common depth of compensation, essentially a uniform base level, where density contrasts even out. Each disagreed

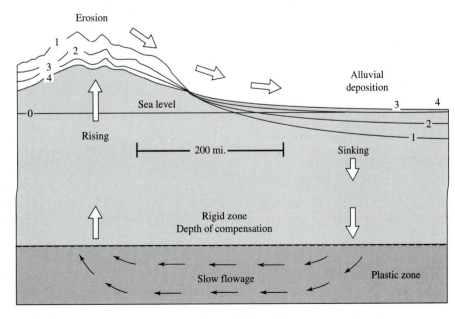

FIGURE 7.3 Diagram illustrating erosion of mountains and isostatic rebound. Numbers 1–4 represent oldest to youngest topography that has been uplifted through isostatic rebound, then eroded off four different times. Each event is then represented by deposition offshore, causing sinking that in turn causes lateral flow in the subsurface, filling in behind the isostatically rising mountains.

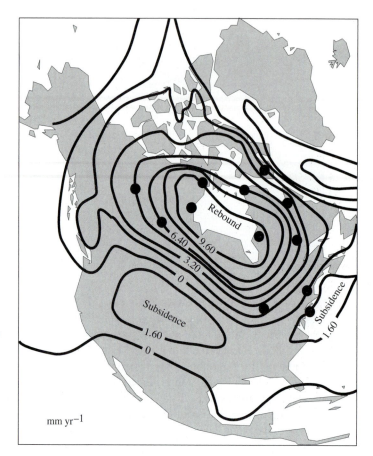

FIGURE 7.4 Canada's isostatic rebound following melting and removal of the last North American ice sheet and subsidence along the margins as the result of lateral subsurface inflow. This process is removing the crustal depression produced by the weight of the ice.

with the other's hypothesis, but we now know that they both were, in part, correct in their basic premises. Mountains do have roots, and there is a common base, the base of the lithosphere.

At the base of mountains, at their roots, temperatures and pressures are high, and the rocks in this region are readily metamorphosed. This produces the regional metamorphic effects discussed in Chapter 6, as well as some partial melting and igneous rock intrusions (Figure 7.5). Erosion and isostatic rebound will eventually bring these metamorphic and igneous assemblages to the surface where they can be studied by geologists (Figure 7.6). The Black Hills region of South Dakota was produced in this way.

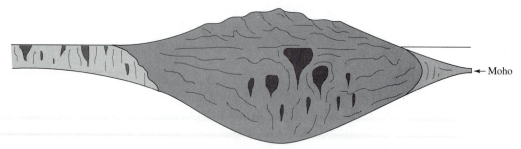

FIGURE 7.5 Mountain cross section illustrating metamorphism and igneous intrusions in the roots of mountains.

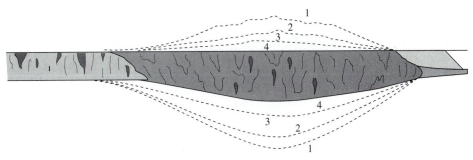

FIGURE 7.6 Mountain isostatic rebound and erosion cross section of the mountain in Figure 7.5, eventually bringing the metamorphosed root zone to the surface. As in Figure 7.3, numbers 1–4 represent oldest to youngest topography that has been uplifted through isostatic rebound, then eroded off four different times.

MOUNTAIN BELTS

Mountains are produced by plate-to-plate collision with a number of variations. When a continental margin encounters a trench, coastal ranges will be produced along the continental margin (Figure 7.7). Examples are the Sierra Nevadas and Cascades, that were both coastal ranges at one time. Now, as the result of several later orogenic events, these mountains lie well into the continental interior (see Chapter 8). As a result of orogenesis and accretion, the North American continent has grown (Figure 7.1) and will continue to grow until it is ripped apart by some rifting event that will produce two or more new continents.

When two continental elements of plates collide (Figure 7.8), mountains, such as the Appalachians along the eastern seaboard of the North American continent and the Atlas Mountains in Morocco, will be produced along the colliding continental margins.

FIGURE 7.7 Illustration of collision between a continent and an ocean and the resulting deformation. This includes andesitic volcanism on the continent and deformation of the near-shore sediments associated with the continent. The Pacific Coast of North America is an example, with the results of collision illustrated in many of the national park areas in this region.

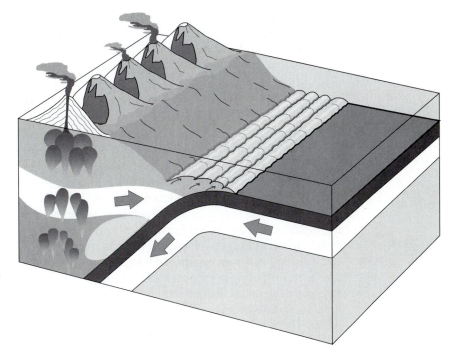

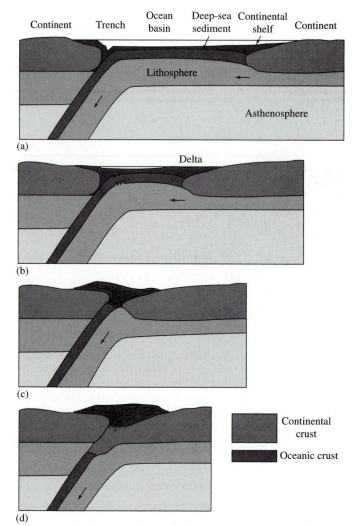

FIGURE 7.8 Illustration of collision between two continents and the resulting deformation. The eastern margin of North America and the park regions associated with the Appalachian Mountains is an example of the results of collision shown here.

On the other hand, island arcs are produced by oceanic plate subduction and ocean crust to ocean crust plate-margin interaction (Figure 7.9), the Aleutians Islands being a good example (Figure 1.11).

Orogenic effects associated with these plate interactions include a number of factors. Magma forms at the core of mountains due to high temperatures and pressures developed deep within them. Also, where subduction is occurring, frictional processes may be important. Magma formation then leads to the emplacement of intrusives and expulsion of extrusives. As the mountains further develop, or as the result of pressures generated during subduction, metamorphic grade increases.

During continent to continent collision, major magnitude folding and thrust faulting is produced from shortening directed toward continental interiors. This may result in the creation of *nappes*, large stacked and overturned folds and thrusts. Associated thrust sheets form from continental materials, from accretionary wedges, and from sediments accumulating in very near-shore *foredeep depressions*. Sediments found in these foredeep depressions include *turbidites* and distinctive black *shales*, together known as *flysch*, and mixed marine deposits. Collectively these sediments are called *molasse deposits*. Turbidites are deposited from high sediment density submarine flows, called

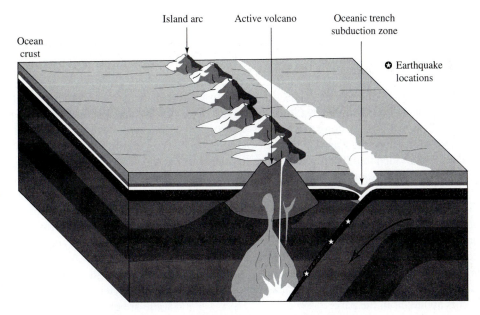

Island arc Active volcano Oceanic trench subduction zone

Ocean crust

⊕ Earthquake locations

FIGURE 7.9 Illustration of collision between two ocean crusts and the resulting deformation. The Aleutian Islands and the associated park regions in Alaska are an example of the results of collision shown here.

turbidity currents (discussed in Chapter 14). The process of collision compresses and displaces accretionary wedges, adding this material to the continental margin by scraping it off subducting slabs. Furthermore, newly uplifted mountains contribute to offshore sediment accumulations because they shed large amounts of sediment through active erosion.

Orogenesis may occur in a series of tectonic cycles of mountain building and erosion following repeated plate collision. In the Appalachian Mountains, for instance, several such cycles are recognized. The first of these occurred in the late Precambrian and is called the *Grenville Orogeny*. Another occurred in the late Ordovician and is called the *Taconic Orogeny*, and a third, the *Acadian Orogeny,* occurred in the Devonian. (In Europe, also involved in the collision, the Acadian is known as the *Caledonian Orogeny*.) The Appalachian mountains we see today are still experiencing isostatic adjustment, but the final orogenic event, called the *Alleghenian Orogeny* by North American geologists, began in the late Carboniferous and climaxed in the Permian. (In Europe the equivalent orogenic event is called either the *Hercynian* or the *Variscan Orogeny*.)

Mountain Belt Types

There are three basic types of mountain chains or belts that develop through different orogenic processes. The first of these, volcanic mountain belts, are composed mainly of basaltic and andesitic extrusions. Examples include island arcs, such as the Aleutian Islands, and continental arcs, such as the Cascades Range in western North America, emplaced by volcanism above the overidden west coast subduction zone. A second type, *fault block* mountain belts, are formed by tensional, normal faulting of large crustal pieces. Excellent examples include the Sierra Nevada, the Grand Tetons, and the mountains in the Basin and Range province. Third, fold and/or tectonic mountain belts result from compressed, folded, crumpled, and thrust-faulted rock sequences. Examples include the Alps in Europe, the Appalachians in North America, and the Himalayas in Asia, all produced during continent-to-continent collision. Fold mountains also result from continent to trench collision, and a good example is the Andes in South America.

EXAMPLE OF PARKS IN VOLCANIC MOUNTAIN BELTS

NORTH CASCADES NATIONAL PARK

Established by Congress in 1968 in northwestern Washington, North Cascades National Park (see Figures 1.12 and 8.4) sits along the central axis of the northern segment of the Cascade Range (Figure P7.1). The park is divided into two segments, one to the north and the other to the south of Ross Lake National Recreational Area (Figure P7.2). When Congress designated the park, they left the recreational area, with its developed areas and man-made lake, intact for recreational purposes. However, roads and bridges in the newly designated park generally are not being maintained and are being reclaimed by natural processes (Figure P7.3). Because of the high precipitation rates, thick vegetation is abundant in the park.

Geology. A wide range of Mesozoic metamorphic rocks are exposed in the park, including Triassic gneisses and schists. Central to the park is the Picket Range (Figure P7.1), containing a granite core. Emplacement of the granites began in the late Cretaceous and continued into the Miocene, as did tectonic activity including thrust and block faulting. Late Tertiary volcanics are exposed throughout the area, as are Quaternary composite volcanoes built from excessive volcanism accompanying western margin subduction below the North American Plate. Due to high precipitation and high altitudes in this region today, there are many glaciers at higher elevations in the park, and some are quite large.

FIGURE P7.1 Picket Range in North Cascades National Park, Washington. (Photo by author.)

FIGURE P7.2 Ross Lake National Recreational Area lying between the north and south segments of North Cascades National Park. (Photo by author.)

FIGURE P7.3 Bridge (no longer used) and vegetation taking over the old roads in North Cascades National Park. (Photo by author.)

EXAMPLES OF PARKS IN FAULT BLOCK MOUNTAINS

 GRAND TETON NATIONAL PARK

Grand Teton National Park (see Figures 1.18 and 8.9) in northwestern Wyoming (see Figures 1.2 and 2.3) was established by Congress in 1929, and in 1950 was enlarged to its present size. Its highest peak is Grand Teton at 13,700 ft (4,197 m).

Geology. The park is composed mainly of Precambrian rocks. Most of these are metamorphic, banded gneisses and a few schists (see Chapter 6) that are very resistant to erosion. The metamorphic rocks are approximately 2.8 billion years old and have been intruded by granite plutons that are approximately 2.5 billion years old. These in turn are intruded by diabase (basaltic) dikes that are approximately 1.4 billion years old (see Figure 5.14). Erosion during the Precambrian left a level surface on which Cambrian sediments were deposited. Those sediments can be seen capping Mt. Moran today (Figure P7.4). The boundary between the Cambrian sediments and the metamorphic rocks that make up Mt. Moran has a special name, a *nonconformity*. Nonconformities are defined by a surface formed by erosion that separates sedimentary rocks above from igneous or metamorphic rocks below (see Chapter 15 for discussion of other types of unconformities).

The Grand Tetons are fault block mountains formed by several events. First, thrust faulting pushed the metamorphic and igneous rocks into position. Second, normal faulting produced a vertical displacement of these rocks of more than 20,000 ft (approximately 6,000 m). The raised block is known as the Teton block. The lowered side of the fault is called the Jackson Hole block (Figure P7.5). Erosion by mass wasting has been modifying the Teton block that in turn has been further modified by alpine glaciation. Deposition of much of the eroded sediments is on top of the Jackson Hole block.

FIGURE P7.4 Mt. Moran in Grand Tetons National Park, Wyoming. (Photo by author.)

FIGURE P7.5 Teton Mountains in Grand Teton National Park, standing above the Jackson Hole depressed fault block. Photo looking toward the south. (Photo by author.)

 GREAT BASIN NATIONAL PARK

Great Basin National Park (see Figures 1.14 and 8.6) was established in 1986 in eastern Nevada. It was developed around Lehman Caves National Monument whose name was changed and area expanded to include Wheeler Peak (Figure P7.6), the highest peak in the Snake Range at 13,063 ft (3,982 m). Lehman Cave was formed by dissolution along fractures in the Cambrian limestone at the site. Locally, the limestone has been metamorphosed to *marble* by granitic intrusions that are now exposed at lower elevations in the park. The park is famous for bristlecone pine trees (see Figure 3.1). Bristlecone pines are used to calibrate ^{14}C dating methods by counting and dating the rings of these trees, some of which are nearly 5,000 years old.

FIGURE P7.6 Wheeler Peak in Great Basin National Park, Nevada. (Photo courtesy of the National Park Service.)

Geology. During the early Paleozoic, deposition of clastic sedimentary rocks, as well as limestones, was followed by uplift and erosion of the area in the late Paleozoic. Beginning in the mid-late Mesozoic and into the Tertiary, emplacement of intrusions metamorphosed some of the Paleozoic sediments. In the Oligo- cene, eruption of lavas and tuffs was followed by the development of Basin and Range topography, as the result of block faulting and uplift in the region. Quaternary alpine glaciation then produced the erosion characteristic of the Southern Snake Range in which the park is located.

BIG BEND NATIONAL PARK

Big Bend National Park (see Figures 1.17 and 8.6), also in the Basin and Range province, was established by Congress in 1944. Located on the Rio Grande in southern Texas, the park includes vast desert and mountain vistas produced mainly by physical weathering processes. The highest peak in the park is Emory Peak at 7,835 ft (2,388 m). The region is the northern extension of the Chihuahuan desert that extends well into Mexico (see Chapter 12). It is the intention of Mexico to create a sister park, just across the Rio Grande (Figure P7.7).

Geology. The park exposes Cretaceous limestones (Figure P7.8) along the Rio Grande, forming spectacular, tilted fault block cliffs and stream-cut canyons. Some of the limestones are highly fossiliferous. Many areas are overlaid by Tertiary volcanics, including tuffs and rhyolite lavas that are mainly Eocene (Figure P7.9). Also exposed are some unusual looking intrusive rocks (Figures P7.10; P7.11), as well as Paleozoic limestones and shales that have been folded and eroded, and some Cretaceous and Tertiary sandstones and shales.

In Pennsylvanian time, the Ouachita Orogeny produced deformation of middle Paleozoic rocks in the park. Later, during late Cretaceous time, the Laramide Orogeny impacted the area, causing thrust faulting and folding. Paleogene volcanism and tectonic activity and Basin and Range block faulting followed by erosion,

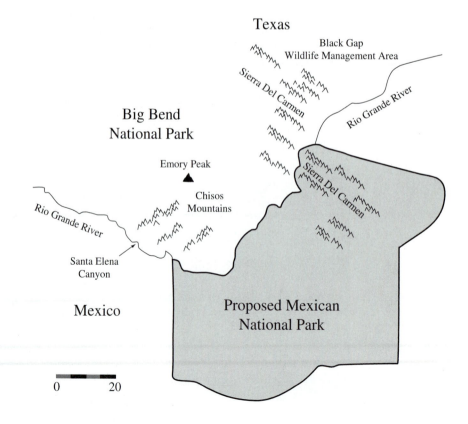

FIGURE P7.7 Map of proposed sister park in Mexico, across the Rio Grande from Big Bend National Park, Texas.

produced the exposures we now see in the park. However, today the area is tectonically stable with no seismic mic activity. In fact, it is so stable that during the Cold War seismographs were located in the region to monitor nuclear testing in the Soviet Union.

Desert landforms dominate the landscape (Chapter 12) with widespread *pediments* formed during weathering by the burial of mountains in their own rock debris. There are a few, but relatively small, *alluvial fans*. These are fan-shaped accumulations of sediment washed out of mountains. The sediment is deposited when water velocity slows as it escapes from the mouths of canyons to spread out over the valley floor (see Chapter 12).

FIGURE P7.8
Limestones along the western margin of Big Bend National Park, Texas. Canyon is Santa Elena Canyon, from which flows the Rio Grande, marking the boundary between Mexico and the United States. Limestones stand above the desert in Texas and Mexico as the result of fault motion. (Photo by author.)

FIGURE P7.9 Sierra Castolon Peak consists mainly of the remnants of volcanic tuffs being eroded in Big Bend National Park. (Photo by author.)

FIGURE P7.10 Unusual looking igneous feeder conduit intruding a pumice layer in Big Bend National Park. These features are often misidentified as petrified trees. (Photo by author.)

FIGURE P7.11 The remnants of an intrusion, called Mule Ear Peaks, in Big Bend National Park. (Photo by author.)

EXAMPLES OF PARKS IN FOLD AND/OR TECTONIC MOUNTAIN BELTS

 ### DENALI NATIONAL PARK AND PRESERVE

Denali National Park and Preserve (see Figure 1.11) was initially established as Mt. McKinley National Park in 1917. The name, Mt. McKinley, replaced the Native American name, *Denali,* meaning "high one," and the name change was a sore point with Native Americans until Congress adjusted the park name in 1980. However, Mt. McKinley, at 20,320 ft (6,194 m), North America's highest mountain (Figure P7.12), retains its English name.

Geology. The park is composed mainly of granites and metamorphic rocks that are covered by snow, ice, and alluvium. Faulting along the Denali transform fault, a very complex "shattered" transform fault sys-tem, has resulted in extreme vertical block displace-ment. Multiple faults have weakened the rocks causing enhanced stream erosion in the area. Exposed in the park are several geological terranes that did not de-velop as part of the North American continent, but in-stead have accumulated along its western margin dur-ing the Paleozoic and Mesozoic as the continent moved westward. In the Tertiary the granites in the park were emplaced, and following uplift in the late Tertiary, the area was extensively eroded by alpine glaciation (Fig-ure P7.13). This was accompanied in the valleys of the park by deposition and accumulation of glacially de-rived sediments.

FIGURE P7.12 Mt. McKinley in Denali National Park and Preserve, Alaska. (Photo courtesy of the National Park Service.)

FIGURE P7.13 Mt. McKinley in background in Denali National Park and Preserve. Alpine glacier is cutting a valley much like Yosemite Valley in California. (Photo courtesy of the National Park Service.)

GATES OF THE ARCTIC NATIONAL PARK AND PRESERVE

Gates of the Arctic National Park and Preserve (see Figures 1.11 and 8.4) was established by Congress in 1980 as one of the many Alaskan parks and preserves. Located in the east-west trending Arctic Rocky Mountains in northern Alaska, north of the Arctic Circle, the park's primary appearance is the result of alpine glaciation. A number of glaciers are still carving the mountains in the park.

Geology. The oldest rocks in the park are Precambrian granites. Also exposed are Paleozoic rocks, mainly near-shore marine and deltaic sediments. During the late Paleozoic and Mesozoic this was followed by tectonic activity, including thrust faulting and metamorphism. Activity continued through the Mesozoic with the accumulation of displaced terranes along the edge of the continent. Uplift and erosion during the Tertiary was followed by intense alpine glaciation during the Pleistocene and into the Holocene. Outwash sediments from melting glaciers are accumulating in the park today along its northern margin (Figure P7.14).

FIGURE P7.14 Northern slope of Gates of the Arctic National Park and Preserve, Alaska. (Photo courtesy of the National Park Service.)

ROCKY MOUNTAIN NATIONAL PARK

Located in the Front Range, just to the north of Denver, Colorado, Rocky Mountain National Park (see Figures 1.18 and 8.9), established in 1915, carries the name of the mighty Rocky Mountains. Its highest peak, Longs Peak at 14,256 ft (4,345 m), is believed to be an uplifted erosional remnant of a *monadnock* (Figure P7.15), a mighty cousin to the similar feature at Mt. Desert Island in Acadia National Park (Chapter 5). Much of the park is represented by glacially carved terrane (Figure P7.16), and many glaciers are still active in the park. Generally, the rocks in the park are Precambrian granites, granite gneisses, and schists, but in the western portion of the park the Never Summer Mountains are the remains of mid-late Tertiary pyroclastic rocks, volcanic ash flows that at one time were quite exten-

sive, but are now heavily eroded. In the northwestern portion of the park is the Little Yellowstone area. Rhyolites in this area have been hydrothermally altered to a yellow coloration typical of such rocks and similar to exposures in Yellowstone National Park (Chapter 11). The color comes from the alteration of iron oxide minerals in these rocks to the mineral *goethite* [FeO(OH)].

At the highest elevations in the park, above the tree line, permafrost is still present today, indicating that the ground remains frozen year around. *Tundra* has developed there, where small, very fragile plants exist in fine ecological balance (Figure P7.17). It is one of the few areas outside Alaska and the Northwest Territories of Canada where tundra can be found on the North American continent.

FIGURE P7.15 Longs Peak in Rocky Mountain National Park, Colorado. (Photo by author.)

FIGURE P7.16 Glacier basin in Rocky Mountain National Park. The feature illustrated here is an eroded dome somewhat like Half Dome in Yosemite National Park. (Photo by author.)

FIGURE P7.17 Tundra preserved in Rocky Mountain National Park. (Photo by author.)

GREAT SMOKY MOUNTAINS NATIONAL PARK

Great Smoky Mountains National Park (see Figures 1.22 and 8.16) was established by Congress in 1934. It is located on the Tennessee–North Carolina border along the crest of the Appalachian Mountains and has its highest point at Clingman's Dome, 6,643 ft (2,025 m). Because of the natural, bluish, hazy mountain beauty of this part of the Appalachian Mountains (Figure P7.18), Great Smoky Mountains National Park has been designated a world heritage site, a place for the world community to cherish.

Geology.   The rocks exposed in the park are mainly Precambrian metamorphic rocks, part of the Blue Ridge province of the Appalachian Mountains (see Chapter 8). There are some sedimentary rocks and intrusives exposed in the park. The Blue Ridge is part of a thick thrust sheet that has been pushed over underlying Paleozoic rocks during the Permian Alleghenian orogenic event that formed the Appalachian Mountains. This thrust sheet was produced by collision between the North American, the European, and African continents, which resulted in the linking of Pangaea.

FIGURE P7.18   The Blue Ridge Mountains in which Great Smoky Mountains National Park is located. (Photo by author.)

 ## SHENANDOAH NATIONAL PARK

Shenandoah National Park (see Figures 1.23 and 8.16) was established in 1926 in north-central Virginia. It includes the crest of the Appalachian Mountains just to the east of the Shenandoah Valley of northern Virginia. The park is long and narrow, and trends northeast-southwest for many miles along the axis of the Appalachian Mountains.

Geology.   Here volcanic rocks are exposed on eroded granites. All are strongly weathered chemically, with the granites taking on a characteristic rounded shape from spheroidal weathering (Figure P7.19). Some Paleozoic sediments are found within the park bound-

aries, but mostly these have been eroded. There are also some Paleozoic limestones in the area and a number of beautiful caves exist just outside the park boundary.

FIGURE P7.19 Granites weathering spheroidally in Shenandoah National Park, Virginia. (Photo courtesy of the National Park Service.)

REFERENCES

BALLY, A. W., and PALMER, A. R., eds. 1989. *Decade of North American geology, DNAG, The Geology of North America.* Vol. A. *The geology of North America; An overview*, p. 619. Boulder, Colorado: The Geological Society of America.

BENNETT, R., ed. 1980. *The new America's wonderlands our national parks,* p. 463. Washington, D.C.: National Geographic Society.

BEUS, S. S., ed. 1987. *Decade of North American geology, DNAG, Centennial field guide.* Vol. 2. *Rocky Mountain section of the Geological Society of America*, p. 475. Boulder, Colorado: Geological Society of America.

CHRONIC, H. 1984. *Pages of stone, geology of western national parks and monuments.* Vol. 1. *Rocky Mountains and Western Great Plains*, p. 168. Seattle: The Mountaineers.

CHRONIC, H. 1986. *Pages of stone, geology of western national parks and monuments.* Vol. 2. *Sierra Nevada, Cascades & Pacific Coast,* p. 170. Seattle: The Mountaineers.

HARRIS, A. G., and TUTTLE, E. 1990. *Geology of national parks.* 4th ed., p. 652. Dubuque, Iowa: Kendall Hunt Publishing Company.

HARRIS, D. V., and KIVER, E. P. 1985. *The geologic story of the national parks and monuments.* 4th ed., p. 464. New York: John Wiley & Sons, Inc.

HAYWARD, O. T., ed. 1988. *Decade of North American geology, DNAG, Centennial field guide.* Vol. 4. *South-Central Section of the Geological Society of America,* p. 468. Boulder, Colorado: Geological Society of America.

HILL, M. L., ed. 1987. *Decade of North American geology, DNAG, Centennial field guide.* Vol. 1. *Cordilleran Section of the Geological Society of America,* p. 490. Boulder, Colorado: Geological Society of America.

SULLIVAN, W. 1992. *Continents in motion, the new Earth debate.* 2d ed., p. 430. New York: American Institute of Physics.

8

United States Geologic Provinces

In Chapter 1 we discussed the national park regions in the United States, established for oversight purposes by the National Park Service. Now we will look at the general geologic provinces of the United States. In the following maps, the numbers refer to the parks list in Table 8.1. In this list the parks are numbered consecutively based on when the park was first established as a national park.

Recent maps by the United States Geological Survey (USGS) from digital data show the striking nature of our country (Figure 8.1). To the west, we see a picture of many mountains, high plateaus, and smooth-bottomed valleys. In the central United States we see a picture of broad, gentle slopes, with river and glacial erosion scars, punctuated here and there by local highs in elevation. Toward the east lies the distinctive Appalachian Mountains, and along the Gulf and Atlantic coasts lies the broad, low-elevation region known as the Coastal Plain.

ALASKA

The Alaska region defines an area that is the result of accreted terranes and complex plate tectonic interactions that created a number of mountain ranges and volcanic arc systems (Figure 8.2). The *Aleutian Range* extends from the Alaska peninsula continental arc, out into the Aleutian *archipelago* arc. It is part of the Pacific "Ring of Fire," so designated because of the volcanoes generated by subduction-related processes, associated with trenches, that extend most of the way around the Pacific Ocean basin. This system of explosive volcanoes, extruding large volumes of andesitic lavas and tuffs, has its Park System representatives in Katmai National Park and Preserve (Chapter 5) in Alaska and a whole series of other parks in other provinces. These generally explosive eruptions produce steep-sided pyroclastic accumulations of volcanic debris that alternate with lavas to form stratovolcanoes or composite volcanoes. They are found all around the "Ring."

The Alaska Range, northeast from the Aleutians, runs east across central Alaska and then swings southeastward into the Yukon Territory. The range is mainly fault con-

TABLE 8.1 NATIONAL PARKS

1.	1872	Yellowstone NP, Wyoming, Montana, and Idaho
2.	1890	Kings Canyon NP, California
3.		Sequoia NP, California
4.		Yosemite NP, California
5.	1899	Mount Rainier NP, Washington
6.	1902	Crater Lake NP, Oregon
7.	1903	Wind Cave NP, South Dakota
8.	1906	Mesa Verde NP, Colorado
9.	1910	Glacier NP, Montana
10.	1915	Rocky Mountain NP, Colorado
11.	1916	Haleakala NP, Hawaii
12.		Hawaii Volcanoes NP, Hawaii
13.		Lassen Volcanic NP, California
14.	1917	Denali NP and Pres., Alaska
15.	1919	Acadia NP, Maine
16.		Grand Canyon NP, Arizona
17.		Zion NP, Utah
18.	1921	Hot Springs NP, Arkansas
19.	1924	Bryce Canyon NP, Utah
20.	1926	Mammoth Cave NP, Kentucky
21.	1929	Grand Teton NP, Wyoming
22.	1930	Carlsbad Caverns NP, New Mexico
23.		Great Smoky Mountains NP, North Carolina and Tennessee
24.	1931	Isle Royale NP, Michigan
25.	1935	Shenandoah NP, Virginia
26.	1938	Olympic NP, Washington
27.	1944	Big Bend NP, Texas
28.	1947	Everglades NP, Florida
29.	1956	Virgin Islands NP, Virgin Islands
30.	1962	Petrified Forest NP, Arizona
31.	1964	Canyonlands NP, Utah
32.	1968	North Cascades NP, Washington
33.		Redwood NP, California
34.	1971	Arches NP, Utah
35.		Capitol Reef NP, Utah
36.	1972	Guadalupe Mountains NP, Texas
37.	1975	Voyageurs NP, Minnesota
38.	1978	Badlands NP, South Dakota
39.		Theodore Roosevelt NP, North Dakota
40.	1980	Biscayne NP, Florida
41.		Channel Islands NP, California
42.		Gates of the Artic NP and Pres., Alaska
43.		Glacier Bay NP and Pres., Alaska
44.		Katmai NP and Pres., Alaska
45.		Kenai Fjords NP, Alaska
46.		Kobuk Valley NP, Alaska
47.		Lake Clark NP and Pres., Alaska
48.		Wrangell–St. Elias NP and Pres., Alaska
49.	1986	Great Basin NP, Nevada
50.	1988	The National Park of American Samoa
51.	1992	Dry Tortugas NP, Florida
52.	1994	Death Valley NP, California and Nevada
53.		Joshua Tree NP, California
54.		Saguaro NP, Arizona

NP = National Park

trolled with the east-west trending Denali transform as its principal fault. The range is cored by batholiths that have been emplaced since the Jurassic. Folding and faulting followed during the late Miocene–early Pliocene, when the Alaska Range was uplifted as the Pacific Plate under-thrust Alaska. The resulting intrusions and orogenesis produced the metamorphic rocks found in the region. Glaciers are large and numerous in the range and reflect abundant moisture from the Pacific. Their effects are quite evident in the landscapes of the region, involving both erosional and depositional features from alpine, *piedmont*, and tidewater glaciation.

PACIFIC BORDERLAND

The Pacific Borderland includes the coastal ranges that extend from Alaska to Mexico. The whole region is unstable tectonically and is geologically complex, due to the fact that the North American Plate is moving westward, overriding segments of the Pacific Plate, the East Pacific Rise, and other subducted plates and plate segments. Mountain building began in the early Tertiary as a result of this plate interaction and continues today. In the latest Tertiary, the coastal ranges were elevated to their present heights. The mountains were heavily glaciated and have been strongly affected by wave erosion along the coast.

FIGURE 8.1 Topographic relief map of the United States. (Modified from USGS data (Thelin and Pike, 1991).)

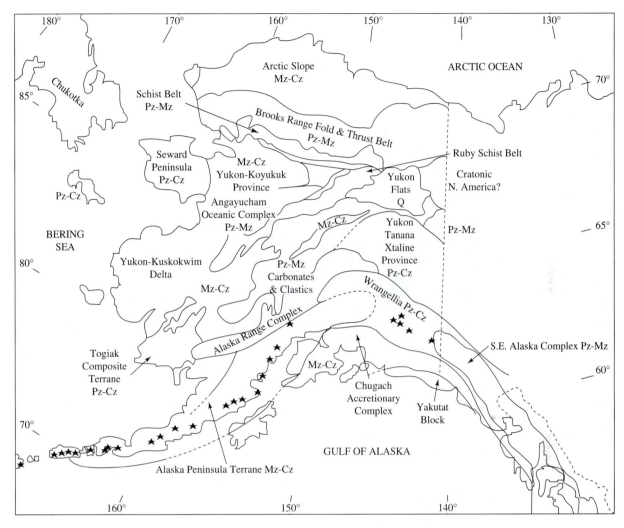

FIGURE 8.2 Geologic provinces in Alaska. Volcanos are represented by (★). PZ = Paleozoic; MZ = Mesozoic; CZ = Cenozoic.

The national parks in this province are Wrangell–St. Elias National Park and Preserve (Figure 8.3; see also Chapter 13) and Glacier Bay National Park (see Chapter 13) in Alaska, Olympic National Park in Washington (see Chapter 14), and Redwood National Park (see Chapter 14) and Channel Islands National Park (see Chapter 14) in California. The latter three can be seen in Figure 8.4.

CASCADE RANGE

The Cascade Range, made up mainly of volcanics, includes the mountains in the northern part of a 1,000-mile-long mountain range extending from the Canadian border into northern California (Figure 8.4). The range consists of a dissected volcanic plateau capped with a number of lofty stratovolcanoes, some that have erupted relatively recently. Examples include Mt. Rainier, which erupted 2,000 years B.P.; Mt. Lassen,

FIGURE 8.3
Wrangell–St. Elias National Park and Preserve in Alaska. (Photo by author.)

which erupted over the period between 1914 to 1921; and Mt. St. Helens, which erupted explosively in 1980. Topography in the Cascades has been strongly modified by alpine glacial erosion.

During the Paleozoic and into the Mesozoic, sediments, eroded from the continental interior to the east, were deposited in a deep basin that lay at that time along the coast of North America. As temperatures and pressures increased at the base of the thickening sediment pile, those materials were metamorphosed, resulting in the development of schists and gneisses. During the Tertiary, volcanics and mudflows were added to the sequence. Then an Oligocene orogenic event resulted in uplift, with accompanying folding, thrust faulting, and volcanism. Erosion to low levels followed. In the Pliocene, following the eruption of the flood basalts, broad upwarping and faulting affected the Cascades province. Erosion left a wide, dissected plateau reaching elevations of 5,000–8,000 ft (1,524–2,438 m) across the province. In the late Pliocene–early Pleistocene, andesitic pyroclastics alternating with lavas were extruded onto the plateau, resulting in the steep-sided stratovolcanoes we see today. The final product has been modified by alpine glaciation to produce the present appearance of each volcano.

There are four national parks in the Cascade Range (Figure 8.4), North Cascades National Park (see Chapter 7) and Mt. Rainier National Park (see Chapter 5) in Washington, Crater Lake National Park (see Chapter 5) in Oregon, and Lassen Volcanic National Park (see Chapter 5) in California.

SIERRA NEVADAS

The Lassen volcanic zone separates the Cascades from the Sierra Nevadas to the south (Figure 8.4). The Sierras are a huge, 650-km-long fault block mountain range, trending north-northwest. Its eastern edge is bounded by the Sierra Nevada fault system, along which the block is tilted up to the west. The southern end of the range is cut off

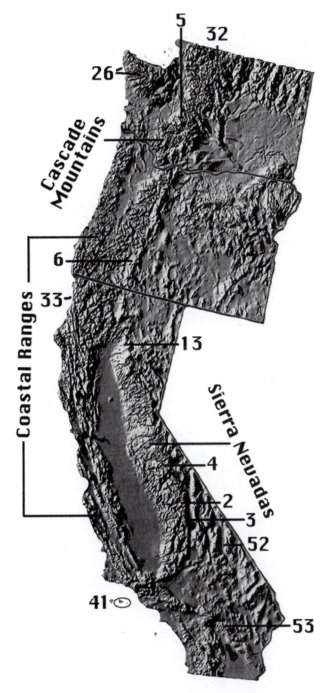

FIGURE 8.4 Map of the Pacific Coast states and the locations of national parks in these areas. Also identified are the Cascade Mountains, Sierra Nevadas, and the Coast Range. Park numbers are from Table 8.1, where parks are numbered based on when the national park was established. National parks identified are: 2. Kings Canyon; 3. Sequoia; 4. Yosemite; 5. Mount Rainier; 6. Crater Lake; 13. Lassen Volcanic; 26. Olympic; 32. North Cascades; 33. Redwood; 41. Channel Islands; 52. Death Valley; and 53. Joshua Tree. Numbers are placed close to, but not immediately over the park so that the park topography is not obscured.

by the Garlock fault, seen in the lower left of Figure 8.4. The Sierra Nevada batholith, a composite intrusive body made up of granites, quartz diorites, and *granodiorites*, is exposed as the central core of the range. During the Jurassic, extensive volcanics were extruded to the east and north of Yosemite National Park (see Figure 8.4; see also Chapter 5). Metamorphic rocks are abundant, some as large xenoliths and others as the result of contact metamorphism caused by heating by intrusives.

In the late Mesozoic, deformation and the generation of granitic magmas occurred, followed by uplift and erosion that unroofed the batholiths. In the late Cenozoic, block faulting, tilting, and volcanic eruptions in boundary areas produced the geology as it exists today. Again, it has been modified by massive alpine glaciation, and today the area is still tectonically active.

Three national parks are located in the Sierra Nevadas of California (Figure 8.4): Yosemite National Park, Kings Canyon National Park, and Sequoia National Park (see Chapter 5).

NW VOLCANIC PROVINCE

To the east of the Cascades extensive flood basalts were erupted, primarily in the Miocene, producing the Columbia Plateau basalts in eastern Washington, northern and eastern Oregon, and western Idaho (Figure 8.4). Other areas in the region are also extensively covered by mid-late Tertiary, Quaternary, and Holocene volcanic materials, including southern Idaho, northwestern California, and parts of northern Nevada and Utah (Figure 8.5). This volcanism extends all the way to Yellowstone National Park, located partly in eastern Idaho, and is part of the volcanic activity associated with the Yellowstone *hot spot*. The distinctive basaltic flows erupted at Craters of the Moon National Monument (C in Figure 8.5; see Chapter 5) are also part of this sequence.

FIGURE 8.5 Map of the Pacific Northwest states. The outlined region includes those areas that were covered by excessive Cenozoic volcanism. "C" indicates Craters of the Moon National Monument and "1" Yellowstone National Park. Numbers are close to, but not immediately over the park so that the park topography is not obscured.

It is believed that underlying Yellowstone is an upper mantle hot spot similar to that which produced the Hawaiian islands and which may be responsible for much of the volcanism in the region. Because Yellowstone is located in a thick continental interior, as opposed to Hawaii, located on a thin oceanic plate, volcanism is very different between the two parks. At Hawaii it tends to be fairly continuous, while at Yellowstone it tends to be episodic in nature, with major eruptions occurring only occasionally. The track of the Yellowstone hot spot can be traced in an arc southwestward from Yellowstone National Park through southern Idaho, with ages of these rocks becoming progressively older with increased distance from the park (Figure 8.5).

Yellowstone National Park (see Chapter 11) is the only national park in the province, but Crater Lake National Park, Mt Rainier National Park, and Lassen Volcanic National Park (see Chapter 5) are all located right on the province's western margin.

BASIN AND RANGE PROVINCE

To the east of the Sierra Nevadas and south of the NW Volcanic province, the Basin and Range province is an area produced by block faulting and tilting during the late Tertiary. In pre-Tertiary time, the province was tectonically active, containing large *thrust sheets,* large slabs of crustal material pushed up over other rocks along low angle reverse faults (see Chapter 7). During thrusting, these rocks were folded and overturned producing a complex pattern of folds. The Basin and Range province extends from southern Oregon to Mexico and from California to southwestern Texas (Figure 8.6). The region is composed of approximately 150 small mountain ranges with basins between them, and the higher mountains are glaciated. Today continental extension is actively pulling the basins apart.

The basins are constantly being filled by sediments eroded from the surrounding ranges because there are no stream outlets and drainage is internal. In some of the basins, large Pleistocene lakes existed that were produced mainly by increased rainfall, but also from runoff as the glaciers melted following the last glacial maximum. Melting occurred approximately 14,000 years B.P., during what has been called the Thermal Maximum; this was accompanied by excessive rainfall. Evidence of ancient shorelines, such as Lake Manley, can be seen in Death Valley National Park (Chapter 12).

Active groundwater circulation has produced several major cave systems in the Basin and Range province. Two important examples include the many caves in the Guadalupe Mountains area of Texas and New Mexico (i.e., Carlsbad Cavern, one of many caves in Carlsbad Caverns National Park discussed in Chapter 11) and in the Big Bend area of Texas. Other isolated caves exist in the region. In addition, most of the area is desert, and physical and chemical weathering have produced the erosional shapes such as jagged mountains with high cliffs, the result of mass wasting. These features will be discussed in Chapter 12.

There are seven national parks in the region (Figure 8.6): Great Basin National Park (Chapter 7) in Nevada; Joshua Tree National Park in California and Saguaro National Park in Arizona; Death Valley National Park, located in mainly in California, with a small area in Nevada (these desert parks are all discussed in Chapter 12); Carlsbad Caverns National Park (see Chapter 11) in New Mexico; Guadalupe Mountains National Park (see Chapter 10) and Big Bend National Park (see Chapter 7), both located in Texas.

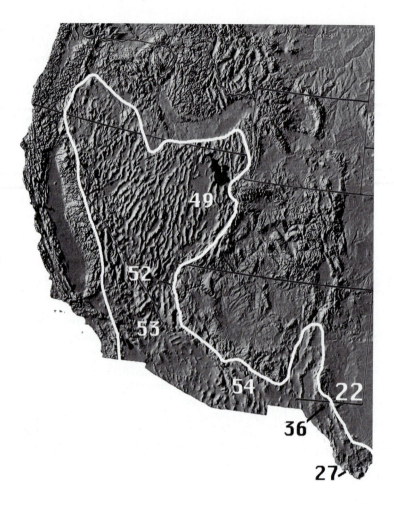

FIGURE 8.6 Map of states containing the Basin and Range province (outline in white). Park numbers are from Table 8.1. National parks identified are: 22. Carlsbad Caverns; 27. Big Bend; 36. Guadalupe Mountains; 49. Great Basin; 52. Death Valley; 53. Joshua Tree; and 54. Saguaro. Numbers are placed close to, but not immediately over the park so that the park topography is not obscured.

COLORADO PLATEAU PROVINCE

Another dry and essentially treeless region, except for canyon interiors and higher elevations, is the Colorado Plateau province (Figure 8.7). Drained by the Colorado River and its *tributaries* the Green, Little Colorado, and San Juan rivers, it is an area of spectacular canyons and deserts. The rocks are mainly flat-lying sedimentary units that range in age from the Precambrian to the Tertiary. Episodic, slow uplift and erosion began at the end of Paleozoic time and continues to the present. Strong uplift during the Pliocene produced broad folding and faulting, bringing the eastern margin of the plateau to elevations of approximately 5,000 ft (1,525 m). To the west elevations rise to over 11,000 ft (3,350 m). The region is capped by thick or resistant strata, consisting of well-cemented sandstones or limestones that persist due to the dry climate and reduced chemical weathering. These resistant units produce scarps and steep walled canyons. *Badlands topography* is also abundant. Badlands topography results from rapid erosion of relatively soft sedimentary beds. When the resistant cap (sometimes sediment held by vegetation) overlying softer sedimentary beds is breached by erosion, the soft interval is exposed and differentially eroded. This yields unusual spires, mounded shapes, and small flat-topped mesas. The last event in the development of the Colorado Plateau was the extrusion of Miocene and younger volcanics, and these rocks are exposed across the region, particularly in the Grand Canyon area. The area is also

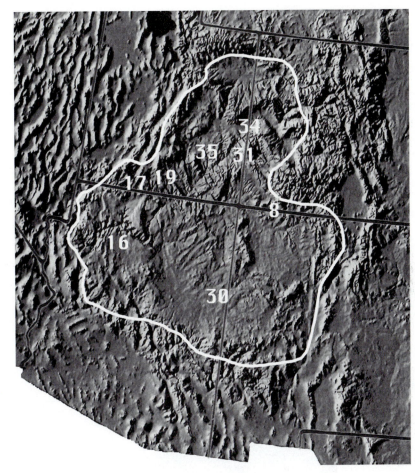

FIGURE 8.7 States of the Colorado Plateau province (outlined in white). National parks identified are: 8. Mesa Verde; 16. Grand Canyon; 17. Zion; 19. Bryce Canyon; 30. Petrified Forest; 31. Canyonlands; 34. Arches; and 35. Capitol Reef.

seismically active, with large numbers of earthquakes being felt along the northwest and southeast margins of the plateau (Figure 8.8).

There are eight national parks in the region (Figure 8.7), Grand Canyon National Park and Petrified Forest National Park, both in Arizona (see Chapter 15); Zion National Park (see Chapter 10); Bryce Canyon National Park (see Chapter 9); Capitol Reef National Park (see Chapter 15); Arches National Park (see Chapter 9) and Canyonlands National Park in Utah (see Chapter 11); and Mesa Verde National Park in Colorado (see Chapter 16).

ROCKY MOUNTAINS PROVINCE

The Rocky Mountains can be divided into four segments, the Arctic, northern, middle, and southern Rockies. These subdivisions extend from Arctic Alaska, through Canada, into western Washington, eastward through Idaho into western Montana, then southeast through Wyoming and southward through Colorado, ending in New Mexico (Figure 8.9).

Mountain building and uplift of the present Rocky Mountains began as the result of the North American Plate's movement against subducting oceanic crust to the west.

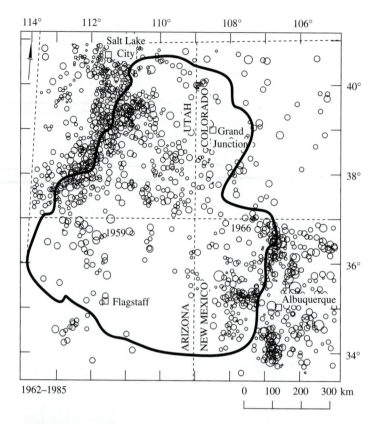

FIGURE 8.8
Earthquakes from 1962 to 1985 in the Colorado Plateau area.

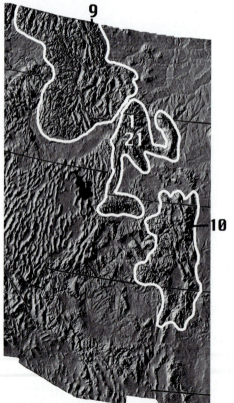

FIGURE 8.9 States containing the Rocky Mountains. Northern, middle, and southern Rockies are outlined in white. National parks identified are: 1. Yellowstone; 9. Glacier; 10. Rocky Mountain; and 21. Grand Teton.

During the *Laramide Orogeny,* toward the end of the Cretaceous Period, as compressive forces intensified, folding, faulting, tilting, and thrusting began to produce the Rocky Mountains. The Laramide Orogeny ended during the Eocene Epoch. Throughout the development of the Rockies, and all other mountain ranges, erosion is a continuing process, sculpting the mountains as they form and destroying them with time. In the case of the Rockies, during the Eocene erosion flattened the mountains, producing a *peneplain,* a flat, eroded plain. Then, *isostatic adjustment* uplifted the peneplain surface, and renewed erosion during Oligocene and Miocene time, dissected the region into new peaks. Extensive volcanism occurred at this time, covering large areas of the southern Rockies. During the Pliocene, another erosional surface called a *pediplain,* developed. A pediplain is a flat surface, similar to a peneplain, but formed in arid or semi-arid environments. It results from coalescence of pediments, each a remnant of an old mountain buried in its own debris, with some segments of the ancient mountain still protruding from the sedimentary cover. Today these old remnants are found as high benches in the Rockies, often with flat tops smoothed by erosion long in the past. One example is Longs Peak in Rocky Mountain National Park (see Figure P7.15).

Isostatic adjustment in the late Pliocene then formed the present day Rocky Mountains, and active uplift is still going on today. The Rockies have been sculpted by alpine glaciation that has occurred from the Pleistocene to the present, resulting in very high erosion rates and characteristic glacial terrains.

In the very north of the chain, the *Arctic Rocky Mountains* extend across northern Alaska from west to east, then, as they swing south through Canada, they become the Canadian Rockies, and extend into northwestern Washington, Idaho, and Montana. The northern and middle Rocky Mountains (Figure 8.9) are made up of several mountain ranges including the Precambrian basement complex Teton Mountains (Figure 8.10). These are fault block mountains characterized by near vertical uplifts that

FIGURE 8.10 View of the Teton Mountains, looking to the west, in Grand Teton National Park in Wyoming. (Photo by author.)

produced asymmetrical anticlines with Precambrian cores. Two major plateaus are located in the region, the Beartooth plateau, a remnant of the Eocene peneplain, and the Yellowstone volcanic plateau, produced by recent volcanism.

The southern Rocky Mountains (Figure 8.9) trend southward through Colorado, ending in New Mexico, and contain the highest mountain peaks in the Rocky Mountain chain, with 52 standing higher than 14,000 ft (4,200 m). These mountains are composed mainly of metamorphic and igneous rocks, with many pegmatites containing economically valuable minerals that have been or are being mined. One of the most striking features in the southern Rockies is the Front Range, including Pike's Peak, standing 14,110 ft (4,301 m), immediately west of Colorado Springs, Colorado. It is cored by the Precambrian Pike's Peak granite and has associated pegmatites with unique occurrences of minerals, such as Amazonite, a beautiful blue-green feldspar. Further to the south lie the Sangre de Cristo Mountains, composed of metamorphic rocks and upturned Paleozoic sediments that were tilted due to motion along the Front Range fault system. The Sawatch Range to the west contains the highest peak in the Rockies, Mt. Elbert at 14,431 ft (4,399 m), and the range is mainly made up of Precambrian rocks intruded by Tertiary granites. The San Juan Mountains, to the west of the Sangre de Cristo Range in southwestern Colorado, are composed mainly of Tertiary volcanic rocks. In New Mexico to the south, the Jemez Mountains are also composed mainly of Tertiary volcanic rocks. The Rockies are constantly being eroded and the debris is filling the down-folded and faulted basins lying between the mountain ranges.

In the southern Rockies after the Laramide Orogeny, evidence of the late Eocene peneplain still remains as 11,500–12,000 ft (approximately 3,500 m) high benches. The higher peaks are resistant rocks called monadnocks that today lie above these high benches at elevations over 12,000 ft (3,500 m). Late Pliocene isostatic adjustment formed the Front Range by normal faulting and during the Ice Ages the Rockies were sculpted by glaciation. In Glacier National Park (see Chapter 13) and Rocky Mountain National Park (see Chapter 7) glaciers are still active today.

There are five national parks in the Rocky Mountains (Figure 8.9): Gates of the Arctic National Park and Preserve (see Chapter 7) in Alaska; Glacier National Park (see Chapter 13) in Montana; Yellowstone National Park (see Chapter 11) in Montana, Idaho, and Wyoming; Grand Tetons National Park (see Chapter 5) in Wyoming; and Rocky Mountain National Park (see Chapter 7) in Colorado.

GREAT PLAINS PROVINCE

The Great Plains (Figure 8.11) east of the Rocky Mountains are composed mainly of sedimentary rocks, with some localized exceptions such as the igneous rocks at Devils Tower National Monument (see Figure 1.3) and the granitic and metamorphic rocks at Mt. Rushmore National Memorial (see Figure 5.3). The sedimentary beds of the Great Plains were upturned in the west by Pliocene tilting during Rocky mountain isostatic uplift. A striking example (Figure 8.12) can be found in the sediments at Garden of the Gods, Colorado Springs (not supervised by the National Park Service). Just to the east of the Front Range, these Paleozoic sedimentary beds flatten out and dip very gently to the east over most of the Great Plains. In the Black Hills of South Dakota, domed during the Laramide Orogeny, glaciation has resulted in major landform changes.

Toward the east from the Front Range a rapid change in moisture from wet to dry occurs over a distance of about 100 miles. From there, for hundreds of miles to the east, the Great Plains are very dry. It was this arid zone that was crossed by early pioneers,

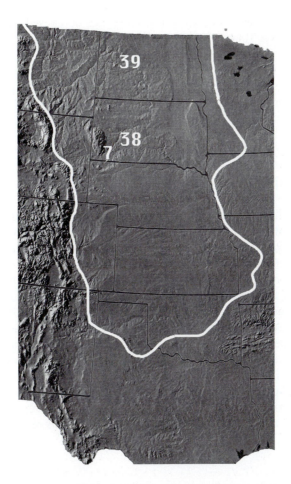

FIGURE 8.11 Great Plains states. National parks identified are:
7. Wind Cave;
38. Badlands; and
39. Theodore Roosevelt.

FIGURE 8.12 Rock fins at Garden of the Gods in Colorado Springs, Colorado. Located just east of one of the southern Rockies Front Ranges, the fins are weathered sedimentary rocks that are dipping nearly vertically at this location. These Paleozoic sediments flatten out with very shallow dips toward the east just a few miles out in the Great Plains to the east. (Photo by author.)

and their trails are now national historic sites composed of stopping sites and trail remnants. Over 100 years later the ruts from the wagons remain to remind us of their passing.

There are three national parks in the Great Plains province (Figure 8.11), Theodore Roosevelt National Park (see Chapter 9) in North Dakota; and Wind Cave National Park (see Chapter 11) and Badlands National Park (see Chapter 9) in South Dakota.

CENTRAL MOUNTAIN PROVINCE

The Central Mountain province (Figure 8.13) is an area where Appalachian Mountain structures surface to the west of the Mississippi River. It includes the Ouachita mountains of eastern Oklahoma and western Arkansas, the *Arbuckle Mountains* in south-central Oklahoma, and the Ozark Mountains in Missouri. In early to middle Paleozoic time, sediments were accumulating in the area, with shales being most abundant, but sandstones, conglomerates, cherts, and limestones also being well represented. During the late Paleozoic, folding and thrust faulting, accompanied by metamorphism in some areas, formed the mountains. The area was twice uplifted and eroded, after which the present valley and ridge topography developed. In some areas Cretaceous igneous intrusions are exposed, and metamorphism altered the shales to slates and the sandstones to *quartzites*.

Hot Springs National Park (see Chapter 11 and Figure 8.13) in the Ouachita Mountains in Arkansas is the only national park in the region. However, at one time there was a national park in southern Oklahoma in this region. Now called the Chick-

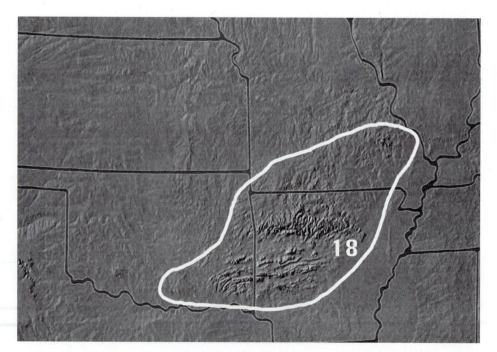

FIGURE 8.13 Central Mountain province states containing the Arbuckle Mountains in Oklahoma, the Ozark Mountains in Missouri, and the Ouachita Mountains in Arkansas. Hot Springs National Park (18) is the only national park in the region.

asaw National Recreation Area it was originally established by Congress as Platt National Park in 1906, and was the result of political levering for one-time Senator Platt of Oklahoma. In 1976 the park was downgraded by Congress to a national recreation area and increased in size.

SUPERIOR PROVINCE

A small segment of the *Canadian Shield*, the Precambrian core of our continent, extends down into the lower 48 states (Figure 8.14). The oldest rocks exposed in the shield in the United States are meta-intrusives and meta-sediments, with younger lava flows covering large areas. These include some pillow lavas indicating extrusion into water. Following extrusion of the lavas, very large mafic intrusions were emplaced. This was followed by more lava flows, some now containing copper ore deposits as copper filling vesicles (*amygdules*) in basalt flows. There are also many iron ore deposits. The iron ores precipitated during the early Precambrian (Archean) in an ocean covering the region, when free oxygen in the water, produced by primitive plants, finally reached sufficient levels to combine with iron to crystallize as iron ores. Later, free oxygen escaping from the ocean into the atmosphere began to oxidize rocks on land, and the age of these oxidized rocks, late Precambrian (Proterozoic), gives an indication of when Earth's atmosphere developed. There are some Paleozoic rocks in the Superior province, around Hudson Bay, and all the exposures have been modified by the glaciation that also formed the Great Lakes.

Only two national parks exist in the region (Figure 8.14), Voyageurs National Park (see Chapter 6) in Minnesota and Isle Royal National Park (see Chapter 5) in Michigan.

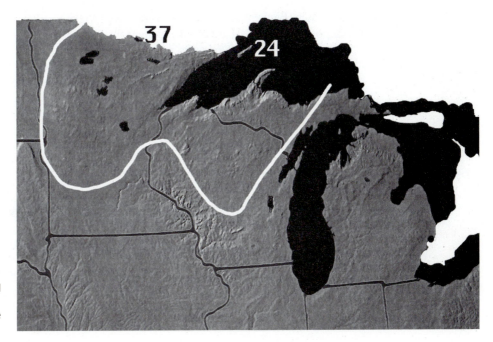

FIGURE 8.14 Superior province states contain the Great Lakes and exposures of Archean rocks. Isle Royale (24) and Voyageurs (37) are the only national parks in the region.

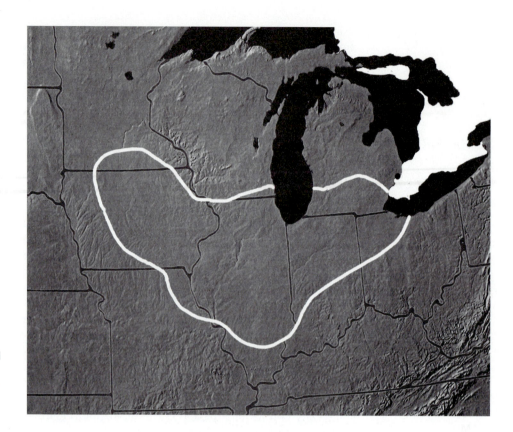

FIGURE 8.15 Central Lowlands geologic province. There are no national parks in the region.

CENTRAL LOWLANDS PROVINCE

East of the Great Plains and south of the Superior province is the Central Lowlands province (Figure 8.15), an area covered by glacial debris and soils formed from glacial deposits. These soils are often excellent for plant growth and make up the productive, glacial deposits–derived soils in Iowa, Illinois, Indiana, Ohio, and other states (see Chapter 13). The Central Lowlands are composed mainly of Paleozoic sediments that are relatively flat lying and undisturbed. Broad up-warps, such as the Cincinnati Arch, and low angle, large down-warps, such as the Illinois basin, characterize the region. There are no national parks in the Central Lowlands province.

APPALACHIAN MOUNTAINS

As the result of collision between the North American, the African, and the Eurasian plates, a major orogenic event, called the Alleghenian Orogeny in the United States, and the Hercynian Orogeny in Great Britain, produced the Appalachian Mountains, a late Paleozoic linear chain of mountains running from Alabama to Canada (Figure 8.16). Associated with this collision was major folding, thrust faulting, and the intrusion of a large number of granitic plutons. This was not the first time that a range of mountains resulted from collision of a plate against eastern North America. At least three other orogenic episodes are recognized, the Grenville event in the late Protero-

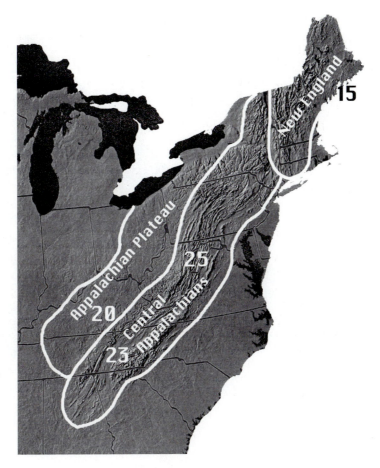

FIGURE 8.16 Eastern seaboard states containing the Appalachian Mountains and associated provinces. National parks identified are: 15. Acadia; 20. Mammoth Cave; 23. Great Smoky Mountains; and 25. Shenandoah.

zoic Eon, the Taconic event during the Ordovician Period, and the Acadian Orogeny during the Devonian Period. All these orogenic events are represented in the rock record in the Appalachian Mountain region. Of course, the implication of these plate collisions is that a whole series of Atlantic Oceans (at least four counting the present Atlantic) developed to the east of North America as the plates moved apart before each collision. And as each new ocean was formed, new oceanic crust was formed, and early phases of the opening Atlantic Oceans resulted in the emplacement of basaltic dikes into the North American continent. These can be seen at many localities today.

The Appalachian Mountains can be subdivided into six provinces. The terminology used here is from Williams (1978).

Coastal Plain

The Appalachian Mountains are a very long, distinct linear trend of geological terrains or distinct northeast-southwest trending zones that can be identified from place to place along the eastern margin of North America. Along the Atlantic and Gulf coastlines is the Coastal Plain, consisting of seaward sloping lowlands (Figure 8.17). It also includes the shallow offshore submerged *continental shelf,* along which sea-level fluctuations have been very important in the past. The Coastal Plain consists primarily of sediments that range in age from Cretaceous along its western border with the next distinct

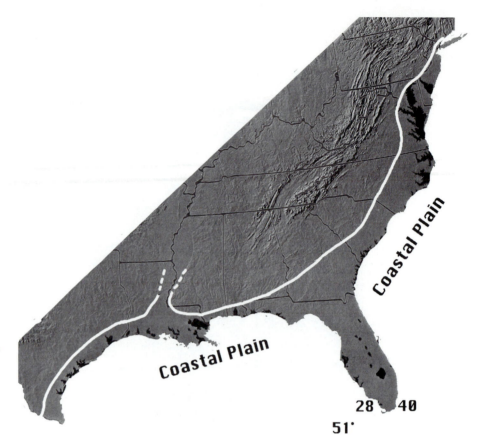

FIGURE 8.17 In the Coastal Plain states there are only three national parks, all located at the tip of Florida. These are: 28. Everglades; 40. Biscayne; and 51. Dry Tortugas.

province, the Piedmont, to Tertiary in the middle zone and Quaternary along the coastline. Large river deltas, like the Mississippi, are important along the Gulf coast. A few coral reefs are present along the Florida coast and are part of the Coastal Plain, and barrier islands are found throughout its extent. The barrier islands resulted from low sea-level stands during the last major glaciation, where beach sands and dunes built up along the shoreline existing at that time. The dunes were stabilized somewhat by vegetation, and when the glaciers melted, rapidly rising sea-level flooded all but the higher elevations (dunes) that built up and now persist as islands along the coast today.

Many forces are responsible for shaping the shore of the Coastal Plain. Big storms, such as hurricanes and northeasters, have strong winds that drive waves against the shore, molding and changing it, often to human displeasure. River runoff brings eroded sediment to add to the coastal materials changing the shoreline. Chemical weathering in the moist, generally warm climate of the coastal regions constantly attacks the rocks, changing their color to red in many areas and eroding and altering these materials. One of the most serious problems facing United States coastal areas is severe erosion caused by our efforts to stabilize the region. This is a very complex problem, but many marine geologists simply recommend that nature be allowed to take its course.

There are three national parks in the Coastal Plain province (Figure 8.17), Everglades National Park, Biscayne National Park (both discussed in Chapter 14), and Dry Tortugas National Park (see Chapter 10) all located in southern Florida.

Piedmont

The Piedmont lies immediately to the west of the Coastal Plain and consists mainly of metamorphic rocks, with some igneous intrusions. It is part of a thick sheet, thrust to the west over Paleozoic sediments. The Piedmont extends into higher elevations up to the Appalachian foothills (Figure 8.16). In some areas it is further divided into zones based primarily on metamorphic grade. No national parks lie in the Appalachian Piedmont.

Blue Ridge

The Blue Ridge includes the high elevations of the Appalachian Mountains (Figure 8.16). It is made up mainly of older Precambrian metamorphic rocks with some sediments and intrusives. A number of major faults emerge at the surface in this region, including the *Brevard Fault Zone,* which has major vertical offsets along it in the southern Appalachians, but dies out as it runs further to the northeast. The Blue Ridge is a high-angle Appalachian thrust sheet, with Precambrian rocks overlying Paleozoic sedimentary units.

The Blue Ridge has two national parks (Figure 8.16), Great Smoky Mountains National Park (see Chapter 7) in North Carolina and Tennessee, and Shenandoah National Park (see Chapter 7) in Virginia.

Valley and Ridge

The Valley and Ridge province consists of a series of valleys created by stream erosion between resistant ridges (Figure 8.16). The rocks are made up of folded and faulted Paleozoic sediments, with many limestones. Caves are abundant in the area. There are no national parks in the Valley and Ridge province.

Appalachian Plateau

To the west of the Valley and Ridge province is the Appalachian Plateau (Figure 8.16). It is composed mainly of relatively flat-lying and undeformed Paleozoic sediments that dip gently to the west. Abundant limestones are present in the region, and the high temperature and rainfall has, in some areas, dissolved much of the limestone, producing a Swiss cheese effect known as *karst topography*. In karst terrains there are abundant caves and many underground streams.

There is one national park in the province (Figure 8.16), Mammoth Cave National Park (see Chapter 11) in Kentucky. Mammoth Cave is the longest cave in the world, with over 300 miles (480 km) of connected passage.

New England Province

The northeastern extension of the Appalachians, beginning at the Hudson River Valley in New York, has a different character from the rest of the Appalachian trend due to climate and glacial effects. The general rock types are the same, with intrusive and metamorphic rocks abundant, but more volcanism occurred in this area than further to the south, and the Appalachian Mountains trend more toward the north in this region (Figure 8.16). Erosion, associated with each orogenic event, has been associated with more physical than chemical weathering. As in the southern Appalachians, several granite monadnocks remain as dome-shaped erosional remnants. One of these is Mt. Desert Island in Acadia National Park (see Chapter 5) in Maine, the only national park in the province.

REFERENCES

BALLEY, A. W., and PALMER, A. R., eds. 1989. *Decade of North American geology, DNAG, The geology of North America.* Vol. A. *The geology of North America; An overview,* p. 619. Boulder, Colorado: The Geological Society of America.

WILLIAMS, H. 1978. *Tectonic lithofacies map of the Appalachian Orogen, Map 1.* St. Johns, Newfoundland: Memorial University of Newfoundland.

Weathering and Erosion of Rocks

Surface and near-surface rocks on Earth are destroyed by weathering through either mechanical or chemical processes. Weathering is active everywhere, but rates vary from place to place depending on several factors. These include availability of moisture, temperature, and topography.

Because chemical and mechanical weathering effects attack some minerals more readily than others, weak rocks are rapidly worn away while resistant rocks remain behind. This *differential erosion* results in deposits that preferentially segregate minerals. While it might not seem obvious to us, global erosion rates are very high. The whole surface area of Earth averages 0.3 m of rock removed every 1,000 years. The evidence of this is all around us. Our streams and dams are clogged with sediment, our fields are constantly eroded causing problems for farmers, and in places our roads are being undercut or are even sliding down hillsides. The same mechanical and chemical processes that break down rocks also destroy man-made structures. These effects cost us a lot of money every year, and the total impact on society is substantial.

MECHANICAL WEATHERING

Mechanical weathering involves those physical forces that cause rocks to disintegrate and to fall and break. These include *frost (ice) wedging, abrasion,* burrowing by organisms, *sheeting,* and heating effects. Frost wedging occurs when freezing water expands in cracks, producing a 9% expansion. This process, however, requires moisture that can penetrate into cracks and freeze and thaw. Abrasion is caused by particles carried in water, wind, and glaciers, like sandpaper scraping away small amounts of rock as the particles pass. Burrowing by animals and plant roots (Figure 9.1) is another important mechanical weathering process, where organisms break rocks apart or disrupt sedimentary layers, bringing grains of rock to the surface where they can be carried off. Another effect, sheeting or unloading, is a result of fracturing parallel to land surface

FIGURE 9.1 Tree roots breaking apart boulder in Devil's Den, Gettysburg National Military Park, Pennsylvania (see Chapter 16). (Photo by author.)

due to erosion, which reduces confining pressure, and therefore unloads or removes the rock load from above. This allows the rock to expand and break, thus creating cracks along which water and roots can penetrate. Whole tops of hills have been known to suddenly pop up, causing an unpleasant sensation for those standing on the hilltop. This has happened due to unloading produced by quarrying operations, where so much rock has been removed that rock expansion and sheeting has occurred. A fifth, minor, element is heat generated by fires, such as that caused by lightning. When rock is heated it expands, sometimes explosively, thus breaking down into smaller pieces. Ancient Native Americans were careful to avoid putting certain types of rocks near their cooking fires because the rocks might explode. Today some quarry operators cut granite by exposing the rock to a flame.

Mechanical weathering is dominant where the climate is cool or cold, and/or where the availability of moisture is low. For example, in desert or near-desert regions such as those found at Arches National Park and many other parks, where moisture availability is low, mechanical weathering, as opposed to chemical weathering, domi-

FIGURE 9.2 Mechanical weathering and erosion of the Navajo Sandstone in Glen Canyon National Recreation Area (see Chapter 11). The cross beds in the sandstone exhibit differential erosion. Photo represents a distance of 6 ft (1.8 m). (Photo by author.)

nates (Figure 9.2); chemical weathering, however, is still important. Likewise, where temperatures are cold, in Arctic regions or at high elevations, mechanical weathering also dominates even though moisture is readily available.

 ## ARCHES NATIONAL PARK

Arches National Park (see Figures 1.18 and 8.7) was established in 1929, originally as a National Monument. It was then upgraded to a National Park in 1971. Located in eastern Utah, it contains Delicate Arch (Figure P9.1), one of the most famous geological features in the Park System.

Geology.   Jurassic rocks dominate in the park. These include two *eolian* sandstones, both with $CaCO_3$ cement. The first is the Entrada Sandstone, discolored red by a coating of hematite (Figure P9.2). Below the Entrada is the Navajo Sandstone (Figure P9.2), also found in Zion National Park and elsewhere. After deposition

of the sandstone units, *salt diapirs* intruded the area. Later uplift caused brittle fractures, then dissolution of the salt led to collapse valleys (Figure P9.1). Still later, in the Pliocene, further uplift produced faulting and folding. Erosion along fracture planes left *fins,* vertical, long thin bands of rock, in the Entrada Sandstone. Differential erosion of the easily dissolved $CaCO_3$ and the process of exfoliation, where the rock breaks off in slabs, formed holes in the fins that were enlarged by mass wasting and further exfoliation. Wind and rain contributed to sculpting the arches we see today (Figure P9.3).

FIGURE P9.1 The famous Delicate Arch, in Arches National Park, Utah, sits just above a collapsed valley (foreground). (Photo by author.)

FIGURE P9.2 The Entrada Sandstone (dark) lying above the Navajo Sandstone (light) in Arches National Park, Utah. (Photo by author.)

FIGURE P9.3 Double Arch in the Entrada Sandstone at Arches National Park, produced by weathering and erosion. (Photo by author.)

THEODORE ROOSEVELT NATIONAL PARK

Theodore Roosevelt National Park (see Figures 1.18 and 8.11) was created by Congress in 1978 in North Dakota, from part of President Theodore Roosevelt's Elkhorn Ranch. It is divided into three sections, large north and south elements, with a small ranch headquarters area lying in between. Both north and south segments of the park are similar in appearance, containing *badland* areas along the Little Missouri River (Figure P9.4). Many of the national parks and monuments contain interesting wildlife, and Theodore Roosevelt National Park is no exception. Both north and south segments have buffalo herds, with the herd in the south being very large. Sometimes in the late afternoon the herd may be seen migrating out of the Little Missouri River basin eastward into the main area of the park (see Figure 1.20), crossing roads and surrounding the vehicles of visitors. Wild animals can be very dangerous, but despite ample warning every year some visitors to the national parks are bitten or gored (see Figure 1.7) by animal residents.

Geology. During the Tertiary, mainly in the Paleocene, clastic fluvial sediments were deposited in the area along with some low-grade coals, primarily lignite, formed from abundant plant material growing along the floodplain (see Chapter 6 for a discussion of coals). Then, in the Pliocene, isostatic uplift elevated the site and stream erosion increased. This was followed by Pleistocene glaciation that changed drainage patterns in the park. Finally, badlands topography developed from erosion, leaving the mounded and unusual forms we see today (Figure P9.5).

Another interesting geological feature in the southern park area relates to lightning strikes during thunderstorms. In the past, fires set by such strikes have set the lignite on fire in areas where it has been exposed by erosion. These fires can burn for hundreds of years, and they result in intense *oxidation* (see "chemical weathering" later in this chapter) at the surface, converting iron minerals in the sediments above the coal beds to natural bricklike materials locally, called scoria. The bright red coloration comes from the oxidation of minerals producing the iron oxide mineral hematite.

FIGURE P9.4 Little Missouri River Valley in the southern segment of Theodore Roosevelt National Park, North Dakota. (Photo by author.)

FIGURE P9.5 Unusual erosional features in the northern segment of Theodore Roosevelt National Park. More resistant materials protect sediments below, and may produce scarps or pedestals, both shown here. (Photo by author.)

 ## BADLANDS NATIONAL PARK

Badlands National Park (see Figures 1.18 and 8.11) was first authorized as a national monument in 1929, then upgraded to a national park in 1978. It is located approximately 50–70 miles east of the Black Hills of South Dakota, in the region where there is a large change in moisture content, drying toward the Great Plains.

Geology. The oldest rocks in the park are Cretaceous marine sediments deposited in a shallow interior sea that covered the region at that time. However, most exposures in the park are Eocene-Oligocene, flat-lying, fine-grained clastic sediments containing some ash beds. These contain *clastic dikes* (Figure P9.6), un-

usual sedimentary formations where sand grains were washed into fractures and then cemented together, usually with calcite, but sometimes with siderite or marcasite. In the Pliocene, uplift occurred as the result of isostatic rebound. This is similar to Theodore Roosevelt National Park. Then, during Quaternary-Holocene time, erosion resulted in badlands formation (Figure P9.7).

An important aspect of the park is the excellent mammal fossil preservation in the Oligocene sediments now exposed by erosion. These include unique giant pigs, sabertooth cats, camels, and other mammals now extinct. Also abundant are plant fossils.

FIGURE P9.6 Clastic dikes exposed in Badlands National Park, South Dakota. (Photo by author.)

FIGURE P9.7 Badlands erosion in Badlands National Park. (Photo by author.)

BRYCE CANYON NATIONAL PARK

Bryce Canyon National Park (see Figures 1.18 and 8.7) was established as a national monument in 1923. Located in southwestern Utah, the park's name and designation were changed to Utah National Park in 1924, then changed again to Bryce Canyon National Park in 1928. Many people have seen pictures of Fairyland Canyon (Figure P9.8), the location of the best known and most unusual, intricate badlands spires (needles or hoodoos) found in the national parks.

Geology. The main rock unit exposed here is the Eocene, Wasatch Formation, composed primarily of calcareous ($CaCO_3$-rich), poorly consolidated *mudstones* and capped by thin sandstones and conglomerates. Figure P9.9 is a north-south cross section along the western margin of the Colorado Plateau through Zion National Park (Chapter 10). The units outcropping at Bryce Canyon National Park (Figure P9.9) are represented along the western margin of the cross section. The mudstones are very rapidly chemically and physically weathered, thus, when the capping unit is breached, erosion quickly forms the interesting hoodoos for which the park is famous.

FIGURE P9.8 Badlands erosion produced needles (hoodoos) in Fairyland Canyon, Bryce Canyon National Park, Utah. (Photo by author.)

FIGURE P9.9 North-south geologic cross section through Zion National Park (Chapter 10), showing Bryce Canyon National Park to the northeast of Zion (modified from a diagram by Peter Coney and Dick Beasley). The upper portion of the section is Tertiary in age and the Wasatch Limestone is exposed at Bryce Canyon. Many of the Mesozoic sedimentary units, such as the Navajo Sandstone and Moenkopi Formation, as well as Paleozoic units, such as the Kaibab Limestone, can be found in many park areas throughout the Colorado Plateau (see Figure 8.7).

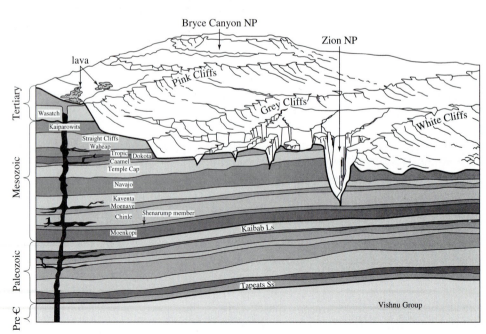

CHEMICAL WEATHERING

Chemical weathering involves those chemical changes causing mineral and rock decomposition. The primary chemical reactions in this process are oxidation, dissolution, hydrolysis, and carbonation. Also important are the climatic factors, availability of moisture and the temperature at which reactions occur. Aiding the process is the surface area or particle size exposed to chemical reactions.

Oxidation is a chemical reaction that takes place when an element or ion combines with oxygen by losing an electron. In the following example the mineral *magnetite* [Fe_3O_4] oxidizes to the mineral *hematite* [Fe_2O_3], and the Fe^{2+} (iron) ions in the magnetite lose an electron, converting to Fe^{3+}. The oxidation reaction involved is the following:

$$4Fe_3O_4 + O_2 \rightarrow 6Fe_2O_3$$

Both magnetite and hematite are manufactured as paint pigments, hematite because of its distinctive red color and magnetite for shades of gray to black. Hematite is also the mineral producing most red coloration in rocks and minerals.

During dissolution of rocks and minerals, the dipolar water molecule loosens mineral bonds by disrupting the electrostatic charges in minerals. Freed ions then go into solution and are carried off, a process called *leaching,* resulting in the removal of minerals. In the following example, table salt, the mineral halite, is broken down into its two constituent ions that then go into solution and can be carried away as groundwater flows in the subsurface.

$$NaCl + H_2O \rightarrow Na^+ + Cl^- + H_2O$$

Carbon dioxide also acts as an important weathering agent by producing carbonic acid [H_2CO_3] when CO_2 goes into solution, a process called *carbonation.* The ions in solution are hydrogen [H^+] and bicarbonate [$(HCO_3)^-$], the hydrogen acting as an acid that readily attacks calcite [$CaCO_3$], the major mineral constituent in limestone and marble. This process has been responsible for the destruction of large amounts of limestone, resulting in the development of karst terrains and caves (see Chapter 11, "Underground Water"), and the continuing destruction of many man-made structures and statues. During the carbonation process, the calcium ion [Ca^{2+}] is carried off in solution. The reactions involved are;

$$CO_2 + H_2O \rightarrow H_2CO_3 \rightarrow H^+ + (HCO_3)^-$$

$$H^+ + (HCO_3)^- + CaCO_3 \rightarrow Ca^{2+} + 2(HCO_3)^-$$

Another important chemical reaction, *hydrolysis,* involves the action of water in forming clays and other minerals. When water is available to the system, from rainfall, groundwater, or other sources, the water can combine with an acid, such as carbonic acid, and attack minerals lying on the surface or in the subsurface, forming new minerals. In the following example, potassium feldspar [$KAlSi_3O_8$], a major constituent of granites, is altered to the hydrated clay mineral known as *kaolinite* [$Al_2O_3 \cdot 2SiO_2 \cdot 2H_2O$] (the equation for kaolinite is written in terms of oxides; it can also be written as [$Al_2Si_2O_5(OH)_4$]). Silica [SiO_2], as well as potassium ions [K^+], are also released and carried off in solution.

$$2KAlSi_3O_8 + 2H_2CO_3 + H_2O \rightarrow 2K^+ + 2(HCO_3)^- + Al_2O_3 \cdot 2SiO_2 \cdot 2H_2O + 4SiO_2$$

A second, simple example of hydrolysis involves the mineral hematite, formed from oxidation. It can be altered by hydrolysis to the iron oxy-hydroxide mineral, *goethite* [FeO(OH)]. The distinctive yellow color of goethite can be readily distinguished in soils and is also manufactured as a paint pigment.

$$Fe_2O_3 + H_2O \rightarrow 2FeO(OH)$$

A general term for several of the oxidized iron oxides and oxy-hydroxides is *limonite*. These minerals are usually formed during soil formation, a process known as *pedogenesis,* and range in color from yellow brown to brown.

WEATHERING PRODUCTS

Exposed surface rocks alter to form soils (Figure 9.3), the most advanced product of the weathering process. Their study is called *pedology*. The zone of weathering, from the surface soils to the fresh rock below, is called the regolith. Soils are often difficult to recognize in the rock record, but if they can be identified, soils become important environmental and climatic indicators. Because of their geological and archaeological significance and their impact on construction, soils have been heavily studied. A number of soil classification systems have emerged. The simplest of these identifies three soil layers or horizons, labeled A, B, and C.

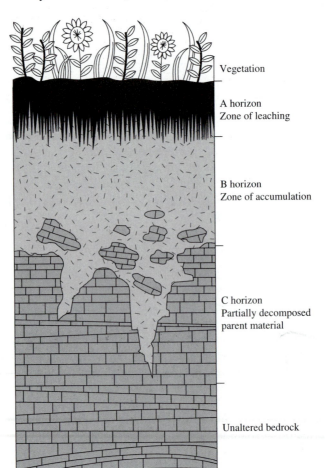

Vegetation

A horizon
Zone of leaching

B horizon
Zone of accumulation

C horizon
Partially decomposed
parent material

Unaltered bedrock

FIGURE 9.3   Soil cross section showing the A–C soil horizons. A is a zone of leaching and zone of plant decay material (humus). B is a zone of accumulation of material derived from A. C is a zone of decomposing parent rock that is partially broken down.

The *A-horizon* includes the topsoil, usually dark in color, with sand and humus as its usual composition. The sand tends to lighten soil color as does clay, but the *humus,* decaying plant material, tends to darken the soil. Rainwater filtering down through the A-horizon dissolves some of the mineral constituents present in this upper layer, and these constituents are then precipitated in lower layers or are carried off in solution.

The *B-horizon* often has some humus, but generally it is dominated by mineral precipitates that are derived from the A-horizon above and form in response to climatic conditions present at the site. There are several distinctive B-type horizons that can develop in temperate to warm-dry climates. Sometimes a nearly impermeable clay horizon may develop in the subsurface, while at other times hard, impermeable *caliche* or *calcrete* [$CaCO_3$] layers develop. Also, dispersed calcite may reprecipitate in the B-horizon. Iron oxides are often abundant in B-horizons, producing typically red or yellow soils, and other materials, depending on climate, often accumulate here.

For example, in tropical B-horizons *laterites* develop. They are very red in color, due to iron oxide residue that has been leached by warm water percolating downward through the soil. This action destroys organic matter (e.g., humus) and silicate minerals, often concentrating hydrated aluminum oxide [$Al_2O_3 \cdot nH_2O$] in the soil as bauxite ore. The small "n" in the bauxite formula represents variable amounts of water that may combine in forming the mineral.

Immediately above the bedrock is the *C-horizon,* which consists mainly of broken, weathering bedrock in the initial stages of destruction. When all soil horizons are fully developed, soils are identified as mature, but more generally, only partial development is identified, and these are classified as immature. Ancient soils are called *paleosols,* and when they are compared to modern soils as an analog, they give an indication of past climates.

Besides soils, other by-products of weathering are formed. For example, *spheroidal* rock represents rocks, often granites, that are rounded during weathering (Figure 9.4). There are many examples of spheroidal rock forms in the national parks, including those found in Acadia National Park (see Figure P5.28), Shenandoah National

FIGURE 9.4 Spheroidal weathering of granite in Mojave National Preserve illustrates the weathering effects on granites in the desert. (Photo by author.)

Park (see Figure P7.19), and Joshua Tree National Park (see Figure P12.4). This weathering occurs along joint surfaces that formed as concentric shells due to unloading during erosion and uplift. The process is called *exfoliation* and causes the rock to peel back like cabbage leaves or onion skins. Large-scale exfoliation domes can be significant geological features and include famous examples such as the rounded side of Half Dome in Yosemite National Park (see Figure P5.16). Mechanical weathering produces mass wasting products such as *talus* deposits, cone-shaped rock debris found at the base of cliffs.

EROSION

Erosion involves the removal and transportation of loose rock materials. There are several ways that material can be carried away from the weathering site. Water removes particles as suspended or dissolved materials in rivers and streams. The *suspended load* includes fine grains that are supported by the turbulence of the stream. Larger particles, known as the *bed load*, are rolled along the bottom as a carpet of material. This carpet moves by frictional forces, as the water and other particles drag these coarser sediments downstream. This mechanism is called *traction transport,* and another term for the bed load materials is a *traction carpet.* All of the particles being moved in streams are collectively called the sediment *load.*

Fluvial sediments are deposited in streams, while *lacustrine sediments* are deposited in lakes. Sediments deposited at the mouths of rivers are called *deltaic sediments* (discussed in Chapter 11). Eventually *terrigenous sediment* reaches the ocean and is deposited in the marine system (see Chapter 14).

Movement by *mass wasting* processes (see Chapter 7), where loose sediment moves downslope driven by the force of gravity, can be important in erosion. This is particularly true in areas of high topographic relief. As the rock debris moves, loose materials collide and break up into smaller and smaller pieces.

Eolian (wind) processes are very important as erosional agents in arid regions. Particles, arcing through the air, eventually hit the ground at high speed, striking other sand grains (Figure 9.5). These collisions cause other grains to bounce into the air along a curved pathway, eventually colliding with still other sand grains lying on the ground that, as a result, are thrown into the air. This process is called *saltation* (a similar process also occurs in streams) and is the method by which small particles are moved downwind. The abrasion produced by these grain impacts causes the frosted appearance associated with dune sands. Eventually sand and silt particles accumulate to form dunes and loess. *Loess* is very fine glacial "flour," the result of wind blowing along the front of ice sheets during periods of major glaciation. These winds pick up the very fine,

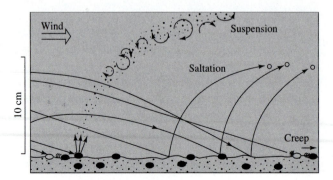

FIGURE 9.5 Diagram illustrating the process of saltation and the movement of sand by wind.

ground-up particles of rock and redeposit them in thick stratigraphic sequences. Major loess sequences are found in Europe, China, Russia, and North America.

Meltwater from glaciers has the effect of breaking up rock by penetrating into fractures and then freezing. Glaciers also abrade the rocks over which they are flowing and carry these materials with them as they move. Eventually this load is deposited and may be carried further by streams. Glacial processes will be discussed in Chapter 13.

CLIMATIC CONTROLS ON WEATHERING

Weathering is controlled mainly by climate, and climate depends on many factors, including temperature and the amount of moisture available. Also important is mineral composition, because composition will determine rates of weathering. In fact, weathering follows the Bowen reaction series (seen in Chapter 5, "Igneous Rocks"), but in reverse; high temperature minerals, such as olivine, weather very quickly, while low temperature minerals, such as quartz, weather very slowly under the same conditions.

Another factor in weathering is topographic relief, controlling how long a mineral will remain in zones of intense weathering. On a hillside, for example, mass wasting processes will remove a mineral exposed at the surface long before it can experience much weathering. But on flat-lands, where rocks remain at one locality for relatively long periods of time, standing water and high temperature can have a significant impact.

Latitude

Temperature is generally latitude dependent (Figure 9.6). At low latitudes, tropical climates have average yearly temperatures between 64–68° F (18–20° C). In tropical

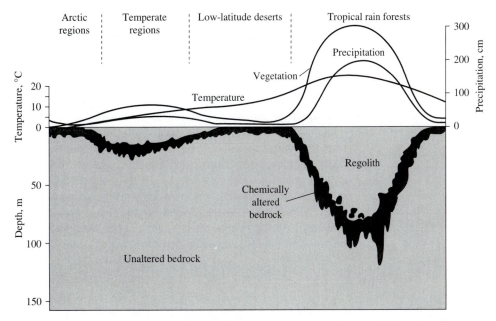

FIGURE 9.6 Chemical weathering depth below the ground surface with changing climate. When temperature is high and water is available, chemical weathering is greatest. But even in temperate climates, if the temperatures are not too cold and if water is available, chemical weathering is still relatively great.

areas that have high rainfall, rain forests develop. Rain forests also develop in cooler, more temperate areas, where rainfall is extremely high, along the Pacific Coast of Washington at Olympic National Park (Chapter 14), for instance. However, where rainfall averages less than 10 in (25 cm) per year, deserts will develop. Clearly, moisture content is critical to climate and weathering.

Further to the north, subtropical climates are characterized by pine forests and *savannas*, grasslands with high rainfall and some exotic plants, such as the Everglades. In areas of moderate rainfall, grasslands, such as those in the Great Plains, develop. When summers are hot and dry, but winters are cool and wet, Mediterranean climates, exhibiting chaparral vegetation, develop.

Cooler temperate climates are typified by grasslands such as those of the Great Plains further to the north in Canada. Forests are deciduous, with trees such as oaks and maples. Subpolar climates are even cooler, with forests dominated by conifers, including spruce, pine, and fir. Tundra is also evident where the near-subsurface is permanently frozen (see Figure P7.17, Rocky Mountain National Park). And the coldest climates of all exist in the polar regions where ice sheets develop. The amount of water claimed by ice determines global sea surface height, and it is of importance to coastal regions. Slight changes in moisture or temperature can dramatically impact global sea levels. This will be discussed further in Chapter 13.

Other Climate Controls

Besides temperature, moisture, and latitude, other important factors controlling climate are ocean circulation, continental distributions, and locality. Ocean circulation patterns bring warm or cold ocean currents to the shores of continents, modifying climate by warming or cooling air masses and by generating weather patterns. For example, effects from the warm Gulf Stream moderate the climate in England, which lies at high northern latitudes, but also generate a large amount of rainfall. Distribution of continents dictate where ocean currents can flow, thus controlling deep ocean and surface circulation patterns. For example, if Central America was removed, ocean currents could flow between North and South America, remaining at low latitudes longer and thus becoming warmer. The whole Earth would be effectively warmed by this change. Instead, ocean circulation today is forced to flow toward the poles where currents become very cold, cooling Earth's climate as they flow back into low latitudes.

A specific locality on a continent has its climate affected by a number of factors. Distance from the ocean is important because as air masses move inland rain reduces the amount of moisture being carried by clouds. Therefore, the farther the air travels from the source of moisture, the ocean, the dryer will be the air and the less rainfall will occur at inland sites. This drying effect is also controlled by the wind pattern over an area. If moist air bypasses a site, then the site will be dry. In part, the distribution of mountains modifies the wind patterns and the moisture available to sites. Generally, *leeward* of mountains (the side of the mountain away from the direction the wind is blowing), rainfall is low, whereas zones of high rainfall are found *windward* of mountain ranges. This will be discussed further in Chapter 12.

REFERENCES

BENNETT, R., ed. 1980. *The new America's wonderlands, our national parks,* p. 463. Washington, D.C.: National Geographic Society.

BEUS, S. S., ed. 1987. *Decade of North American geology, DNAG, Centennial field guide.* Vol. 2. *Rocky Mountain Section of the Geological Society of America,* p. 475. Boulder, Colorado: Geological Society of America.

BOGGS, S. 1994. *Principles of sedimentology and stratigraphy.* 2d ed., p. 800. New York: Macmillan College Publishing Company.

CHRONIC, H. 1984. *Pages of stone, geology of western national parks and monuments.* Vol. 1. *Rocky Mountains and Western Great Plains,* p. 168. Seattle: The Mountaineers.

CHRONIC, H. 1988. *Pages of stone, geology of western national parks and monuments.* Vol. 4. *Grand Canyon and the plateau country,* p. 158. Seattle: The Mountaineers.

HARRIS, A. G., and TUTTLE, E. 1990. *Geology of national parks.* 4th ed., p. 652. Dubuque, Iowa: Kendall Hunt Publishing Company.

HARRIS, D. V., and KIVER, E. P. 1985. *The geologic story of the national parks and monuments.* 4th ed., p. 464. New York: John Wiley & Sons, Inc.

10

Sedimentary Rocks

CLASTIC SEDIMENTARY ROCKS

Clastic (detrital) sedimentary rocks are formed from the pieces of other rocks (sedimentary, igneous, or metamorphic). After erosion, they are usually transported and deposited at some other locality, and finally undergo *lithification.* Besides clastic sedimentary rocks there are also chemical sedimentary rocks, such as cave dripstone deposits, and biologically produced sediments, mainly limestones.

After being broken down, the particles produced are carried away and deposited elsewhere. The processes of erosion and transportation have a major impact on the particles that are being moved. The primary effects are to concentrate, sort, and segregate minerals so that the final sedimentary deposit is often very distinctive and diagnostic of the environment of deposition, source of material, climate, and other factors. Figure 10.1 illustrates the complex *terrigenous* (continental as opposed to marine) sedimentary depositional environments. Clastic particles may be eroded and transported by glaciers, streams, waves, and/or wind. These materials may then be deposited at the margins of glaciers, in and along streams, in lakes, in dunes, in river deltas, or ultimately in the marine environment. Lithification may follow deposition, or the particles may continue to be moved following initial deposition.

Particles are concentrated and shaped as they are moved along by streams toward their final depositional site. As the result of the abrasion of one particle against another, the grains are rounded and become more spherical in shape. The term sphericity is used to show how closely a particle approximates a sphere. As the grains are carried along in streams, large grains tend to settle to the bottom, roll along, or drop out of the current when stream velocity slows, such as when a stream rounds a bend in the river. Grains with cleavage planes, such as feldspars, are more easily broken down into smaller grain sizes, while quartz is much harder to break because it has no cleavage. The result of these and other factors is that grains are *sorted* by size, with larger particles accumulating closer to the source from which grains were initially derived and finer grains accumulating farther from the source. Particles are also segregated by min-

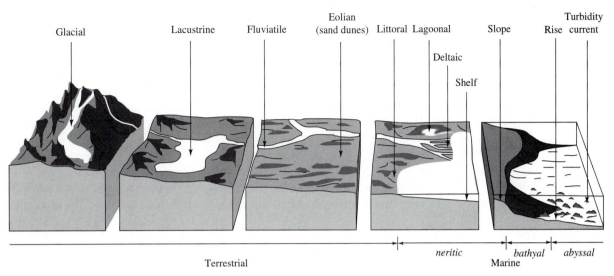

FIGURE 10.1 Cross section of a continent illustrating the terrigenous sedimentary environments and extending into the marine environments.

eral type (composition), as the result of differing densities. Quartz is lighter than magnetite, thus equal-sized grains of quartz are carried farther than are the denser magnetite grains that drop out sooner. Most grains, silt, and clay particles, are carried in suspension.

The effects of rounding, sorting, and segregating sand grains can be seen on many beaches, where most of the grains of sand are about the same size and are made primarily of quartz. On Daytona Beach, in Florida, these effects have produced a very fine quartz sand that packs so well that in places, and at certain times, cars can drive on it as though it were a paved road. It can become so smooth and hard that it was used for many years to set the world's land-speed record.

Deposition

Deposition of grains may occur in either the marine or the terrestrial environment. For those rocks that are ultimately preserved, deposition usually occurs where the sediment can eventually be lithified. In the case of a stream, sediment usually is deposited when the water velocity slows, such as when it enters a lake, or the ultimate base level, the ocean. Deposition along the length of a stream may also occur. There are some places in the national parks where such sediments are preserved and are important. For example, dinosaur bones are found in fluvial sediments at Dinosaur National Monument (Chapter 15).

Wind-blown deposits are important in some areas. These are the result of erosion by wind and later deposition by wind as wind velocities slow, dropping the silt or sand being carried. Great Sand Dunes National Monument has dunes made up of loose sand that is being deposited in this way. Zion National Park, Capitol Reef National Park (Chapter 15), and other national parks and monuments in the southwestern part of the United States contain rocks that represent vast areas of lithified sand dunes.

GREAT SAND DUNES NATIONAL MONUMENT

Great Sand Dunes National Monument (see Figure 1.18) was set up in 1932. It is located in a cul de sac along the eastern margin of the San Luis Valley (Figure P10.1) at the base and just to the west of the Sangre de Cristo Mountains in southern Colorado. Located here are some of the world's highest sand dunes, attaining a height of 700 ft (213 m) or so (Figure P10.2). After struggling for some time to reach the crest of the tallest dunes, visitors are often startled by the high winds driving sand particles painfully into their faces.

Geology. The San Luis Valley (Figure P10.1) is a depressed fault block that has been faulted down and filled by sediments. At one time a lake also filled the valley, but this has since evaporated, leaving behind evaporite deposits. Sands, eroded from San Juan Mountains to the west, including quartz but mostly feldspars, have been carried to the east by strong winds. As the rising wind encounters the Sangre de Cristo Mountains, velocity slows a bit and the sand is dropped at that point. In addition, magnetite eroded out of the Sangre de Cristos has been added to the dunes by swirling winds along their leading edge. Further migration of the sand is retarded by intermittent stream flow washing out of the Sangre de Cristos (Figure P10.3).

FIGURE P10.1 San Luis Valley, Colorado, containing Great Sand Dunes National Monument. The San Juan Mountains to the west are the main source of the sand, which is piled up at the base of the Sangre de Cristos to the east. The sand dunes can be seen in this map.

FIGURE P10.2 The summit of the highest sand dune in Great Sand Dunes National Monument. The wind is constantly reforming the crests of these large dunes, and the surface sand is always in motion, either by saltation or small scale slumping. (Photo by author.)

FIGURE P10.3 Rain in the Sangre de Cristos brings intermittent stream flow to Great Sand Dunes National Monument, washing away the eastern-most dune sands. (Photo by author.)

Diagenesis

Diagenesis involves post-depositional changes, including dissolution, *cementation,* and *replacement* of minerals. Material being dissolved may be carried completely away, changing the composition of the sediment, or it may reprecipitate as coatings on other grains, or be deposited in pore spaces between grains, as cement. The result is often the formation of new, *authigenic* minerals. Unstable minerals may recrystallize during diagenesis, changing the original character of the sediment.

Lithification

Lithification involves the formation into rock of deposited loose detrital material. Diagenesis begins to alter this material, and as more and more sediment accumulates, pressures in thickening sedimentary piles cause compaction and dewatering (desiccation) of the material. Compaction plays a significant role in the development of sedimentary rocks, where grains are squeezed together as sediment overburden increases. Fine-grained sedimentary rocks, such as shales, are formed mainly by compaction. Eventually coarser grains are cemented together, thus filling the pores between grains, producing a rock. These processes form the final sedimentary rock.

SEDIMENTARY ROCK TYPES

Clastic sedimentary rocks are classified by the size, shape, and composition of particles contained within the rock. Particle size is a result of abrasion, density, or shape sorting. As particles are transported downstream, coarse deposits are formed near the source, where stream velocities are high. Finer deposits, farther from their source, are deposited from moderate velocity streams. Suspended, very fine deposits are found long distances from their source and are deposited from low-velocity flows. Composition of abraded deposits is, in part, a function of how easily the particles being transported are broken down. Quartz, for example, is resistant to abrasion because it has no cleavage planes and because it is also relatively hard. It persists in many environments and often accumulates on beaches. Feldspars, on the other hand, are rapidly destroyed by physical abrasion, because they exhibit high cleavage-plane breakage, and by chemical alteration, because they are susceptible to chemical attack.

Lithification affects large grains differently that it affects small grains. Large grains are cemented by silica, carbonate (e.g., calcite—$CaCO_3$), or other cements (e.g., iron oxides), while fine grains may be lithified simply by compaction. Grain size determines the five clastic sedimentary rock types. These are, from coarse to fine grained: conglomerates, containing the coarsest particles; sandstones, sand grains from $\frac{1}{16}$ to 2 mm in size; siltstones, particles from $\frac{1}{256}$ to $\frac{1}{16}$ mm in size; shales and mudstones, containing grains less than $\frac{1}{256}$ mm in size. *Conglomerates* include the coarsest particles, ranging from gravel to boulders. Such deposits may contain a wide variety of clast sizes. Sometimes they are fairly well sorted, having only one general size of particles, such as gravel, and sometimes they are poorly sorted, containing the full range of particles.

Sandstones are usually well-sorted, cemented sands. In addition to the size classification, composition is also used to further differentiate sandstones. For example, if the sandstone is composed mainly of quartz, it is called a *quartz arenite;* if it is composed of quartz and feldspar, an *arkose;* or if it contains some coarse fragments, a *lithic sandstone*; and finally, if it contains some clay, it may be called a *greywacke,* often poorly sorted. Wind-blown, cemented dune sands, such as the Navajo Sandstone exposed at Zion National Park, are often bound together with calcite cements. There are extensive examples of such sandstones found in the Colorado Plateau region (Figure 10.4).

 ## ZION NATIONAL PARK

Zion National Park (see Figures 1.18 and 8.7) was first established as a National Monument in 1918. Located in southwestern Utah it was upgraded to national park status the following year. In 1937, a rather strange thing happened; a second Zion National Monument was created. Then in 1956, Congress decided to consolidate them both into one Zion National Park. It is a place of rare beauty, containing unusual rock formations, deep wooded canyons, and steep cliffs (Figure P10.4).

Geology. The rocks in Zion are mainly Mesozoic in age with some Cenozoic lavas also exposed within the park. The oldest rocks are from the Moenkopi Formation, also found in many areas of the Colorado Plateau region (Figure P9.9). But the most significant and extensive unit exposed in the park is the Jurassic Navajo Sandstone (Figure P10.4). It is composed of lithified dune sands (eolian) with crossbeds (Figure P10.5) that formed as grains slid down the leeward edge of the dunes. The steeper face of the dune (*leeward* side) generally lies in the migration direction of the dune, the direction toward which the wind is blowing. Typical frosted grains are observed in the Navajo Sandstone, the result of natural sand blasting from quartz grain collisions. $CaCO_3$ has cemented the quartz grains together, forming a sandstone that, in some places, is up to 2,200 ft (670 m) thick. But because the cement is $CaCO_3$, the Navajo Sandstone is very easily weathered and therefore is quickly destroyed if not protected by more resistant overlying beds. At Zion, the Navajo Sandstone is topped by more resistant units that in many areas form a cap, resistant to erosion. Pliocene uplift caused differential cracking of the sandstone because calcite has excellent cleavage and is easily broken. These breaks then became zones of rapid weathering and erosion leaving interesting patterns such as those observed on the face of Checkerboard Mesa (Figure P10.6).

FIGURE P10.4 Shapes produced by erosion of the Navajo Sandstone, after capping units have been breached in Zion National Park. (Photo by author.)

FIGURE P10.5 Lithified cross-beds in the eolian Navajo Sandstone (Checkerboard Mesa), Zion National Park. Sand grains are cemented by calcite [$CaCO_3$]. (Photo by author.)

FIGURE P10.6 Fracture dissolution on Checkerboard Mesa in Zion National Park. (Photo by author.)

Siltstones are composed of small, well-sorted silt particles. Compaction has squeezed them together in such a way that they adhere to each other and form a sedimentary rock. Some siltstones also contain cemented silt-sized grains. *Shales* and *mudstones* are composed of silt and finer, flat clay grains. These are the most abundant sedimentary rock and may be composed of thin *laminae* (layers) that result from compaction of fine, clay-sized particles that form sheets (resembling sheet silicates, see Figure 2.10). These grains adhere to each other, producing a very *fissile rock* that is easily broken along the layering. Sometimes these layers contain fossils, due to the burial, compaction, and preservation of the remains of plants and animals that died and settled to the bottom along with the accumulating sediments. This usually happens in a lake where very fine-grained sediments have accumulated and are undisturbed. Florissant Fossil Beds National Monument, in the mid-Colorado Rocky Mountain Front Range, is one such locality, well known for its fine, delicate fossils. Another similar locality, known for its outstanding fish fossils, is Fossil Butte National Monument. It is discussed in Chapter 15.

FLORISSANT FOSSIL BEDS NATIONAL MONUMENT

Florissant Fossil Beds National Monument (see Figure 1.18), near Florissant, Colorado, was fashioned from two tourist attractions in 1969 and sits in the mid-central Colorado Front Range. It lies immediately to the west of Pikes Peak and to the west of Colorado Springs, Colorado. The monument is well known for petrified trees (Figure P10.7), other plants, fish, and, most important, 1,200 species of insects recovered from the shale beds at the locality.

The land now within the boundaries of the monument includes what were two private tourist establishments, Colorado Petrified Forest to the north and Pike Petrified Forest to the south. Other land has also been incorporated within the monument, including a large portion of the fossil-containing shales that were deposited in "Lake Florissant." Self-guided and ranger-guided tours allow visitors to see a number of spectacular petrified tree stumps exposed within the park.

Geology. During Eocene/Oligocene time, a broad valley existed in the area. This was dammed by lava, forming Lake Florissant. The climate at the time was subtropical, and palms and sequoias grew in the swampy areas surrounding the lake. For many years, fine-grained sediment filled the lake, incorporating within the accumulating sedimentary layers the fossils

FIGURE P10.7
Petrified sequoia (Big Stump) at Florissant Fossil Beds National Monument, Colorado. Saw blades (broken ends are visible in the photograph) were used in an attempt to cut up the stump so that it could be more easily moved to a World's Fair. The attempt failed and the stump was never moved. (Photo by author.)

that are preserved today (Figure P10.8). Abruptly, the area was covered, first by mud flows, 10–20 ft (3–6 m) thick, and then by lavas that capped the mud flows. These volcanic eruptions cut off the upper parts of the trees that were sticking above the mudflows, but in the mud their lower portions were protected. There, *petrification (petrifaction)* occurred by slow replacement of the wood by SiO_2. The area was further covered by sediment, and the muds, fine silts, and clays in the lake were compacted, forming shales. Then, during the Pliocene, isostatic uplift of the Rockies caused increased erosion that then exposed the beds.

Today in the monument a number of petrified trees have been excavated, and some shale exposures,

granites surrounding what was once the lake, and volcanic tuffs, can be viewed by visitors. At the boundaries of the monument, just outside government land, there are areas where the curious, with permission, can dig for fossils. Paleontologists, working in new exposures in the monument, are currently looking for new fossils and the information they contain. Fossils still lying buried in the ground, hidden within shale layers, will provide exciting new information about Earth history. In addition to the 1,200 species of insect and 50 species of spider, the Lake Florissant sediments have already yielded some mammal and bird fossils, as well as 8 or 9 species of fish.

FIGURE P10.8 Plant fossils in shale beds from Florissant Fossil Beds National Monument. (Photo courtesy of the National Park Service.)

CHEMICAL SEDIMENTS

Chemical sediments are chemical precipitates or organic sediments that result from natural or biological precipitation, or evaporation in special environments. Shallow basins, where evaporation is rapid and salts are concentrated, are areas where supersaturation and eventually precipitation of mineral *evaporites* occurs. In fresh water the primary evaporites include *halite* (NaCl, common table salt, also called rock salt), *gypsum* ($CaSO_4 \cdot 2H_2O$), and its dehydrated equivalent *anhydrite* ($CaSO_4$). In relatively shallow, warm marine waters, such as those around the Bahamas or found in the Florida Keys at localities such as Dry Tortugas National Park, calcite as lime mud is a commonly produced mineral, as are some iron oxides, carbonates, and sulfides; these, however, are accessory minerals found in relatively small quantities.

DRY TORTUGAS NATIONAL PARK

Dry Tortugas National Park (see Figures 1.22 and 8.17) was established by Congress in 1992. It is located at the end of the chain of coral reef islands known as the Florida Keys, 68 miles to the west of the town of Key West, Florida. Access is only by boat or seaplane. It was first established in 1935 as Fort Jefferson National Monument on the largest key in the Dry Tortugas, Garden Key (Figure P10.9). No fresh water is available in this region of the Keys, thus the name Dry Tortugas (sea turtles).

Geology. The park is made up of seven small keys built from coral reefs as wave action broke up the coral ($CaCO_3$). Included as part of the park are the surrounding staghorn reefs, small patch reefs, and very small barrier reefs. As the corals die and are broken up by wave action, carbonate sand is piled on reef tops and builds up as sediments. Plants then anchor the sand and form soil. These processes are actively occurring today.

History. The Dry Tortugas have an interesting history. The site was discovered by Ponce de Leon in 1513. He recognized its strategic importance for controlling the eastern end of the Gulf of Mexico, but Spain never occupied the site. The United States included the islands as part of the state of Florida, and after the War of 1812 decided to fortify Garden Key. Construction of Fort Jefferson began in 1846 (Figure P10.10) but was never completed, and no shot was ever fired in anger from or at the fort. It was last used as a coaling station in 1908.

Fort Jefferson was the largest Civil War fortification built in the United States and was used as a prisoner-of-war camp. But it was clear early in the war that the invention of rifled cannon made forts of this type obsolete. (See the discussion of Fort Pulaski National Monument in Chapter 16.) Dr. Samuel Mudd was the most famous person to be imprisoned here. He had helped John Wilkes Booth by setting his leg, broken after Booth shot President Lincoln. But Dr. Mudd did not know who Booth was and, as a doctor, was obliged to help him. Nevertheless, he was convicted as a conspirator and imprisoned at Fort Jefferson. Dr. Mudd was eventually pardoned by President Johnson in 1869 after he saved the fort from a yellow fever epidemic in 1867, but he was never exonerated and died in 1883. His ancestors continued to try to have his name cleared, until he was finally exonerated by President Carter in 1978.

FIGURE P10.9 Fort Jefferson on Garden Key, a carbonate reef, in Dry Tortugas National Park, Florida. (Photo by author.)

FIGURE P10.10
Entrance into Fort Jefferson, Dry Tortugas National Park. The fort served as a prison during the Civil War. (Photo by author.)

Chemical sediments that are formed by biological precipitation are usually marine (originate in the ocean) and composed of calcium carbonate ($CaCO_3$), forming shell material, often microscopic, and reefs. This shell material may be calcite or *aragonite,* a *polymorph* of calcite (the two minerals have the same chemical composition, but different crystal structure). Aragonite is relatively unstable and will often recrystallize to calcite.

The varieties of calcite formed by biological activity are lithified to form *limestones,* such as those found in Guadalupe National Park, Texas. Limestones can be very generally classified (Figure 10.2) as *micrite,* with or without larger grains; *sparite,* made up of calcite cement with little or no micrite; or *coquina,* where shells are loosely cemented together. Two of the national monuments in Florida contain forts made from quarried coquina (Figure 10.3), Castillo de San Marcos National Monument and Fort Matanzas National Monument. These are discussed in Chapter 16.

Limestones also have distinctive textures, depending on their constituents. For example, *skeletal* textures represent limestones containing microscopic plant and animal remains, while *microcrystalline* textures contain very fine crystals of calcite. An interesting texture is caused by *oolitic* (spherical particles like small pearls) limestones. In cave pools it is common to find "cave pearls," very round, finely layered calcite speleothems, similar to individual oolites. These form by very slight natural motion of the pearl that allows the calcite to precipitate on all sides as fine layers.

FIGURE 10.2 Types of limestone textures, from aphanitic limestone, or micrite where no clear fossils or structures are distinguishable, to limestone very rich in bioclasts.

Aphanitic Limestone

Limestone

Microcrystalline

Oolitic

Coquina

FIGURE 10.3 Coquina limestone blocks, dominated by abraded bioclasts that have been lithified, (indicative of beach erosion), were used in the construction of Castillo de San Marcos National Monument and Fort Matanzas National Mounument, Florida. (Photo by author.)

Other chemical sediments include *dolostone,* composed mainly of the mineral dolomite $[Ca,Mg(CO_3)_2]$, which is similar to calcite but has magnesium in the crystal lattice. Dolostones tend to be more resistant to erosion than are limestones, and geologists tell the difference in the field by dripping a drop of dilute hydrochloric acid on the rock. Limestones rapidly fizz (effervesce), but dolostones bubble only slightly.

Silica $[SiO_2]$ is another chemical sediment that may result from the accumulation and lithification of biological shells, such as the development of *chert* from *radiolarian* and *diatom* shells. These plankton are usually found living in marine surface waters. Silica may also have a nonbiological origin. The addition of water to silica produces the mineral *opal* $[SiO_2 \cdot nH_2O]$, often found associated with biological materials, and also found in the petrified tree stumps at Florissant Fossil Beds National Monument.

GUADALUPE MOUNTAINS NATIONAL PARK

Guadalupe Mountains National Park (see Figures 1.17 and 8.6) was established in 1972. Located in west Texas, the park is jointly administered with Carlsbad Caverns National Park, and the headquarters is in Carlsbad, New Mexico. Today, the two parks have a common boundary. The boundary extension by Congress was designed to protect caves within the parks from mining operations outside the parks that affect groundwaters flowing through the region.

Geology. As with Carlsbad Caverns National Park, Guadalupe Mountains National Park is composed of Permian reef, skeletal limestones, and the park contains many caves, none that are visited by Park Service tours, but many that are regularly visited by caving groups. In Texas, the leading edge of this limestone forms a clif, and rising above the plains below is El Capitan (Figure P10.11). At 8,085 ft (2,464 m) elevation, El Capitan can be seen for 100 mi (160 km) in Texas. The park also contains Guadalupe Peak, the highest point in the state of Texas, at 8,749 ft (2,667 m). It is not as distinct as El Capitan because it lies farther back along the ridge formed by the reef limestones and is dominated by El Capitan. The limestones in the park form limestone exposures typical of desert weathering (Figure P10.12).

FIGURE P10.11 The Permian limestone cliffs of El Capitan Mountain in Guadalupe National Park, Texas. El Capitan can be seen from a great distance. (Photo by author.)

FIGURE P10.12 Desert weathering typical of Guadalupe National Park, forming limestone cliffs with debris lying along the flanks. Along with abundant cacti, there are also tree-lined canyons where the changing leaves produce beautiful fall colors. (Photo by author.)

REFERENCES

Bennett, R., ed. 1980. *The new America's wonderlands, our national parks,* p. 463. Washington, D.C.: National Geographic Society.

Beus, S. S., ed. 1987. *Decade of North American geology, DNAG, Centennial field guide.* Vol. 2. *Rocky Mountain Section of the Geological Society of America,* p. 475. Boulder Colorado: Geological Society of America.

Blatt, H., Middleton, G., and Murray, R. 1980. *Origin of sedimentary rocks.* 2nd ed., p. 782. Englewood Cliffs, New Jersey: Prentice-Hall, Inc.

Boggs, S. 1994. *Principles of sedimentology and stratigraphy.* 2nd ed., p. 800. New York: Macmillan College Publishing Company.

Chronic, H. 1984. *Pages of stone, geology of western national parks and monuments.* Vol. 1. *Rocky Mountains and western Great Plains,* p. 168. Seattle: The Mountaineers.

Chronic, H. 1986. *Pages of stone, geology of western national parks and monuments.* Vol. 3. *The desert southwest,* p. 168. Seattle: The Mountaineers.

Chronic, H. 1988. *Pages of stone, geology of western national parks and monuments.* Vol. 4. *Grand Canyon and the plateau country,* p. 158. Seattle: The Mountaineers.

Harris, A. G., and Tuttle, E. 1990. *Geology of national parks.* 4th ed., p. 652. Dubuque, Iowa: Kendall Hunt Publishing Company.

Harris, D. V., and Kiver, E. P. 1985. *The geologic story of the national parks and monuments.* 4th ed., p. 464. New York: John Wiley & Sons, Inc.

Hayward, O. T., ed. 1988. *Decade of North American geology, DNAG, Centennial field guide.* Vol. 6. *South Central Section of the Geological Society of America,* p. 468. Boulder, Colorado: Geological Society of America.

Neathery, T. L., ed. 1986. *Decade of North American geology, DNAG, Centennial field guide.* Vol. 6. *Southeastern Section of the Geological Society of America,* p. 457. Boulder, Colorado: Geological Society of America.

11

Rivers, Streams, and Underground Water

RIVER SYSTEMS

Running water is the most important erosional agent on Earth, ultimately flowing to the sea via stream flow, confined within banks in stream valleys. Generally, river systems develop in well-ordered *drainage basins* (Figure 11.1) that are separated by distinct boundaries or divides. These are high points from which water flows into a basin on either side. River systems are part of the *hydrologic system,* whose main water source is precipitation in the form of snow, rainfall, or dew. After precipitation, the water is shed by *overland flow,* sheet surface flow, or runoff that is regulated by vegetation. Vegetation, in turn, is controlled mainly by rainfall, rock or soil type, fires, elevation, and topography.

Tributaries bring flowing water into streams overland by collecting and funneling surface runoff to the main stream trunk (Figure 11.1), with excellent examples found in Canyonlands National Park. Some of the water penetrates into the ground by infiltration, feeding *groundwater* that, in turn, also flows, eventually reaching streams via springs. We see a clear decrease in the number of tributaries further downstream, but the tributary length increases. The stream gradient, or *slope* (Figure 11.2), decreases exponentially downstream, from the head (source) of the stream to its mouth, where it empties into the ocean. The *main trunk* of the stream is the central core of the transporting system. Downstream, trunk channels become deeper and wider, and valley size increases in proportion to stream size, with corresponding increasing height of the containing valley walls (Figure 11.3).

When rivers flow into lakes, into dry basins, or into the ocean, the *dispersing system* of *distributaries* dumps the sediment load and funnels the water through a network of channels. It is from distributaries that deposited sediment produces deltas. These will be discussed later in this chapter.

FIGURE 11.1   Mississippi River drainage basin, the largest in North America.

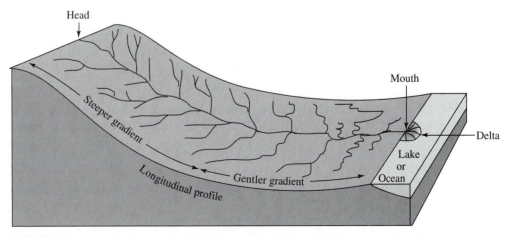

FIGURE 11.2   Longitudinal profile of a stream, representing the change in gradient from steep to gentle along the length of a stream from its source (head) to its mouth.

FIGURE 11.3   The Mississippi River channel from Effigy Mounds National Monument, Iowa. Effigy Mounds, burial mounds built, from 500 B.C. to about 1,300 A.D., in the shape of animals, lie on the western bluffs overlooking the western channel of the Mississippi River at this point. The eastern edge of the valley can be seen in the haze at the far right in the photograph, beyond an island separating the eastern and western channels of the river. (Photo by author.)

 CANYONLANDS NATIONAL PARK

Canyonlands National Park (see Figures 1.18 and 8.7) was established by Congress in 1964. Located in eastern Utah, the park contains canyons that were eroded by the Colorado River and its major tributary, the Green River. In Canyonlands these tributaries merge with the main trunk and the canyons developed here (Figure P11.1) are well representative of desert stream erosion patterns. In the past, funding for the park has been less than adequate. As a result, for many years the visitor center was located in a mobile home.

Geology.   During the Paleozoic a great deal of sediment was deposited in the area. This began in the Cambrian and Ordovician Periods, when sediment was deposited in the then-offshore Cordillera passive margin. In the later part of the Ordovician and into the Silurian, uplift and erosion occurred, followed by further deposition from Devonian to Mississippian time. During the Pennsylvanian (the oldest rocks found exposed in the park), the Uncompahgre uplift occurred, as well as the formation of the adjacent Paradox Basin, with large accumulations of salt as evaporite deposits. This was followed by Permian deposition, uplift, and erosion.

In the Mesozoic, deposition of a series of sandstones occurred followed by deposition of the eolian Entrada Sandstone that caps the area. Fracturing then occurred, the result of broad folding and the emplacement of salt intrusions. Differential erosion then produced needles, arches, and other badlands features as the capping Entrada Sandstone was breached. Recent major block-fault movement has also affected the area. Today stream erosion has produced broad stream valleys in the area (Figure P11.2)

It is argued by some geologists that Canyonlands National Park contains one of the most spectacular meteorite impact structures in the world (Figure P11.3). Known as upheaval dome, the impact crater probably occurred toward the end of the Cretaceous and, theoretically, is one of a series of bolide impacts that caused the extinction of many organisms on Earth, including the dinosaurs. The feature was initially 5–6 mi (8–9 km) across, but has since been heavily modified by erosion. For many years geologists interpreted the feature as being due to uplift from salt intrusion below, but now the impact hypothesis seems very appealing to many.

FIGURE P11.1 Typical stream canyon erosion in desert areas, from Canyonlands National Park, Utah. (Photo by author.)

FIGURE P11.2 Stream channel eroded by the Colorado River at Canyonlands National Park. (Photo by author.)

FIGURE P11.3
Upheaval dome, believed to be a bolide crater created in the latest Cretaceous, in Canyonlands National Park. (Photo courtesy of the National Park Service.)

STREAM FLOW

Stream flow results in erosion and the removal of a large amount of sediment from the continental interior, transporting this material to lakes and ultimately to the sea. In order to conduct its work, streams must flow at a relatively fast velocity, governing the stream's capacity to erode and transport sediment. Much of the erosion occurs during floods as the result of excess rainfall, and river valley floors may often be flooded as streams overflow their banks. The area flooded is called the *flood plain* and is the beneficiary of fine sediment, silt, and mud accumulation that creates fertile farmland, which is replenished by periodic flooding. Generally the water is moving as a chaotic *turbulent flow,* represented by high velocity, disturbed movement, and circular flowing eddies that support grains in suspension (Figure 11.4). However, along the very margins of streams there is a narrow band where *laminar flow* occurs, represented by low velocities and parallel flow lines with no between-layer mixing.

Water volume in streams is controlled by the amount of surface runoff, as well as input from groundwater seepage, an important source of water volume. Those streams that continue to flow year-round, even during times of drought, are called *permanent streams,* because they have a constant water supply. Some *intermittent streams* dry up seasonally because the groundwater seepage is absent or too low to maintain the stream flow.

Further from the source, the slope of the land, and therefore the *stream gradient* changes. The head-to-mouth profile of streams, known as the *longitudinal profile* (Figure 11.2), shows steep gradients near the source and flat slopes near the mouth. Such longitudinal profiles result from erosion by the stream and its tributaries, flattening the Earth's surface as more and more water is collected and erodes the land surface down-

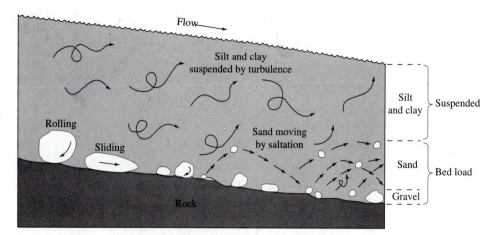

FIGURE 11.4 Stream flow with turbulence and particles carried by the flow. Larger particles are rolled or they slide along the bottom, while the smallest particles are carried in suspension. Sand is moved mainly by saltation.

stream. Channel size and shape are produced by these erosion effects, with some elements in the stream retarding flow by increasing the *channel roughness.* Items such as boulders act to retard flow near the source, but these are slowly eroded. Farther from the source sediments become finer in streams and in stream deposits, such as in alluvial fans (Figure 11.5). Ultimately stream channels erode to a semicircular shape, providing the lowest surface area and least resistance to flow for the stream. Streams also change downstream in other ways. For example, in the eastern United States, where rainfall is relatively high, the amount of water carried by a stream, its *discharge,* increases because more and more new water is fed into the stream by tributaries. Stream channel width and depth, as well as velocity, also increase downstream.

Tributaries develop a number of distinctive patterns that are usually controlled by the underlying rock over which they are flowing. Shapes may be *dendritic*—treelike (Figure 11.6a)—in character due to the erosion of massive, uniform, flat-lying strata where resistance to erosion is everywhere relatively similar. *Radial stream patterns* develop by erosion outward from a common central high point in elevation, such as the center of a volcano (Figure 11.6b). Sometimes *rectangular stream patterns* exist where fractures in underlying rock controls the erosion pattern with near right-angle bends following bedding planes of steeply inclined strata. Erosion follows joints, faults, or foliation planes. A fourth, the *trellis stream pattern,* is developed when erosion is controlled by underlying bedrock along zones of different weakness. The pattern is rectangular, but with very long tributaries such as those commonly seen in the Appalachian Mountains. An extension of the trellis pattern often produces a *water gap,* where rivers flow through a gap in a ridge after broad erosion of an area has *superposed* (superimposed) the stream onto buried structures that are exposed during erosion (Figure 11.6c). The ridge remains due to its resistance to erosion, the surrounding rocks having been eroded away. Tributary patterns are best seen in high-altitude photographs and can often be easily recognized on topographic maps.

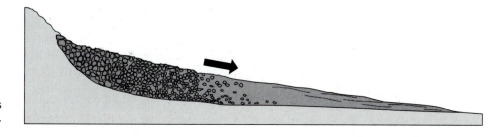

FIGURE 11.5 Decreasing grain size of sediment in alluvial fans further from the source.

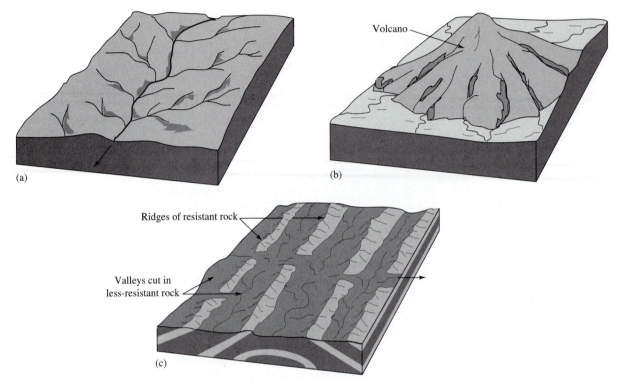

FIGURE 11.6 Erosion patterns controlled by the geology or topography; (a) dendritic, (b) radial, and (c) rectangular or trellis patterns (see text).

Eventually streams reach their *base level,* the lowest level to which a stream can erode. Local base levels are found as streams flow into lakes, but these are only temporary. Continued erosion will, in time, fill the lake with sediment and the stream will then breach the lake and continue to erode until it reaches its final base level, the ocean. Then sediment is dumped along continental margins where it accumulates until tectonic processes bring the sediment back to the surface. Base levels can change by the removal of dams or obstructions after which streams will erode away sediments already deposited, or by change of sea level, either through continental uplift or through ice volume buildup or depletion. Coastal effects during these base level changes can be dramatic, with increasing erosion by streams as sea level falls, or choking streams with accumulating sediment as sea level rises.

RIVER SEDIMENTS

The sediment load can be carried by streams for long distances before being deposited, or it can be deposited very near the source. Fine-grained particles are carried in suspension (Figure 11.4), while sand and gravel is carried as bed load, rolling or bouncing along the bottom of the stream. The sand moves by the process of saltation, discussed in Chapter 9 (Figure 9.5).

Alluvial fans are a special type of stream-flow deposit. Sediment is washed rapidly down and out of the mouths of mountain canyons by intermittent streams, and the sediment load is then dropped due to a sudden decrease in water competence as the slope of the land flattens out. Such fans contain large fragments of eroded material deposited in fairly thick fan shapes near the source (Figure 11.7). These sediments be-

FIGURE 11.7 Alluvial fan along the eastern margin of Death Valley National Park. (Photo by author.)

come thinner and finer away from the source (Figure 11.5). Typically alluvial fans are found in arid mountainous regions and alluvial fans are abundant in many of the national parks. For example, roads into Death Valley National Park (Chapter 12) and Great Basin National Park (Chapter 7) are built on very large alluvial fans.

The stream channels on the surface of alluvial fans, as well as those resulting from glacial outwash, show a criss-cross complex pattern (Figure 11.8). The deposits associated with these streams are called *braided stream deposits*. They are deposited from

FIGURE 11.8 Braided stream deposits, developed on alluvial fans and also found in glacial outwash sediments. These materials are relatively coarse grains, sand sized and larger. Light areas represent bar deposition of relatively coarse-grained sediment.

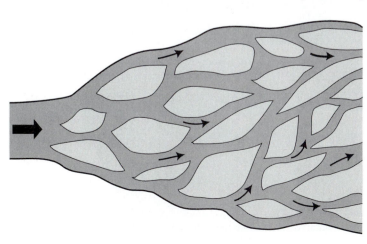

water flowing with moderate to high energy; the sediment contained within them is made up almost entirely of sand and gravel. This produces a complex pattern of *inter-bedded* coarse- and finer-grained sediments.

Stream sediments are constantly being deposited and re-eroded as the stream continues to flow toward the sea. In some areas *stream terraces* (Figure 11.9) can be observed that are remnants of former stream sediment levels, now cut through and eroded by the stream that originally deposited them. These terraces may be found in pairs, on both sides of a channel, or as an *unpaired terrace,* found only on one side. Stream terraces are generally found associated with meandering rivers (Figure 11.10), where streams flow back and forth across the river valley in gentle arcs, due to the low to moderate energy of the flowing water in this stable river form. Meandering streams exhibit solitary channels and gentle downstream slopes. They are constantly eroding in meander bends, undercutting where the energy of the stream directly impacts the bank and depositing where the stream slows as it moves around the bend (Figure 11.11). These deposits, *point bars,* are points where sediment is dropped inside a bend, resulting in

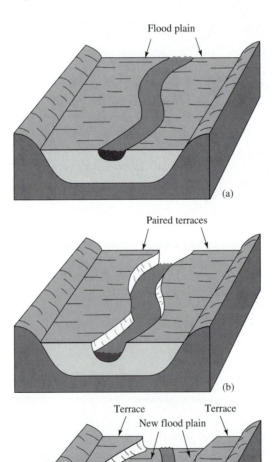

FIGURE 11.9 The development of stream terraces by erosion where (a) erosion through a flood plain (b) cuts a deep channel (c) that eventually widens and a new flood plain develops. High elevation areas above the flood plain but within the river valley, are terraces.

FIGURE 11.10 The Rio Grande, a meandering river flowing along the southern border of Big Bend National Park, Texas, between Texas and Mexico. (Photo by author.)

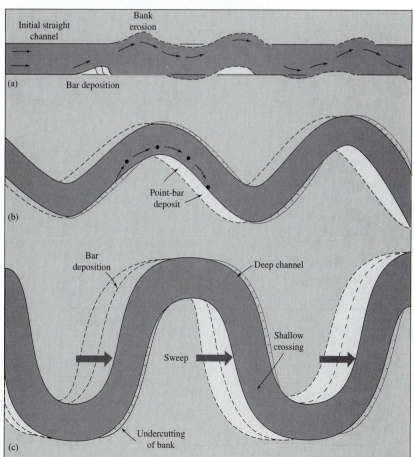

FIGURE 11.11 The development of a meandering stream, illustrating erosional patterns and deposition of point bars from the (a) initial stages in stream development, through the (b) intermediate stream state, to (c) the equilibrium state.

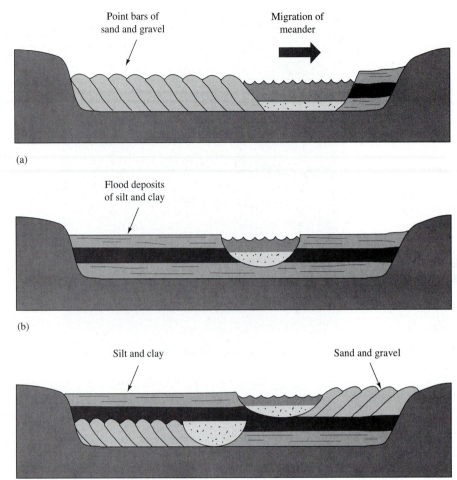

FIGURE 11.12 Cross section of stream deposits illustrating how cross-beds develop in point bar deposits (a–c). These are intermixed with fine sand and clay deposits as a result of the meander, erosion, and depositional patterns of streams.

cross-bedded, sand-sized sedimentary deposits (Figure 11.12). It is in point bars that dinosaur bones were deposited, preserved, and now exposed at Dinosaur National Monument (Chapter 15). In some cases, like at Natural Bridges National Monument, meanders undercut resistant strata, producing natural bridges.

 ## NATURAL BRIDGES NATIONAL MONUMENT

Natural Bridges National Monument (see Figure 1.18) was established in 1908 in southeastern Utah to protect some unique erosional features forming along the eastern edge of what is now Lake Powell.

Geology. Deposition during the Permian, of the thick, massive Cutler Formation has produced the dom-

inant rock unit in the park. Meandering streams have undercut sandstone units in the Cutler on both sides, and in places the meanders have broken through to form natural bridges (Figure P11.4). These initial openings have been enlarged by weathering, exfoliation, and mass wasting to form the stone bridges present in the park today.

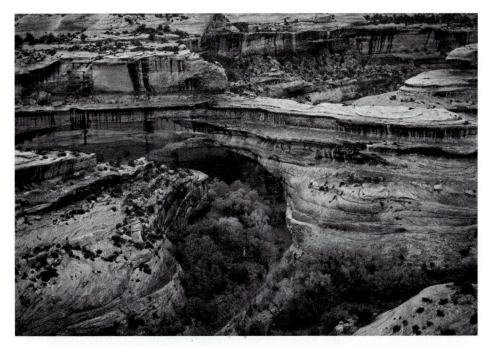

FIGURE P11.4 Natural bridge at Natural Bridges National Monument, Utah, where stream erosion has undercut the Permian Cutler Sandstone. The bridges have been enlarged by exfoliation and mass wasting. (Photo by author.)

During flooding, a stream overflows its banks, depositing sediments outside the stream channel. This causes *natural levees* (Figure 11.13), low ridges of fine sand, to form. These levees then become the upper banks of streams, containing the water at times when discharge is low. Floods, short periods of exceptional high water that produce the *overbank deposits*, leave mud accumulating in backswamps (Figure 11.14; see also Figure 11.13) behind natural levees. The backswamps may be quite large and are very important in agriculture. This is because stream flooding brings rich topsoil to the floodplain, that part of a stream valley that is periodically flooded, and dissolves and carries off salts that precipitate during the rest of the year. Flooding also increases the valley size due to lateral erosion of valley walls, and fills in the valley floor formed, thus widening and flattening the terrain affected by the stream.

Characteristic residual features are produced as streams meander through stream valleys, including *cutoffs* and *oxbow lakes* (so called because they resemble the "U" shape of an ox harness) that result from narrow neck, meander-bend cutoff (Figure 11.15). Such features are *meander scars* left behind by the passing eroding and

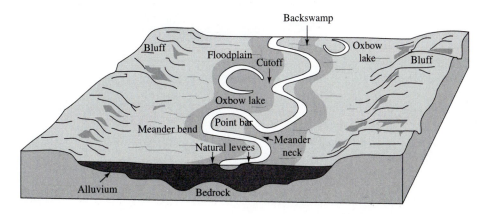

FIGURE 11.13 Diagram of stream morphology.

FIGURE 11.14
Backswamp environment associated with an oxbow meander scar at Congaree Swamp National Monument, South Carolina. (Photo by author.)

depositing stream. Another result of this meander process is lateral channel migration as streams move back and forth across the valley floor. In cross section, a stratigraphic sequence is produced where sediments get finer upwards as the lateral migration progresses and finer and finer sediment is deposited (Figure 11.12) until the stream moves on (Figure 11.11).

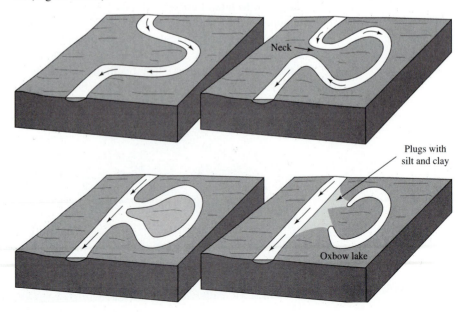

FIGURE 11.15
Diagram illustrating the creation of oxbow lakes from meandering streams.

GEOMORPHOLOGY

Geomorphology is the study of landforms. Streams are the dominant force in the development of most continental landforms, although other controls on geomorphology are also important. Valleys, cut by streams within a drainage basin, continue to modify the landscape as erosion and deposition occur. Relatively sharp and steep forms represent recent activity, and these landscapes are said to be youthful. On the other hand, after long erosion and deposition the landscape becomes flattened and mature. Extremely old and flat landscapes are identified by the term *peneplain,* "almost a plane." The development of these landscapes is also controlled by rock strength, known as *erodibility*, with resistant rocks yielding steep valleys and weak rocks in wide valleys.

The whole system can be *rejuvenated* and produce youth from old age, due to abrupt tectonic changes. If, for example, the surface gradient increases, the result is increasing velocities and increased erosion of sediment. Rejuvenation may also result from local or regional base-level changes, such as sea level falling due to continental ice-sheet build-up. Or it may be due to climatic changes caused by changes in wind patterns. This, in turn, can cause increases or decreases in rainfall with corresponding changes in runoff and erosion, which, in turn, is controlled by other factors such as changes in vegetation. Factors that increase runoff, such as overgrazing or overpaving, may also control local changes in geomorphology.

DELTAS

A large amount of sediment is deposited at river mouths as streams flow into standing water (base levels, i.e., sea level) and current velocity abruptly slows. The sediment accumulates to the level of the standing water surface as river deltas (Figure 11.16). This becomes the top of the delta sediments, the rest of the sediments being brought in by the stream are distributed outward and then down along the leading edge of the accumulating sediment pile. Deltas are generally triangular in shape, due to distributaries (Figure 11.16) spreading sediment mainly outward from where the river hits the shoreline, but also to either side in reduced quantities. The coarsest particles (sand) are deposited closest to the source, forming the upper delta surface called the *delta* plain. These beds, called *topset beds* (Figure 11.16), are nearly horizontal and locally crossbedded. Overflow from distributary channels during floods forms levees and shallow shoals, as well as marshes that form in shoal areas where very fine clay is deposited. Intermediate grain sizes (silt and clay) are deposited in the middle distances forming

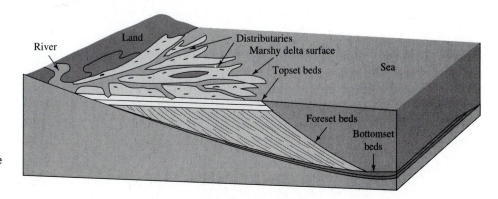

FIGURE 11.16 Surface and cross section of a delta.

delta fronts or *forset beds* that accumulate at steep bedding angles in thick beds (Figure 11.16). Farthest from the source, deposited on the lake or sea floor, are very fine clays that form *prodelta* or *bottomset beds,* with shallow bedding angles and thin layers (Figure 11.16).

As the result of the deposition of sediments, deltas *prograde* (grow or migrate) seaward forming a sequence of beds from the top, coarsest-grained topset beds to the bottom, finest grains, forming bottomset beds. In some river-dominated deltas, such as the Mississippi River delta, sediment deposition rates are so high that it dominates over oceanic wave effects, which are working to disperse and erode the sediment. River-dominated deltas form a series of sediment lobes, which are the result of sediment distribution from place to place in the delta (Figure 11.17). In the case of the Mississippi River, the total sediment distribution looks a little like a bird's foot and is therefore called a "bird's foot delta" (Figure 11.17). Accumulating deltaic sediments build up in thick stacks, and the sediment pile then begins to sink due to compaction and isostatic pressure. This allows younger lobes in these deltas to accumulate on older lobes as time progresses.

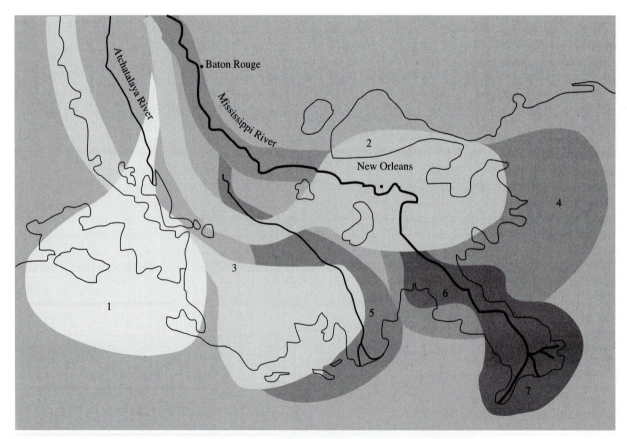

FIGURE 11.17 The Mississippi River "Bird's Foot" delta, so named because it is shaped somewhat like a bird's foot. Also illustrated are older lobes of sediment deposition, the present lobe is labeled "7" and the oldest lobe, associated with the Atchafalaya River, is labeled "1."

JEAN LAFITTE NATIONAL HISTORICAL PARK AND PRESERVE

Jean Lafitte National Historic Park and Preserve (see Figure 1.17) was established in 1978 to include parts of the city of New Orleans, Louisiana, and two other elements along the Mississippi River and on the extensive Mississippi River delta. The Barataria portion was the pirate Jean Lafitte's refuge, an island developed along a distributary channel of the delta. Chalmette, on the Mississippi River where it flows over deltaic sediments south of New Orleans, was the scene of the last battle (1814) of the War of 1812, where Jean Lafitte's pirates helped General Andrew Jackson defeat the British. Chalmette was first designated as a historic park in 1907, and Barataria and the French Quarter in New Orleans have since been added to the park.

In the case of wave-dominated deltas, such as the Nile River delta in Egypt, oceanic wave effects dominate the sedimentary pattern. Distinct small lobes are destroyed and the sediment spreads out along the shore, while the broader distributaries form a single active lobe. The sand tends to form into beaches and transverse sand bars. Most coastal plain deltas are wave-dominated.

A third type of delta, *tidal deltas,* forms in lagoons and on the ocean side of tidal *inlets* as tidal flow enters from the ocean (Figure 11.18). Slowed ocean water dumps its sediment load that then forms obstructions across the mouths of tidal inlets, making navigation through these channels difficult. Sediment obstructions, in turn, break the tidal flow into *tidal channels,* much like the distributaries in river deltas. Examples of tidal deltas are numerous in the National Park System, primarily in the national seashores, including Cape Hatteras National Seashore and Canaveral National Seashore, both discussed in Chapter 14. Matanzas Inlet in Fort Matanzas National Monument (Chapter 16) also has a tidal delta.

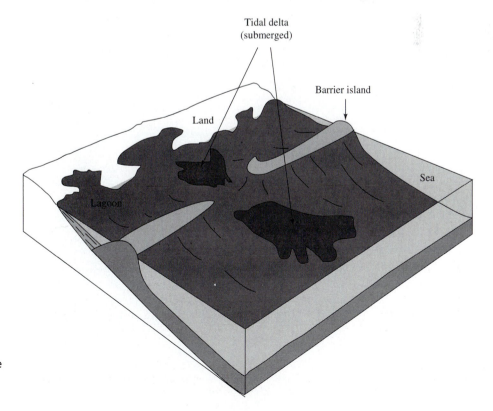

FIGURE 11.18 Tidal delta where sediment builds seaward and landward of erosion inlets in barrier islands. Tidal delta structures are similar to river deltas (Figure 11.16).

MANIPULATING RIVERS

Humans manipulate rivers by building dams or by channeling flow to prevent flooding. We do this for recreation, energy, water storage, navigation, and other purposes. But building dams creates an artificial base level where several unwanted things happen. Lakes formed behind dams end up being areas of high evaporation, therefore areas of high water loss, often in places where water is scarce. Lakes created where lakes did not exist naturally can also experience unusual water loss by infiltration. The production of a new base level causes the dammed stream to drop its sediment load, rapidly filling in and shallowing the lake, and reducing the dam's ability to store water and reduce flooding. Many of our national recreation areas are created around lakes created by man-made dams.

GLEN CANYON NATIONAL RECREATION AREA

Glen Canyon National Recreation Area (see Figure 1.18) was established in 1972. Located in southeastern Utah, it is centered around Glen Canyon Dam (Figure P11.5) and Lake Powell, a 186-mile-long lake behind the dam (Figure P11.6). One of many dams on the Colorado River, the Glen Canyon Dam is one of the world's highest. To the north of Lake Powell, the beauty of the Colorado River valley can still be seen (Figure P11.7).

FIGURE P11.5 Glen Canyon Dam creating Lake Powell in Glen Canyon National Recreation Area, Utah. The dam is built into the Navajo Sandstone. (Photo by author.)

Because it is built into the Navajo Sandstone, an eolian sandstone with a $CaCO_3$ cement holding the grains together, there are some problems with the dam. Calcite cements tend to dissolve in fresh water, and the water in Lake Powell is dissolving the cement; thus the sandstone is unstable. This creates problems for the dam, including leakage and weakening of the support-ing rock. Calcite also has excellent cleavage, which adds to the problem because the Navajo Sandstone breaks easily. Infiltration and evaporation has signifi-cantly reduced the downstream discharge in the Col-orado River and contributes to the water problems in California and Mexico.

FIGURE P11.6 Lake Powell in Glen Canyon National Recreation Area, Utah. (Photo by author.)

FIGURE P11.7 Colorado River valley to the north of Lake Powell in Glen Canyon National Recreation Area. (Photo by author.)

LAKE MEAD NATIONAL RECREATION AREA

Lake Mead National Recreation Area (see Figure 1.14; Figure P11.8) was established in 1964. It lies along the border between southeastern California and southern Nevada, centered on the Colorado River. It was the first national recreation area and is named after Lake Mead, which was created when Hoover Dam was built. (Initially named Boulder Dam, it was renamed several times. An early book by Zane Grey, titled *Boulder Dam,* was written, in part, about construction of the dam.) Also included in the area are several other lakes, such as Lake Mojave, behind Davis Dam further downstream along the Colorado River.

Geology.   In places, the Colorado River has cut deep canyons into the rocks through which it passes. The best example is the Grand Canyon, but other striking canyons have also been cut. South of Hoover Dam, the Colorado River has left a jagged scar as it cut through metamorphic rocks below the dam (Figure P11.9).

As in many lakes created by dams in the western United States, evaporation and infiltration is also a problem in this area, significantly reducing water in the the Colorado River. Concentrated salts due to high levels of evaporation have made the Colorado saline by the time it enters Mexico, and is a source of contention between the United States and Mexican governments. Sediment is filling Lake Mead (and all the lakes) and its effective life as a lake is reduced as a result.

FIGURE P11.8 Lake Mojave in Lake Mead National Recreation Area. Here, a swimmer is jumping from volcanic cliffs along the western margin of the lake. (Photo by author.)

FIGURE P11.9 The Colorado River below Boulder Dam in Lake Mead National Recreation Area cuts through metamorphic rocks that anchor the dam. (Photo by author.)

Cities have significant effects on runoff and stream flow patterns. Because building, paving, and reducing vegetation increase runoff, increased flooding results downstream. Furthermore, infiltration is decreased, thus the water table is lowered in many areas because it is not being recharged. These effects are opposite to those needed to support cities.

GROUNDWATER

Groundwater is a significant resource, in many areas providing the only readily available source of water for drinking and irrigation. Therefore it is very important to understand this resource. Water penetrates into the ground from precipitation and collects in the subsurface in pores and fractures in the bedrock and regolith. When we consider the total amount of water on the Earth, 97.6% is contained in the oceans and marginal seas. The next largest percentage is contained in ice, mainly as part of the Antarctica and Greenland ice sheets. Less than 1% is in groundwater, originating mainly from precipitation, with a tiny amount coming from magma. Even though the percentage of groundwater is very low relative to the total water on Earth, groundwater still makes up considerably more water by volume than is contained in lakes and streams.

Most of the groundwater is located near the surface of the Earth, mainly above 750 m depth. Some water is unavailable to us because water adheres to grains and is not free to move. But the rest is moving because the groundwater surface slopes toward the nearest stream and gravity drives it "downhill." This subsurface slope is known as the *hydraulic gradient*. Figure 11.19 is a simple schematic that illustrates where the water resides in the subsurface. Water from precipitation penetrates or *infiltrates* into the ground and percolates through the soil, a zone of relatively low permeability in which some of the water is retained by molecular attraction. This relatively water-free zone is called the *zone of aeration*. Groundwater eventually reaches the *zone of*

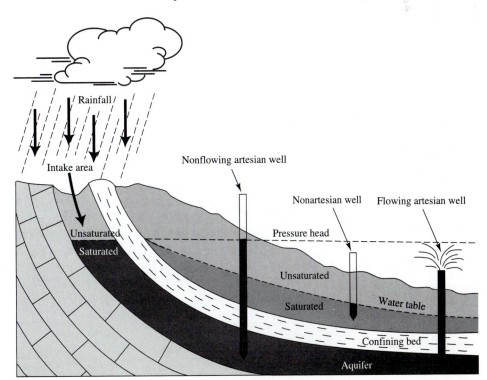

FIGURE 11.19
Diagram illustrating recharge and flow of aquifers in permeable layers between impermeable beds in the subsurface and the generation of artesian wells.

saturation, where just about everything is wet and water fills the fractures and pores in the rock. The *water table* is defined as the upper surface of the zone of saturation (Figure 11.20). A small amount of water seeps upward from the water table by *capillary action* (molecule to molecule movement) above the water table, and the region that contains this water is called the *capillary fringe.* In order to maintain the water table at some level, there must be recharging, usually by rainwater, but pumping of water for irrigation, for drinking water or for other purposes, such as cooling nuclear reactors, can seriously deplete the groundwater and significantly reduce the level of the water table.

Groundwater is constantly moving, driven by gravity to flow slowly downslope by percolation through pore spaces. This movement is governed by *porosity,* the percentage of pore spaces in the bedrock or regolith, and *permeability,* the capacity of the rock for transmitting fluids through which the water is flowing. Essentially, groundwater movement is controlled by the size of openings and their continuity. Those zones in the subsurface that contain a large amount of slowly flowing groundwater are called *aquifers.* Such underground water movement pathways are contained between nonporous (impermeable) strata (Figure 11.19), and may appear in unusual places. Figure 11.21 shows water flowing from a hole in the ceiling of a cave passage. A caver bumped his head, knocked off a cave formation and opened a hole to an aquifer immediately above the passage, and the water from the aquifer now forms a spring.

Springs are places where natural groundwater seeps to the surface. They are due to pressure variations caused by one end of the water table sitting at a higher elevation than the surface of the ground where a spring is generated. This creates a *pressure head* that causes seepage along aquifers at topographic lows. When seepage occurs naturally above the aquifer, rather than where it intersects the surface, we call this *artesian flow.* It results from pressurized aquifer flow being released through a break in the containing impermeable roof. Such breaks many be natural (faults) or artificial.

Springs and wells are prominent pathways bringing water from aquifers to the surface. Ordinary wells have been drilled to intersect the water table and require pump-

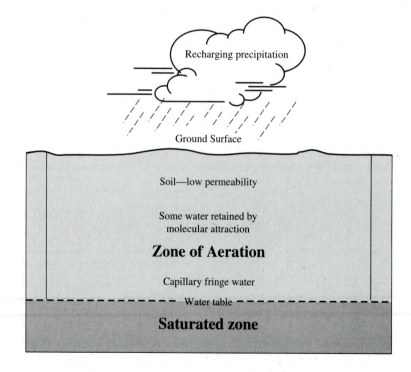

FIGURE 11.20
Subsurface diagram showing the location of water, including the water table, zone of aeration, and capillary fringe.

FIGURE 11.21 A spring in a cave produced when a caver bumped his head, opening a hole to the aquifer above. (Photo by author.)

ing to bring the water to the surface. Drawing the water up out of the ground creates a *cone of depression* that results in the lowering of a local water table. This depression will recover slowly when left alone, as the groundwater percolates back to fill the cone.

THERMAL SPRINGS

Thermal springs (Figure 11.22) are natural springs where the water is hotter than the temperature of the human body. Groundwater is heated within the Earth, usually by

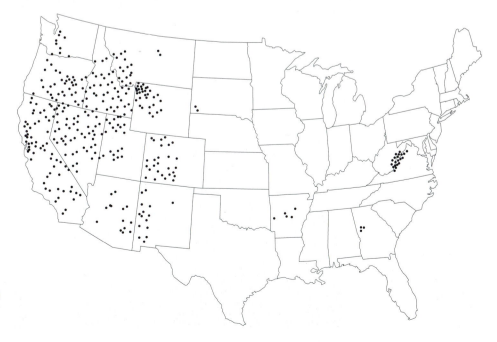

FIGURE 11.22
Location of thermal (hot) springs in the United States.

interaction with magma (particularly in the western United States, such as at Yellowstone National Park), or by deep penetration, geothermal gradient heating (more common in the eastern United States, such as at Hot Springs National Park). In some places, where subsurface conditions are right, *geysers* form. Steam pressure may cause heated water to periodically burst violently from the ground.

YELLOWSTONE NATIONAL PARK

Yellowstone National Park (see Figures 1.18, 8.5, and 8.9), located mainly in Wyoming with a small portion of the park located in Idaho and Montana, was established in 1872 as the first national park in the world. It set a precedent for worldwide preservation of natural wonders. Because of its spectacular and unique character it has recently been designated a world heritage site. Old Faithful (Figure 1.1) is probably the most famous geyser in the world and is typical of the varied and continuing geothermal activity in the park. Boiling water, mud pots, beautifully colored pools, and geysers (Figure P11.10) that erupt at different intervals make the park a fairyland of geologic wonders. These features are driven by a convecting water system interacting with hot magmas lying below Yellowstone.

There have been some interesting controversies surrounding the park. A few years ago serious fires threatened much of the park, but Park System administrators argued that in nature, no one would step in to stop the fires. So they decided to let them burn. In the end the destruction was so extensive that firefighters were brought in to save some of the park areas, and it has taken years for the trees to recover (Figure P11.11). However, opening up the forest has provided new grasslands with increased food for grazing animals in the park.

FIGURE P11.10 Norris Geyser basin at Yellowstone National Park. (Photo by author.)

Geology. Yellowstone lies on a broad undulating volcanic plateau, between 7,500 and 8,000 ft (2,286 and 2,438 m) high. The park area is seismically active (Figure P11.12), and had an earthquake in 1959 that measured 7.5 M on the Richter Scale (see Table 4.1). The volcanic materials exposed at Yellowstone are extensive, and all intrusive-extrusive relationships are found here. Most of the rocks exposed in the park are Eocene in age or younger, with the youngest rocks approximately 10,000 B.P. Some of the volcanic eruptions were so violent that ash from Yellowstone has been identified as far away as Georgia in the southeastern United States.

During the Pliocene, block faulting produced two important mountain ranges, the Gallatin Range in the northwestern part of the park and the Teton Range just to the south. Then in the Pleistocene, *rhyolitic* lavas, the most common in the park, were erupted. These have since been eroded and exposed by the Yellowstone River along the Grand Canyon of the Yellowstone (Figure P11.13). Here the rhyolites have been yellowed (goethite has been produced in abundance) by hydrothermal alteration, chemical alteration by hot groundwater. Today there are beautiful examples of the hydrothermal alteration taking place in the subsurface, leaving travertine at Mammoth Hot Springs (Figure P11.14) and quartz at most other geysers (Figure P11.15), as groundwater or steam, carrying dissolved ions, reaches the surface and these ions precipitate. (*Travertine* is a finely crystalline, calcium carbonate precipitate, usually forming a layered or fibrous structure.) As in many areas of high elevation in the western United States, glaciation has modified the geomorphology of the park.

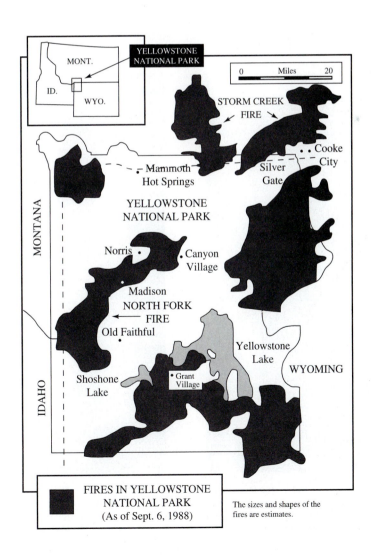

FIGURE P11.11 Area burned in 1988 at Yellowstone National Park.

FIRES IN YELLOWSTONE NATIONAL PARK (As of Sept. 6, 1988)

The sizes and shapes of the fires are estimates.

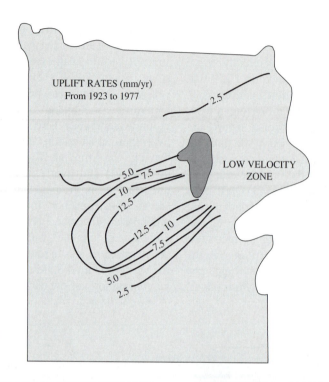

FIGURE P11.12 Uplift rates at Yellowstone National Park over a period of approximately 50 years. In places, the uplift rate has been greater than 15 mm/yr. Seismic low velocities indicate a magma chamber in the sub-surface in the park and it is associated with the area of fastest uplift.

FIGURE P11.13 Grand Canyon of the Yellowstone containing Lower Falls at Yellowstone National Park. The walls of the canyon are rhyolites discolored yellow as the result of hydrothermal alteration. (Photo by author.)

FIGURE P11.14 Mammoth Hot Springs in the northern part of Yellowstone National Park, where large amounts of travertine have precipitated. (Photo by author.)

FIGURE P11.15   Castle Geyser demonstrates the silica mineral precipitation associated with most geysers in Yellowstone National Park. (Photo by author.)

 ## HOT SPRINGS NATIONAL PARK

Hot Springs National Park (see Figures 1.17 and 8.13) was established in 1921 in southwestern Arkansas. It is the site of 47 hot artesian springs that emerge along faults and joints (Figure P11.16). These hot springs were thought to be medicinal so the site has been visited for many years by people looking for physical improvement. Today, 45 of the springs are capped and piped into bathhouses for visitors to the park. Only 2 springs remain natural. They are heated to elevated temperatures in the subsurface (due to the geothermal gradient), with the average spring temperature being 143°F (62°C). Recently, the temperature and flow rate has been slowly declining.

Geology. Hot Springs National Park is located mainly on the nose of a large plunging anticline (Figure P11.17). This fold and corresponding plunging syncline are part of many large compressive features formed as the Ouachita Mountains were uplifted, and the sediments in the area were folded, tilted, and metamorphosed. One interesting geological feature in Hot Springs is the quarries of *novaculite* (Figure P11.18), a very pure variety of quartz, called *chert* (SiO_2), that is often pure white and is mined for whetstones, millstones, etc.

FIGURE P11.16   One of two hot springs at Hot Springs National Park, Arkansas, that is still relatively natural. (Photo by author.)

HOT SPRINGS NATIONAL PARK

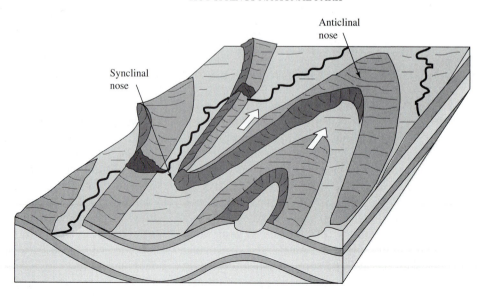

FIGURE P11.17 Diagram showing the nose of the plunging anticline, and associated plunging syncline, on which Hot Springs National Park is located.

FIGURE P11.18 Badly fractured novaculite exposure in Hot Springs National Park, from which whet stones were quarried. The severe fracturing finally made it economically infeasible to continue quarry operations. (Photo by author.)

WATER QUALITY

One significant aspect of groundwater is water quality. Water quality is defined based only on its intended use. For example, if the application is for cooling a nuclear power plant, then cool temperatures are critical. Mineral particles and solutes, often a consideration in water quality, are of consequence in this application only if they might plug coolant lines in the plant. Other water quality concerns are organic material, usually bacteria. But appearance is usually not a concern, unless the intended use is for drinking, and then it is a concern only if someone complains.

Wastewater purification is necessary for reclaiming this natural resource and can be accomplished by a number of means. Water can be purified by percolation through medium porosity material, where such mechanical filtering tends to remove bacteria that adhere to grains as the water passes. Chemical oxidation can destroy bacteria as well, or bacteria can be consumed by other organisms, a process that is being used more and more for water purification.

There are a number of groundwater problems resulting from human use or abuse, and these are becoming crucial as population increases. The worst problems are pollution by toxic or organic constituents due to invasion of waste or organisms from landfills, septic tank leakage, industrial pollution, and other causes (Figure 11.23). Saltwater invasion is another problem in coastal areas, where groundwater pumping and

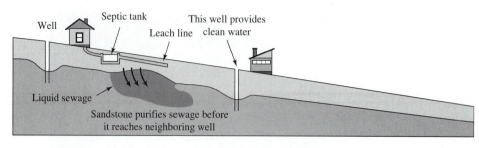

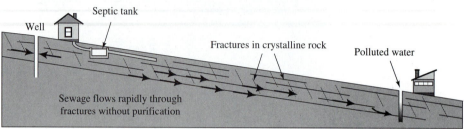

FIGURE 11.23
Pollution problems associated with groundwater contaminating wells. In some cases pollution plumes may be purified before reaching the well, but in others pollution may move rapidly in the subsurface, causing contamination.

lowered pressure cause salt invasion. There are also examples where accidental salt-water recharge of groundwater has destroyed aquifers. Land subsidence due to pumping is another problem. In some cases, when combined with tectonic and sea-level changes, this can be a very serious matter. For example, the first story of many buildings in the city of Venice, Italy, is underwater, the result of groundwater pumping and sinking. In Los Angeles, pumping oil out of the ground caused sinking, until flooding became a potential problem. The oil is still pumped out, but it has been replaced by salt-water recharging.

CAVERNS

Caverns are created primarily by chemical weathering of limestone by groundwater, which dissolves the calcite and carries the calcium ions away. The groundwater then carries and provides the calcium ions for reprecipitation as cave formations. The following dissolution reactions have already been discussed in Chapter 10, where carbon dioxide combines with water to form carbonic acid that attacks the calcite.

$$CO_2 + H_2O \rightarrow H_2CO_3 \rightarrow H^+ + (HCO_3)^-$$

$$CaCO_3 + 2H^+ \rightarrow H_2O + CO_2 + Ca^{2+}$$

As a result of these reactions, dissolution along joints occurs (Figure 11.24). When the water given off by the reaction, now rich in CO_2 and calcium, drips from the ceilings of caves or flows into pools, evaporation and degassing of slight amounts of CO_2 into the air concentrates the calcium. This then crystallizes as calcite, producing *speleothems,* a general term for all secondary formations found in caves. The speleothems produced include some familiar to many people, such as *stalagmites* that grow up from the cave floor and *stalactites* that are found hanging from cave ceilings (Figure 11.25). When these grow together they produce continuous *columns.* Many other calcite speleothems have also been identified, as well as speleothems that have formed from minerals other than calcite, including gypsum and quartz.

Dissolution of limestones by groundwater can result in the removal of vast amounts of rock, with the result being an unusual surface topography known as *karst*

FIGURE 11.24
Limestone dissolution along a fracture forming a cave passage. (Photo by author.)

FIGURE 11.25
Speleothems, including stalactites, stalagmites, and columns, in Carlsbad Caverns National Park, New Mexico. (Photo by author.)

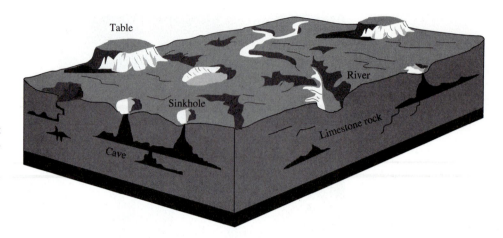

FIGURE 11.26
Schematic of a karst terrain showing the development, as a result of limestone dissolution, of depressions, tables, sinkholes, and caves. Karst topography is an indication that much limestone has been removed.

topography (Figure 11.26). This consists of closely spaced *sinkholes*, large solution cavities that may open to the air, and indicates large destruction of subsurface carbonate rock. Karst terrains contain many rounded mounds or angular and very rough limestone blocks. Small lakes and ponds are also common. Figure 11.27 illustrates the groundwater level in a cave where dissolution is occurring along the water-air interface, thus increasing the size of the cave. Because groundwater rises and falls, dissolution occurs over a range of elevations in caves. Rainwater flowing into caves from the surface erodes surface sediment that is then deposited within the cave. This can often make the caving experience more difficult (Figure 11.28).

Very large sinkholes may develop in karst areas, and sometimes these may be 200 m deep or more. Figures 11.29 and 11.30 illustrate two such sinkholes, one in North America (Figure 11.29), and the other in the Moravian Karst National Park in the Czech Republic, Europe (Figure 11.30). Note the difference in caving styles, United States cavers generally prefer ropes for deep sinkholes, while European cavers prefer ladders.

FIGURE 11.27
Groundwater level standing in a cave. Dissolution occurs along the air-water interface. (Photo by author.)

FIGURE 11.28 Caves are natural storm drains. Here a caver searches for his pack, somewhere in the mud in front of him. The mud is sediment that has been eroded and washed into the cave from the fields above. (Photo by author.)

FIGURE 11.29 Caver descending into a deep limestone sinkhole in the United States; some of these sinkholes or "pits" are greater than 600 ft (185 m) deep. Note the caver is suspended from a rope. (Photo by author.)

FIGURE 11.30   Caver climbing a ladder in a deep limestone sinkhole in a cave in the Moravian Karst National Park in the Czech Republic, Europe. Many nations have turned cave and karst areas into national parks. (Photo by author.)

MAMMOTH CAVE NATIONAL PARK

Mammoth Cave National Park (see Figures 1.22 and 8.16) was authorized by Congress in 1926. Located in Kentucky, Mammoth is by far the world's longest cave with more than 330 mi (531,069 m) of continuous passage mapped so far. Because of its length and long history of use, it has been designated a world heritage site. Mammoth Cave has a lunchroom and an elevator that allows handicapped access. It also has a special, paved cave passage that is reserved for handicapped visitors.

History.   The cave has been used for a very long time. Over 3,000 years ago Native Americans were using the cave. Evidence of this includes bundles of reeds used as torches and a mummified body, found lying under the rock that was the cause of death. More recently, the cave was a source of sodium nitrate that was mined during the War of 1812 and used in gunpowder manufacture. Today, *cavers* (cave explorers and researchers)

are actively studying in the cave system, researching (Figure P11.19) and mapping. The Park Service insists that as tours progress through the cave, cavers hide themselves until the tour group passes. In some places in the cave, off in the distance, there may be many pairs of eyes watching as these tours pass. One caver tells the story of the time he was climbing just above a tour passage when the lights came on. He froze in place, just a few feet over their heads, as the people in the party passed beneath. At the very back of the tour was a father dragging his young daughter along behind him. She was looking all over, clearly a curious child, when she spotted the caver immediately above her. "Daddy, Daddy!" she cried and pointed, but her father just dragged her along, urging her not to be such a laggard.

Crystal Cave, once thought to be independent from Mammoth Cave, is now part of the Mammoth Cave system. It is the site of the famous Floyd Collins

FIGURE P11.19 The late Dr. Victor Schmidt (right above) leading an expedition with other researchers to collect sediment samples in Mammoth Cave National Park, Kentucky, for paleomagnetic research. His research yielded a long history of changes in the Earth's magnetic field direction recorded in Mammoth Cave sediments. (Photo by author.)

tragedy that occurred in 1923. Collins was an explorer in the region, looking for new caves that were pretty and accessible so that they could be opened to tourists. He would go into these caves for days at a time and explore on his own, sometimes blasting openings in narrow passageways so that he could get through to see the cave beyond. It is on the Floyd Collins story that the National Speleological Society (NSS), a national organization of cavers, bases their safety techniques, because Collins appears to have done everything wrong. In modern caving it is strongly recommended that cavers use three sources of light, that they notify someone where they are going and when they will return, and that they go caving with at least two other people. And by all means, never blast in a cave! Collins broke all the rules, and when returning through a blasted passageway in Crystal Cave, a rock dislodged and trapped his boot. He could move, but couldn't pull his boot through the hole. When they finally found him days later, it was clear that he had a real problem, and when word got out that he was trapped, a carnival atmosphere developed outside the cave. A movie about the experience was made about a reporter who covered the story and won a Pulitzer Prize. Collins finally died needlessly of exposure, a couple of weeks after he was trapped, and provides to this day an example of what not to do when exploring caves.

Geology.   Mammoth Cave is formed in a Mississippian limestone, that has as its capping unit a Mississippian sandstone. The sandstone creates a unique and unusual flat roof over some of the passages. The cave developed by dissolution at the water table, following slow uplift of the area during the late Tertiary and Quaternary. As the level of the water table dropped, cave dissolution progressed, leaving a complex, multilevel cave system. Work on sediments deposited in the cave has indicated that early materials were deposited more than 2 million years B.P.

CARLSBAD CAVERNS NATIONAL PARK

Carlsbad Caverns National Park (see Figures 1.17 and 8.6) contains some of the most beautiful and famous caves in the United States. It was established as a National Monument in 1923 and then upgraded to park status in 1930. Located in southeastern New Mexico, the park contains more than 40 caves with Carlsbad Cavern being the best known. Carlsbad is the ninth deepest cave in the United States, at 1,028 ft (313 m) and the twentieth longest at 19.1 mi (30.7 km). The park area is designed for self-guided walking tours and is well lighted. In places in the main part of the cave, not too far from the lunchroom (Figure P11.20), one can see the remains of old exploration activity. For example, the National Geographic Society conducted a cave survey in the 1930s and visitors can still see a metal ladder, used at that time, hanging down into a hole in the Big Room. From above, the ladder still looks intact, but from below it can be seen that it is broken and stops many meters above the floor of the cave.

We can only hope no child will ever attempt to climb down! The Big Room is easily accessible to wheelchairs by elevator. In summer, visitors leave behind the 100°F plus heat of the New Mexican desert when they descend to the lunchroom, 750 ft (230 m) below the surface, where the temperature is 58°F!

Carlsbad Cavern is also known for its bats and bat cave area. During the day the bats nest in the cave, but at night they fly out and down into the valley below the cave to feed.

There are also raccoons and ring-tailed cats living in the cave, feeding on food dropped by visitors in and near the lunchroom. In the morning, park rangers find trash cans knocked over and garbage scattered about by racoons searching for food. There is a story about a graduate student who, while working with his professor in the cave (Figure P11.21), was sent to the lunchroom on an errand. As he approached, he heard a loud commotion that turned out to be three raccoons

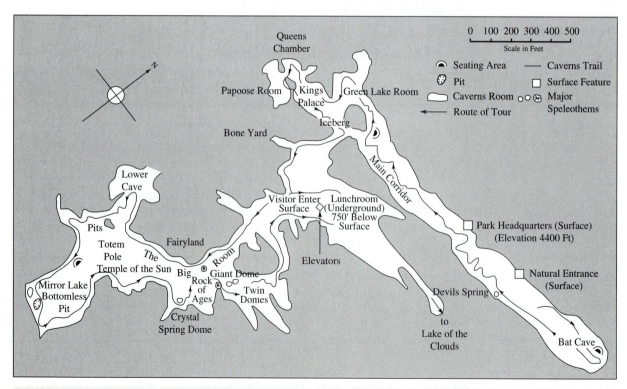

FIGURE P11.20 Carlsbad Cavern map, Carlsbad Caverns National Park, New Mexico. Indicated on the map is the Lunch Room and Lower Cave where the Texas Toothpick is located (see Figure P11.21).

FIGURE P11.21
Researchers drilling the "Texas Toothpick," a large stalagmite in Lower Cave, Carlsbad Cavern, Carlsbad Caverns National Park. The calcite cores recovered were used in dating and climate analysis. (Photo by author.)

fighting over some scraps of food on the floor. When he came into the room, they appeared to ignore him and just kept fighting, but he was sure they were eyeing him as a very large scrap of food. Very slowly and quietly he backed away and disappeared into the darkness.

Slaughter Canyon Cave, at one time called New Cave, another cave in the park, can be visited on a lantern tour and gives a better idea of what caving in a "wild" cave is like. It was in Slaughter Canyon Cave that some of the movie *King Solomon's Mines* was filmed, and light bulbs burned out during filming were cast under the formation called the Clansman, shown in the movie (Figure P11.22). The Park Service has not removed them, indicating that they are now part of the history of the cave. The Park Service renamed the cave because of the discovery of Lechuguilla Cave, the newest cave in the park.

Lechuguilla Cave is now the deepest cave in the United States at 1,565 ft (477 m). It is also longer than the Carlsbad Cavern with 30.897 mi. of mapped passage. The cave is filled with spectacular, unique, and beautiful speleothems, but because of its fragile nature will probably never be open to general tours.

Geology. Carlsbad Caverns National Park developed in a Permian Reef system similar to the Great Barrier Reef in Australia. The cave limestone is made up of shells and coral skeletons that have been recrystallized into a limestone. The limestone has been fractured and the caves have developed as the result of dissolution along fractures. Secondary calcium carbonate and gypsum deposition has produced the speleothems present in the cave. Some of the columns are gigantic.

FIGURE P11.22 The Clansman, a large stalagmite in Slaughter Canyon Cave, Carlsbad Caverns National Park. The Clansman appears in the movie *King Solomon's Mines,* filmed in the 1950s. (Photo by author.)

 ## WIND CAVE NATIONAL PARK

Wind Cave National Park (see Figures 1.18 and 8.11) was established in 1903. Located in the Black Hills of southwestern South Dakota, it is the fourth longest cave in the United States with over 70 mi (110 km) of mapped passage (tenth longest in the world). An interesting fact about caves is that they "breathe" due to barometric (atmospheric) pressure changes, causing air to move in and out of their mouths. The original entrance to Wind Cave is very small, and the name derives from the "wind" blowing in and out of the entrance. The average yearly temperature is maintained within a cave, and caves are very humid. Thus, in the winter, as barometric changes occur, water vapor plumes can be seen rising into the air from cave entrances as the warmer moist air is blown out into the colder external air. Cavers in many parts of the country use this winter phenomenon to locate unknown caves.

FIGURE P11.23
Boxwork structure in Wind Cave at Wind Cave National Park, South Dakota. The limestone in which the cave was eventually formed was first fractured, and the fractures filled with a calcite slightly more resistant to dissolution than the limestone matrix. Dissolution of the matrix when the cave formed left behind the box-shaped beds. (Photo by author.)

Geology. Wind Cave is developed in a Mississippian limestone. Interestingly, many areas in the United States—northeast, southeast, and west—have major cave systems also developed in Mississippian limestones. Mammoth Cave is one example, and there are many others. While there are caves in limestones of other ages, in no age group are they so abundant as in the Mississippian. Dissolution of the limestone probably began in the late Tertiary and continues today.

Wind Cave is well known for its unique *boxwork structure,* quite evident in the cave (Figure P11.23). It results from limestone that has been finely fractured; the fractures filled with calcite, then the limestone dissolved back, leaving the calcite boxes behind.

 JEWEL CAVE NATIONAL MONUMENT

Jewel Cave National Monument (see Figure 1.18) was established in 1908. Also in the Black Hills of southwestern South Dakota, it is the second longest cave in the United States, with over 80 mi (130 km) of mapped passage and the fourth longest cave in the world.

Geology. As with Wind Cave, Jewel Cave is developed in a Mississippian limestone. Its unique character includes calcite crystals shaped like jewels, thus the name (Figure P11.24). It is also important because it is a very long cave system and therefore provides an excellent laboratory for cave process study.

FIGURE P11.24   Calcite "jewels" for which the cave is named in Jewel Cave National Monument, South Dakota. A long, thin calcite speleothem called a "soda straw" is also seen in the photo. (Photo by author.)

One of the many problems in all the parks is vandalism, but in the parks' caves the problem is very serious. When serious vandalism occurs in the National Park System, the Federal Bureau of Investigation (FBI) is called to investigate. In Carlsbad Cavern, park rangers have documented the destruction and theft of thousands of speleothems, some quite unique, now lost to the world. An example of this crime is illustrated in Figures 11.31 and 11.32, a before and after shot of the destruction of an unusual speleothem known as an "angel wing" (not in Carlsbad Cavern). As theft and destruction of this type continues, our country's natural beauty is diminished.

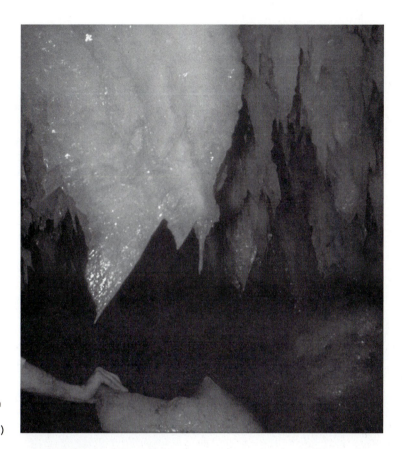

FIGURE 11.31 Caver observing a spectacular white calcite speleothem known as an "angel wing." (Photo by author.)

FIGURE 11.32 Caver pointing to the calcite "angel wing" shown in Figure 11.31 after it was destroyed by vandals. (Photo by author.)

REFERENCES

BENNETT, R., ed. 1980. *The new America's wonderlands our national parks,* p. 463. Washington, D.C.: National Geographic Society.

BEUS, S. S., ed. 1987. *Decade of North American geology, DNAG, Centennial field guide.* Vol. 2. *Rocky Mountain Section of the Geological Society of America,* p. 475. Boulder, Colorado: Geological Society of America.

BOGGS, S. 1994. *Principles of sedimentology and stratigraphy.* 2nd ed., p. 800. New York: Macmillan College Publishing Company.

CHRONIC, H. 1986. *Pages of stone, geology of western national parks and monuments.* Vol. 3. *The desert southwest,* p. 168. Seattle: The Mountaineers.

CHRONIC, H. 1988. *Pages of stone, geology of western national parks and monuments.* Vol. 4. *Grand Canyon and the plateau country,* p. 158. Seattle, Washington: The Mountaineers.

HARRIS, A. G., and TUTTLE, E. 1990. *Geology of national parks.* 4th ed., p. 652. Dubuque, Iowa: Kendall Hunt Publishing Company.

HARRIS, D. V., and KIVER, E. P. 1985. *The geologic story of the national parks and monuments.* 4th ed., p. 464. New York: John Wiley & Sons, Inc.

HAYWARD, O. T., ed. 1988. *Decade of North American geology, DNAG, Centennial field guide.* Vol. 6. *Southeastern Section of the Geological Society of America,* p. 457. Boulder, Colorado: Geological Society of America.

NEATHERY, T. L., ed. 1986. *Decade of North American geology, DNAG, Centennial field guide.* Vol. 6. *Southeastern Section of the Geological Society of America,* p. 457. Boulder, Colorado: Geological Society of America.

12

Deserts

DESERTS AND ARID BASINS

Deserts, such as Death Valley, and arid basins make up 25% of the total world lands. Arid climates are defined as having low rainfall, less than 10 in (25 cm) per year, and generally as having high evaporation rates and high temperatures. Deserts are places of noticeably high winds, the result of heating and convective air mass movement.

Desert regions generally lie between 20° and 30° north and south latitude (Figure 12.1) and are produced by atmospheric effects—the circulation of sun-warmed air. Hot temperatures at the equator warm the air, which expands and rises; as it rises the sun-warmed air begins to cool, dropping much of its moisture as rain. As the air continues to rise and cool, it is deflected toward the poles. As it cools it gets heavier and eventually begins to sink, and as it does so it is deflected back toward the equator (Figure 12.2). These air masses are deflected to the right in the northern hemisphere and to the left in the southern hemisphere as the Earth rotates under the moving air. This deflection, called the Coriolis effect, is responsible for breaking air flow into belts and producing the westerly and easterly winds at different latitudes. At 30° latitude, the descending air warms. This descending warm air is relatively dry and holds what moisture it contains, resulting in reduced precipitation as the air mass returns to the equator.

In the western United States deserts are generally found to lie in the Basin and Range province. These include four major desert regions: the Great Basin in the north; the Mojave, mainly in California (see Death Valley National Park and Joshua Tree National Park as examples); the Sonoran, extending from California and Arizona into Mexico (Organ Pipe Cactus National Monument and Saguaro National Park are examples); and the Chihuahuan, extending over parts of Arizona, New Mexico, Texas, and Mexico (Figure 12.3).

Types of Deserts

Some deserts are controlled by global air mass circulation, the best example being the Sahara in North Africa, an extensive region that is starved for moisture (Figure 12.1).

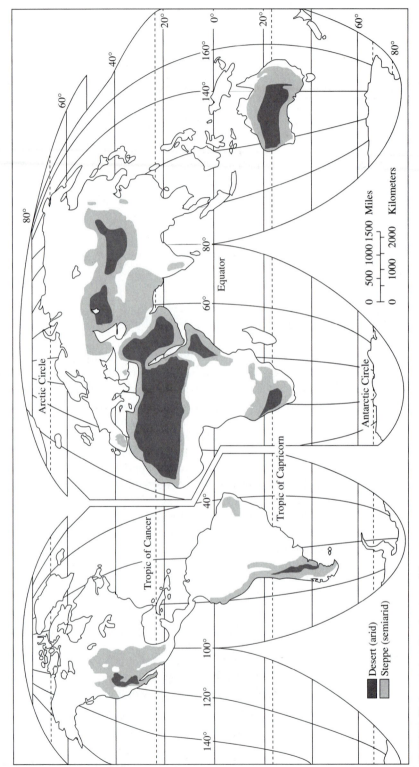

FIGURE 12.1 Global distribution of deserts.

Desert (arid)
Steppe (semiarid)

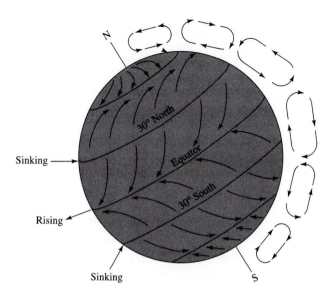

FIGURE 12.2
Atmospheric air mass circulation.

The image we have of the Sahara is one of vast sand-dune fields, but, in fact, most areas have been swept free of sand, and *desert pavements* of small rocks are all that remain. Deserts also develop in continental interiors that are very hot in the summer and that receive dry, cold, continental air in winter. Another area where deserts commonly develop is in the lee of mountain ranges, such as the deserts of the southwestern part of

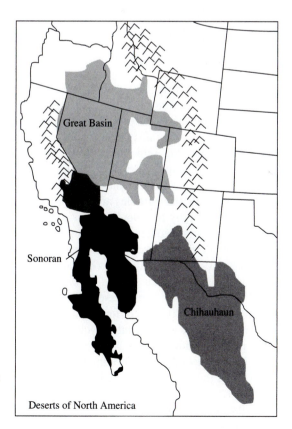

FIGURE 12.3 Deserts in the western United States and Mexico.

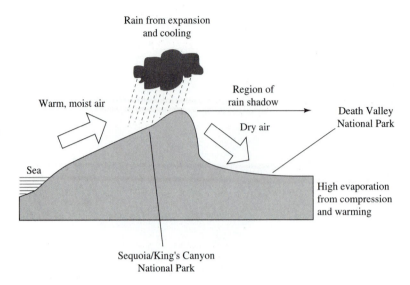

FIGURE 12.4 Diagram showing how mountains create rain shadows that help cause the development of deserts. Areas of high rainfall occur on windward sides of mountains, in this example at the location of Kings Canyon/Sequoia National Parks. Death Valley National Park, on the leeward side of the mountains, is in the rain shadow.

North America. Here, moist air moving in from the Pacific Ocean encounters the Sierra Nevadas and rises over the windward slope (Figure 12.4). As the air rises it cools and drops rain on the windward slopes. The giant sequoias, like those at Sequoia and Kings Canyon National Parks (Chapter 5), grow in this moist region. Then, after the air mass has crested the mountain, it descends on leeward slopes as dry air, warming and holding its little remaining moisture, causing zones of reduced rainfall that develop into deserts.

DESERTS—WEATHERING AND EROSION

Weathering in deserts is primarily mechanical, although chemical weathering can be very important. Desert weathering produces coarse particles and creates steep slopes because mass wasting is the dominant process. Erosion of particles is relatively rapid, with flash floods from occasional rains a major agent in carrying particles away. Wind erosion results in *deflation*, the removal of loose, relatively small particles (Figure 12.5). Deflation slows and may even stop due to the development of a rock armor covering the surface. Such desert pavements are made up of pebbles that remain as the stable ground cover after smaller particles are blown away.

Wind transport is a highly turbulent process, carrying very fine particles in suspension, supported by turbulent eddies, but moving most of the transported mass along the ground as a bed load. This zone is less than a meter high, where sand grains arc into the air and move through saltation/collision processes (discussed in Chapter 9, see Figure 9.5). This produces characteristic *frosted grains* that can be recognized in lithified materials and is diagnostic of eolian processes. This operation easily forms sand particles into dunes, but the suspended load of fine sand, silt, and clay is transported for very long distances, and drops out when wind velocities fall. Fine particles carried by the wind act as sand paper, abrading rocks lying on the desert floor and producing *ventifacts,* rocks that have polished greasy looking surfaces that are pitted or *fluted.* These rocks develop facets in the direction facing the wind and are separated in the rock by sharp edges (Figure 12.6).

Wind deposits include *dunes,* piles of sand grains, usually quartz, controlled by obstacles, such as vegetation, that cause wind velocities to slow and sand grains to

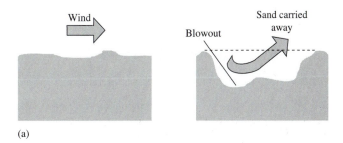

(a)

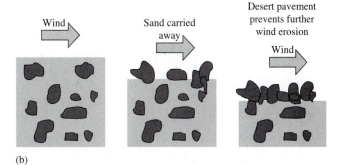

FIGURE 12.5   Wind erosion causing deflation (a, b) and the development of an armored pavement (b).

(b)

FIGURE 12.6   Photo of a desert ventifact formed by wind erosion and an armored pavement like that in Figure 12.5 (b). (Photo by author.)

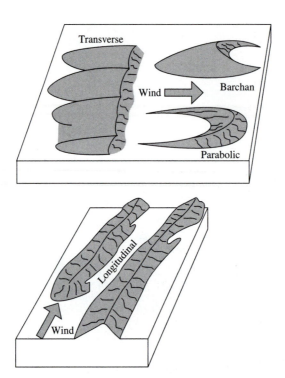

FIGURE 12.7   Dune shapes illustrating the wind direction and the common types of dune found in deserts.

collect. Sometimes the dunes are made from other minerals, such as the gypsum dunes found at White Sands National Monument. Dunes are cross-bedded as the result of the way in which grains are usually deposited along the leading edge of the dune, first by blowing up the low-angle windward slope and then by cascading down the leeward slip face. The leeward face represents the steep slope of the dune where sand grains lie on about a 34° slope, the angle of repose of most sand grains. As more grains move over the crest of the dune, this angle steepens, causing grains to avalanche down the slope.

Accumulation on the leeward face produces cross-beds as the sand dune migrates in the direction the wind is blowing, thus giving an indication of the prevailing wind direction. Dunes take on many shapes, including *barchan, parabolic,* and *transverse* forms (Figure 12.7), which are useful in determining wind direction. (Barchan dunes have even been identified on Mars.) Longitudinal dunes, however, develop parallel to the prevailing wind direction (Figure 12.7), and are more difficult to interpret. They have representatives in Death Valley (Figure 12.8). With some exceptions, the prevailing wind direction in ancient, lithified dunes can be determined using cross-bed orientation.

DESERT WATER

In the desert, the water table generally lies deep below the surface, near-surface water having been evaporated away. Streams, on the other hand, cause the greatest amount of erosion in deserts, not wind. Desert streams are usually intermittent, seeping into the ground or rapidly disappearing as the result of runoff and evaporation. Groundwater is too deep for most streams to be maintained by spring-fed flow. However, some mountain-fed, long, permanent rivers, such as the Nile River in Egypt and the Colorado River in the western United States, do continuously provide water to desert regions.

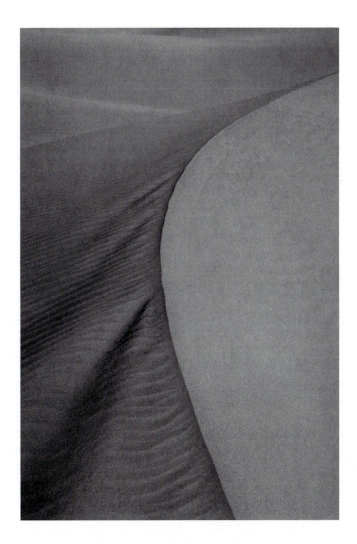

FIGURE 12.8
Longitudinal dune in Death Valley National Park like that seen in Figure 12.7. The ripple marks near the dune crest give an indication of the wind direction, parallel to the crest of the dune. (Photo by author.)

Flash floods occur in the deserts, producing sheets of water that flow over flat valley floors and rush down steep-sided, flat-bottomed canyons. These are important geological agents and can be very dangerous to the unsuspecting traveler. In many desert regions drainage runs into the interior of basins, thus the water never reaches the sea, but infiltrates or evaporates from the basin. Lakes formed in this manner are called *playa lakes*. They are short lived, only receiving water periodically. Evaporation results in salts accumulating as *evaporite* or *playa deposits*.

Rain in mountains surrounding deserts produces braided streams, where coarse particles are carried by high energy flow, and deposition of these materials leaves cross-stratified sediments. Braided streams produce a complex network of channels that are found on most alluvial fans (Figure 12.9).

DESERT LANDSCAPES

Desert landscapes are dry types, with sparse vegetation. Wind erosion is important in producing dune fields, and deflated desert pavements, but stream effects are evident in

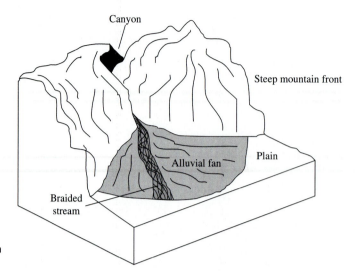

FIGURE 12.9 Braided stream on an alluvial fan in the desert. Many examples can be found in national park areas.

all deserts, as well. For example, the development of large alluvial fans, fan-shaped accumulations of sediment washed out of mountains and accumulating at the mouths of canyons, is an indicator of intense, high energy, intermittent stream action. In the desert, alluvial fans (Figure 12.9) form from periodic rainfall that causes flash floods to rush down mountain canyons, eroding as it goes along. When velocities slow as the water escapes from the mouth of the canyon, it deposits the material as it spreads out on the valley floor. When alluvial fans from several neighboring canyons grow very large, they begin to overlap or coalesce into one very large apron of outwash material, called a *bajada*. Excellent examples are found in Death Valley National Park.

As bajadas grow they accumulate along the base of the mountain ranges from which they are derived. Pediments represent the remnants of those mountains buried in their own talus (Figure 12.10). *Bolsons* are enclosed flat basins, with no drainage out-

FIGURE 12.10
Pediment in the Sonoran Desert, where the mountains are being buried in their own debris. (Photo by author.)

FIGURE 12.11   Playa lake in Death Valley National Park. (Photo by author.)

let, that are being filled by accumulating alluvium. Lying in these basins are playas, dried up lake beds that are periodically flooded and vegetation free (Figure 12.11), flat-floored areas usually covered by thin, fine-grained, flat-lying sediments and extensive evaporites (salts). The major ions making up these evaporites include chlorine, sulfur, magnesium, potassium, calcium, and sodium, producing a large range of mineral compositions. Playas are also known as *salt pans* and *alkali flats*.

 ## DEATH VALLEY NATIONAL PARK

Death Valley National Park (see Figures 1.14 and 8.4) was proclaimed as a national monument in 1933. It is located mainly in the southeastern California *Mojave Desert* (Figure 12.3), extending a little into southwestern Nevada. It became a national park in 1994. Hot and dry, Death Valley contains all the classic desert landscape forms. Even some of the desert geology termi-

nology was developed here. The park ranges in elevation from a high point at Telescope Peak, 11,049 ft (3,368 m), to a low point of 282 ft (86 m) below sea level. The best time to visit is in the late fall, and it is a great place to hike, in part because hikers don't need to adjust to high elevations, often required when hiking in other areas.

FIGURE P12.1 Satellite photograph of Death Valley in Death Valley National Park. Photo shows large bajada along the western (upper portion in photo) margin of the valley and discrete alluvial fans along the tectonically active eastern margin. Between lie playas.

Geology. Death Valley (Figure P12.1) is part of the Basin and Range province, where block faulting of the valley floor in the late Tertiary, began to produce the basin. Motion continues today, with the east side moving down faster than the west, basically as a rotating fault block. The result yields large bajadas extending all along the western, relatively stable margin of the valley, with discrete alluvial fans developing along the eastern margin (Figure P12.1). There has been some volcanism, with Holocene steam eruption craters present in the northern end of the park, where interaction of magma and groundwater produced violent explosions.

The rocks found in surrounding mountains range in age from the Precambrian to the Tertiary. Many of these rocks have been eroded and redeposited within the valley. Some interesting and beautiful examples exist, where this material has been eroded, redeposited, eroded again, and then polished. Some examples are found in Mosaic Canyon (Figure P12.2).

At one time Death Valley was filled by Lake Manly, a Pleistocene lake that was 90 mi (145 km) long and 585 ft (180 m) deep. Evidence of its presence can be seen by wave cut shorelines relatively high on the mountain slopes. Water in the lake evaporated during the Thermal Maximum, approximately 11,000 years B.P., leaving behind thick salt playa deposits (Figure P12.3).

There are also quartz (generally) sand dune fields in the park, the site of several movies and TV commercials. And the hydrated mineral ulexite [$NaCaB_5O_9 \cdot 8H_2O$], a unique mineral with very unusual optical properties (some have called it "TV rock"), was first identified here.

FIGURE P12.2
Photograph of a wall in Mosaic Canyon, Death Valley National Park. Light tan blocks of dolomite have been eroded and deposited, then exposed and polished as water rushed through the canyon and out onto the valley floor. (Photo by author.)

FIGURE P12.3 Thick playa salt deposits in Death Valley National Park. (Photo by author.)

JOSHUA TREE NATIONAL PARK

Joshua Tree National Park (see Figures 1.14 and 8.4) was established as a national monument in 1936 in the southeastern California Mojave Desert. It became a national park in 1994. The park has many spectacular geological features, but is best known for its Jurassic granite outcrops, weathered in unusual rounded shapes as the result of exfoliation processes (Figure P12.4; see also Chapter 9 for a discussion of weathering). The stark beauty of the park, like that in Death Valley, has been used by Hollywood as the site of a number of western movies. It was originally proposed as a monument, in part, to protect the Joshua trees that grow here (Figure P12.5). Uplift and subsequent Basin and Range faulting during the mid-late Tertiary produced the mountains and basins now present in the park. Very late in the history of the park volcanism added to the scenery. The area is still tectonically active.

Geology. The oldest rocks found in Joshua Tree are Precambrian gneisses exposed in the Little San Bernardino Mountains along the southwestern boundary of the park. These rocks were intruded by Precambrian granites and some ancient, slightly altered sedimentary rocks are also exposed in the region. No Paleozoic rocks are exposed in the park. During the Mesozoic the granites that show some of the spactacular shapes found in the park were intruded and the area was uplifted. Erosion then carved away the land until gentle uplift, then Basin and Range block faulting in the middle Cenozoic, produced the general basin and range topography that characterizes the park. Late in the Tertiary deserts developed in the region as the result of uplift of the Sierra Nevada to the west, which cut off the flow of moist air to the region. A small area of the park contains lavas that have been erupted relatively recently.

FIGURE P12.4 Skull rock in Joshua Tree National Park, California, typical of the unusual, rounded granite spheroidal weathering forms found in the desert. (Photo by author.)

FIGURE P12.5 Joshua tree in Joshua Tree National Park. In the background are weathered granites. (Photo by author.)

ORGAN PIPE CACTUS NATIONAL MONUMENT

Organ Pipe Cactus National Monument (see Figure 1.14) was established in south central Arizona in 1937. Located in the center of the Sonoran Desert (Figure 12.3), it contains large exposures of Miocene volcanics (lavas and ash flow tuffs), as well as extensive bajadas and pediments. The monument was primarily conceived to protect the unusual varieties of cactus present in the region (Figure P12.6), but the rock exposures, as in many desert areas, are spectacular.

Geology. Most of the rocks exposed in the park are Mesozoic and Cenozoic in age, although there are some small outcrops of Precambrian granites located in the monument. There are no Paleozoic rocks exposed. Mesozoic rocks in the monument are represented mainly by gneisses and schists, and these have been intruded by Jurassic and Cretaceous granites. Tertiary granites are also exposed in the monument. In the Miocene, Basin and Range faulting and explosive volcanism produced the abundant lavas and tuffs that form many of the ridges and dominate the geology.

FIGURE P12.6 An organ pipe cactus growing in Organ Pipe Cactus National Monument, Arizona. In the background are volcanic ash flow tuffs, abundant in the monument. (Photo by author.)

SAGUARO NATIONAL PARK

Saguaro National Park (see Figures 1.14 and 8.6) was established as a national monument in south central Arizona in 1933, and was later upgraded to national park status in 1994. Located on the eastern edge of the Sonoran Desert (Figure 12.3) to the east of Organ Pipe Cactus National Monument, it was formed to protect thick Saguaro forests (Figure P12.7) from the rapid encroachment of civilization. It is divided into two parts, Saguaro west, just to the west of Tucson, Arizona, and Saguaro east, the larger of the two segments. In the west there are good examples of bajadas at the foot of the Tucson Mountains, and exfoliated granites are well exposed. To the east of Tucson in the Rincon Mountains, the park offers excellent examples of high desert

FIGURE P12.7 Saguaro cacti growing in the western segment of Saguaro National Park, Arizona. In the background are rounded, eroded granites. (Photo by author.)

plant communities. The rocks have been darkened, burned by the sun, their surface covered by desert varnish.

Geology. Throughout the Paleozoic the region was relatively stable, with periods of marine sedimentation followed by erosion, while the Mesozoic was mainly a time of erosion. Then in the Cretaceous, as the North American plate began to override the East Pacific rise, the area experienced uplift and some deformation. This was accompanied by granite intrusions, including the Amole Peak–Wasson Peak granite, multiple other intrusions, and rhyolite extrusions. Following this compressive phase, in the Middle Tertiary a tensional phase leading to the development of Basin and Range topography began. The area continued to be deformed and igneous activity also continued. In late Tertiary time, thinning of the crust and gravity sliding of decoupled crustal blocks resulted in unloading of the crust and isostatic uplift. This was accompanied by faulting and erosion, producing the shapes in the mainly granite mountains in the park.

During the Pleistocene, as global climate got cooler, rainfall in the area increased. This increased erosion of the mountains and sedimentation in the basins. Then as temperature again increased, lakes dried up, producing playa deposits and desert conditions.

WHITE SANDS NATIONAL MONUMENT

White Sands National Monument (see Figure 1.17) was established in 1933 in New Mexico. It contains the largest gypsum [$CaSO_4 \cdot 2H_2O$] dunes in the world, and is located in the Chihuahuan Desert. The dunes are extremely white and behave and look very much like snow drifts, drifting across park roads and periodically being plowed back by the Park Service (Figure P12.8).

Geology. The gypsum crops out in the San Andres and Sacramento Mountains to the west and east (respectively) of the Tularosa Valley, which contains White Sands. It is then dissolved and carried into Lake Lucero, a playa lake to the southwest, where evaporation allows reprecipitation of the gypsum as *selenite,* the mineral name for gypsum crystals. Capillary action, bringing gypsum saturated groundwater to the surface around the lake, also contributes to the gypsum. Mechanical weathering then breaks the selenite down into very small crystals that are blown into dunes and migrate as dune fields, with cross-bedding developing as they move.

FIGURE P12.8 To keep the white gypsum sand from covering the road into White Sands National Monument, New Mexico, the Park Service has the road cleared using a snow plow. Movement of the sand is from left to right in the photograph. (Photo by author.)

REFERENCES

BENNETT, R., ed. 1980. *The new America's wonderlands, our national parks,* p. 463. Washington, D.C: National Geographic Society.

BEUS, S. S., ed. 1987. *Decade of North American geology, DNAG, Centennial field guide.* Vol. 2. *Rocky Mountain Section of the Geological Society of America,* p. 475. Boulder, Colorado: Geological Society of America.

BOGGS, S. 1994. *Principles of sedimentology and stratigraphy.* 2nd ed., p. 800. New York: Macmillan College Publishing Company.

CHRONIC, H. 1986. *Pages of stone, geology of western national parks and monuments.* Vol. 3. *The desert southwest,* p. 168. Seattle: The Mountaineers.

HARRIS, D. V., and KIVER, E. P. 1985. *The geologic story of the national parks and monuments.* 4th ed., p. 464. New York: John Wiley & Sons, Inc.

HILL, M. L., ed. 1987. *Decade of North American geology, DNAG, Centennial field guide.* Vol. 1. *Cordilleran Section of the Geological Society of America,* p. 490. Boulder, Colorado: Geological Society of America.

13

Glaciation

GLACIERS

Today, less than 10% of Earth's land surface is covered by glaciers (Figure 13.1), but in the past this figure has been much greater. Glaciers are part of the water cycle and are the result of accumulated snow compacting into ice, with continuing recrystallization as the system compacts, a complex process (Figure 13.2). If excess ice/water is available, snow and ice will build up until the weight creates enough pressure that the ice will deform and begin to flow. During times of insufficient rainfall, glaciers begin to shrink due to *ablation,* a process combining melting and/or *sublimation.* Sublimation results when ice goes directly to the gaseous state (water vapor).

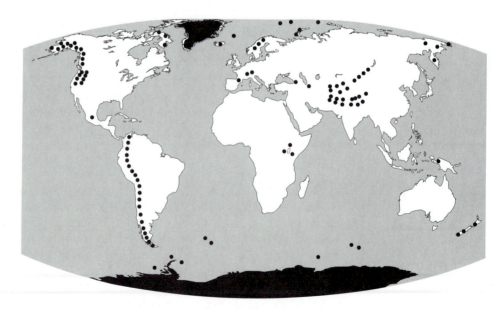

FIGURE 13.1
Distribution of glaciers on Earth today.

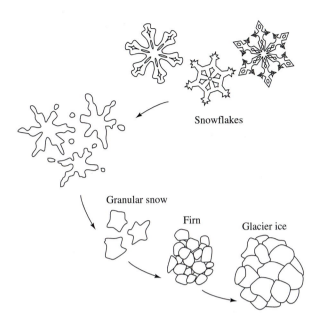

Snowflakes

Granular snow

Firn

Glacier ice

FIGURE 13.2
Snowflakes being
converted to glacier ice.

The growth or decline of glaciers depends on very slight climate changes that determine the renewal rate of snow due to the availability of moisture. The accumulation of ice in large amounts on land means that water normally contained in the ocean becomes tied up on land, thus causing global sea level changes (Figure 13.3). In the past, sea level stood nearly 100 m lower than it does today, and at times, sea level was even higher than it is today. Sea level lowering results in the exposure to the air of low elevation coastal regions (areas along continental shelves) and, in turn, drops the global base level, allowing renewed erosion. The impact of sea level change is enhanced by isostatic rebound of continental shelves as the weight of ocean water is removed when sea level falls. One of the interesting effects associated with lower sea level is that terrestrial plants and animals migrate out onto exposed continental shelves. Archaeologists have discovered mastodon and woolly mammoth fossils (Pleistocene animals) located in areas that are currently flooded (Figure 13.4), but during the last glacial maximum were available to animals for grazing.

Erosion during times of past low sea-level stands has produced deep submarine canyons that cut through the continental shelves along the eastern and western seaboards of North America. It also caused the build-up of sand dunes on continental shelves that were the foundation of what are now barrier islands along the eastern and southern coastlines of North America. These are seen in the national seashore areas (see Chapter 14).

Types of Glaciers

Several types of glaciers have been identified and classified by *glaciologists,* geologists that study glaciers and their effects. *Cirque glaciers* are very small glaciers that occupy *cirques,* convex cavities cut into the side of mountains by mechanical weathering. When water filling rock fractures freezes, it expands and breaks up the rock, allowing the water to penetrate farther inside. As the process continues, a little bowl-shaped depression develops on the mountainside where cirque glaciers form and continue to expand. When cirques erode three or more sides of a mountain, a mountain-sized rock spire called a *horn* is formed such as the example in Figure 13.5 in Glacier National

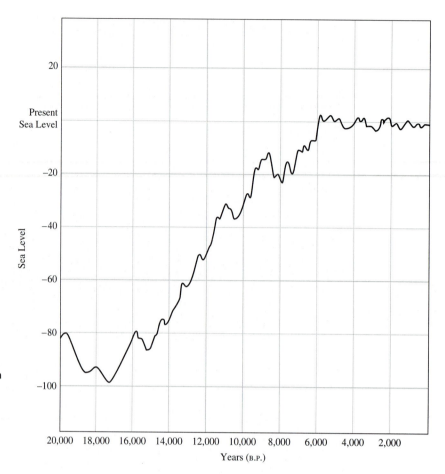

FIGURE 13.3 Global sea level changes during the past 20,000 years. In the past sea level was nearly 100 m lower than it is today. Years uncalibrated carbon 14 ages.

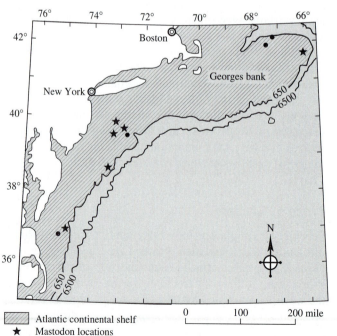

FIGURE 13.4 Area of the continental shelf along the North Atlantic U.S. coastal plain that was exposed during the last major glaciation. Mammoth and mastodon fossils have been found in this area and indicate that these animals were living there during low sea-level stands.

FIGURE 13.5 Glacially carved horn in Glacier National Park, Montana. (Photo by author.)

Park. The best known example of a horn is the Matterhorn in Switzerland. Sometimes, rather than a horn, cirques form thin, saw-toothed ridges called *arêtes*.

As cirque glaciers expand they coalesce into larger glaciers that eventually begin to flow into the valleys below. Such *valley glaciers* are continuously fed by cirques, and their flow carves the characteristic U-shaped valleys, so evident in glacial terrains (note the U-shape in Figure 13.5). Further coalescence of valley glaciers produces the large *piedmont glaciers* that develop on lowlands at the base of mountains. Wrangell-St. Elias National Park and Preserve has the largest piedmont glacier in the United States. As piedmont glaciers flow into the ocean, blocks of ice break off their leading edge, (*calve*) forming icebergs.

WRANGELL-ST. ELIAS NATIONAL PARK AND PRESERVE

Wrangell-St. Elias National Park and Preserve (see Figure 1.11) was first established as a national monument in 1978, but was upgraded in 1980. Located in southeastern Alaska just a few miles north of Glacier National Park, Wrangell-St. Elias is the largest park in the system with an area of 13,188,000 acres. The park shares a common boundary with Kluane National Park in Canada. Wrangell-St. Elias has been designated as a world heritage site, in part because of the shared border with Canada that crosses the summit of Mt. St. Elias, rising 18,008 ft just 23 miles from the sea (Figure P13.1). Lying below this spectacular mountain is the largest piedmont glacier in the United States, the

Malaspina Glacier, and elements of the glacier can be seen feeding into the bay below Mt. St. Elias (right side in Figure P13.1). Abundant icebergs calve off the front of the glacier to produce a constant grinding noise in the bay, and many icebergs with bizarre shapes can be seen floating past or stranded on the beach by the tide (Figure P13.2).

The park is filled with evidence of glacial activity. Many active glaciers, hanging valleys, cirques, aretes, striations, and *glacial till* (jumbled, unsorted sediment dropped by glaciers; see Figure P13.1) are abundant in the park. Bald eagles and varieties of bears frequent the park and can be seen going about their

FIGURE P13.1
Mt. St. Elias in Wrangell-
St. Elias National Park
and Preserve, Alaska. The
foreground is composed
of glacial till. (Photo by
author.)

FIGURE P13.2
Completely natural ice
"bird" that floated
ashore and remained on
the beach when the tide
went out. (Photo by
author.)

daily activities. High energy erosion along the coast is constantly undercutting the pine trees, leaving their trunks strewn along the beach (Figure P13.3; similar effects are seen in Figure 1.13, Olympic National Park).

Access to Wrangell-St. Elias National Park is by boat, float plane, or helicopter, and there is a small landing strip for wheeled aircraft. Before becoming a park, the area was extensively logged (Figure P13.4) with some very large, old-growth trees being the prime target of these activities. Most of the lumber was floated offshore to waiting Japanese freighters and shipped directly to Japan. Apparently this was illegal, and the practice has been stopped now that the area is a national park!

Geology. A wide variety of rocks and ages of rocks exist in the park, due in part to the accretion of displaced terranes in much of Alaska and the Western Cordillera. The coastal areas are composed mainly of late Tertiary and Quaternary till. This till was deposited in thick sections offshore and has since been lithified, uplifted, and is being eroded by active glaciation in the park.

FIGURE P13.3 Erosion along the Pacific coastline in Wrangell-St. Elias National Park and Preserve. (Photo by author.)

FIGURE P13.4 Logging in Wrangell-St. Elias National Park and Preserve before it became a national park. (Photo by author.)

Very large ice sheets may develop that are extremely large glaciers of irregular shape. Iceland has some smaller ice caps, small ice sheets that can grow into much larger continental ice sheets of such great size and thickness that they can cover mountain ranges. Today continental ice sheets cover most of Greenland to a thickness of approximately 3 km, and in Antarctica ice is, in places, more than 4 km thick. Antarctica contains 66% of the world's fresh water, all located over the South Pole in a gigantic continental ice sheet!

As glaciers flow, the ice motion over topographic irregularities causes the brittle upper surface to bend and crack, producing crevasses. Internal flow or creep of the ice causes crystals to deform and allows the glacier to slide or slip along the ground, a process called *basal slip.*

GLACIATION

The process of glaciation actively erodes, transports, and deposits rock, leaving behind distinctive traces of the glacier's passing. As an erosional process, glaciation causes frost wedging, the result of water freezing in rock fractures and forming cirques high on the side of mountains where glaciers often get started. Rocks, rock fragments, and blocks of rock are *plucked* or quarried from the bedrock as the glacier passes over (Figure 13.6). Also, striations, grooves scraped into the bedrock (Figure 13.7), are evidence of abrasion caused by rocks held in the ice of the passing glacier and point in the direction of its travel.

Glacial valleys, often reshaped stream valleys (young stream valleys are V shaped because they are eroded quickly), develop a characteristic U shape in cross section. Some valleys, called *hanging valleys,* can be seen far above a main valley floor, where side, tributary glaciers entered to coalesce with a main valley glacier (these can be seen forming from glaciers in Glacier Bay National Park; Figure P13.5). After melting, evidence of these smaller glaciers remains as U-shaped hanging valleys. A cirque is likely to be found at the head of these valleys. *Fjords* are glaciated valleys or troughs that are now partly submerged by seawater (Figure P13.5). They are produced as the result of glacial erosion, usually during low sea-level stands during times of expanded glaciation, and these valleys are later submerged as sea level rises. Other examples in Alaska are found at Wrangell-St. Elias National Park and Preserve (this chapter), Kenai Fjords National Park (this chapter), and Lake Clark National Park and Preserve (Chapter 5).

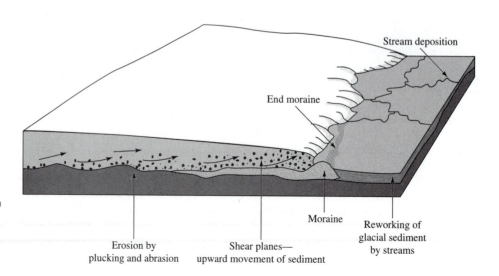

FIGURE 13.6   Diagram illustrating some of the effects of glaciation including plucking and deposition of moraines.

Stream deposition

End moraine

Moraine

Reworking of glacial sediment by streams

Erosion by plucking and abrasion

Shear planes— upward movement of sediment

Ice flow direction

FIGURE 13.7 Striations left by glaciers in Acadia National Park, Maine (Chapter 5). Ice movement direction is given. (Photo by author.)

 ## GLACIER BAY NATIONAL PARK

Glacier Bay National Park (see Figure 1.11) was initially proclaimed in 1925 as a national monument. Located in the extreme southeastern part of Alaska (Figure 1.11), it was upgraded in 1980 to park status, at the same time that many areas in Alaska were also so designated. The park consists of extensively glacially carved coastal ranges, with many examples of active valley glaciers (Figure P13.5). Erosion has resulted in the development of fjords along the park's coast. Access to the park is by boat, float plane, or helicopter.

Geology. As at Wrangell-St. Elias National Park, a wide variety of rocks and ages of rocks exist in the park, due in part to the accretion of displaced terranes. The real character of the park, however, comes from glacial erosion of Tertiary and Quaternary glacial drift that was deposited in thick sections offshore and has since been lithified, uplifted by the Coastal Range Orogeny, and is being eroded by the active glaciers in the park.

FIGURE P13.5
Coalescing glaciers in Glacier Bay National Park, Alaska. (Photo courtesy of the National Park Service.)

 GLACIER NATIONAL PARK

Glacier National Park (see Figures 1.18 and 8.9; see also Figure 13.5) was established in 1910 in Montana. It is a mountainous park that has been sculpted by glacial action. Even today there are 50 small but active glaciers (Figure P13.6). The park extends to the United States–Canadian border and has been combined with Canada's Waterton Lakes National Park in Alberta as the International Peace Park. Glacier is located in the Rocky Mountains, and its highest point is Mt. Cleveland at 10,468 ft.

FIGURE P13.6 One of 50 small glaciers in Glacier National Park, Montana. (Photo by author.)

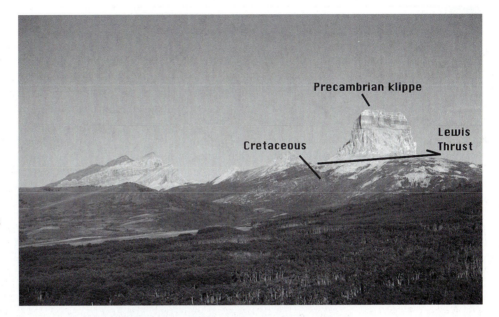

FIGURE P13.7 Chief Mountain, a klippe in Glacier National Park, where Precambrian rocks have been thrust over Cretaceous rocks along the Lewis thurst. Erosion then isolated Chief Mountain. (Photo by author.)

Geology. The Lewis Thrust, a major thrust fault separating flat-lying Proterozoic Belt Supergroup rocks above from folded Cretaceous rocks below, has thrust the Proterozoic rocks, mainly limestones, out onto the plains beyond the mountains. Chief Mountain, a limestone remnant of the thrust sheet known as a *klippe,* stands alone beyond the eastern extent of the mountains (Figure P13.7). Erosion after thrusting has destroyed most of the rocks above the thrust plane, except for the klippe.

The park has exposed in it flat-lying clastic sedimentary beds and limestones that have been intruded by basaltic sills and dikes, and locally some pillows were extruded and can be seen. In Rising to the Sun Mountain the limestone beds have been intruded by a large sill that has baked the upper and lower margins of the limestone, producing *contact marbles* (Figure 5.15). Then, during the Pleistocene, glaciation produced many classic glacial landforms in the park, including hanging valleys, arêtes, horns (Figure 13.5), cirques (Figure P13.8), and striations.

FIGURE P13.8 Cirque in Glacier National Park. (Photo by author.)

KENAI FJORDS NATIONAL PARK

As were many parks in Alaska, Kenai Fjords National Park (see Figures 1.11) was originally a national monument established in 1978 and was later upgraded to national park status in 1980. Located in southern Alaska (Figure 1.11), the coastal fjords (Figure P13.9) in the park give it its name. It also contains the Harding Icefield, one of four major glaciers in the United States.

Geology. During the Mesozoic, a range of sediments were deposited offshore, including siltstones, sand-stones, limestones, tuffs, and even some lavas were erupted producing pillow forms. Deformation and further deposition during the late Mesozoic produced the metasediments dominant in the park. Later in the Tertiary, further intrusions, including large granite bodies, were also emplaced. As with all the coastal parks in Alaska, late Tertiary and Quaternary till has been lithified (producing *tillite*), uplifted, and sculpted by glaciation.

FIGURE P13.9 Fjord in Kenai Fjords National Park and Preserve. (Photo courtesy of the National Park Service).

Glacial Transport and Deposition of Sediment

Glaciers transport a great deal of sediment and are significant in sculpting many of the national parks. The material being eroded is carried by wind, scraped, carried and pushed directly by the glacier, and washed out by meltwaters flowing from the glacier. Glacial sediments in the rock record are an important indicator that the climate at the time of their deposition was cold.

The material carried and deposited from wind is a very fine-grained and well-sorted sediment called loess. It is composed primarily of silts with some clay and fine sand-sized particles and, once deposited, is resistant to erosion, tending to form cliffs due to molecular attraction between particles. Loess is usually not stratified and is a heavily *bioturbated* (reworked by burrowing organisms), productive soil. Vast thicknesses of glacial debris accumulated in states such as Iowa, Illinois, Ohio, Nebraska, and Kansas, where soil developed on glacial material supports the agriculture in this part of the United States. Thick loess accumulations also exist in a band across Europe, in countries such as France and Germany, as well as in Asia, where very thick loess sequences are known in China.

As sediment-carrying glacial ice melts, it deposits this material, usually as an unstratified, unsorted jumbled mass called glacial till (Figure P13.1). Because it lacks any organization, these deposits are difficult to study and many geologists ignore them with a shrug and a statement like, "You can't work with that stuff!" In fact, the larger clasts in glacial till can be very instructive, for example indicating the various rock outcrops captured by the glacier along its path.

Along the margins of glaciers, large quantities of rock, often boulder sized, are deposited in *moraines* (Figure 13.8). These are found in several forms and are the result of constant glacial movement; erosion breaks up rocks lying at the margins and under glaciers and the glacier carries this material to the front and sides of the advancing or receding ice flow. For example, *ground moraines* are widespread but relatively thin accumulations of glacial debris called *drift,* and have a gentle, undulating topographic surface. Moraines result from rocks being deposited as a glacier rapidly receded. *End moraines* form as ridgelike glacial margin accumulations. When they accumulate as *lateral* (along the sides of glaciers) *moraines* they give an indication of flow direction (Figure 13.8). At the end of glaciers, *terminal moraines* mark the forward extent that a glacier moved before it melted back. At some localities, such as in Ohio, terminal moraines are hill-sized topographic features representing a relatively stable glacial front that melted at the same rate it was moving forward, resulting in large accumulations of drift being brought to the leading edge of the glacier and dumped, much like a conveyer belt carries and dumps material. *Medial moraines* are ridgelike accumulations of drift formed between two merging valley glaciers (Figure P13.5).

Melt waters are constantly flowing from glaciers, even in the winter, and the sediments transported look very much like those deposited from streams. This material,

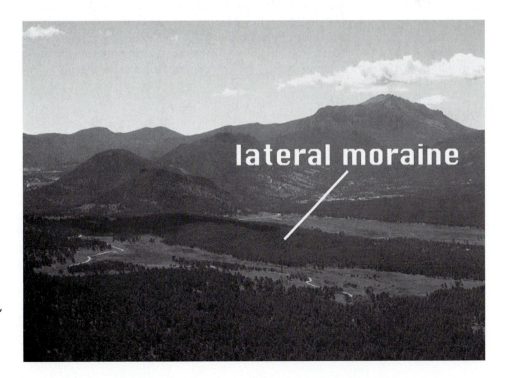

FIGURE 13.8 Lateral moraine in Rocky Mountain National Park, where glacial drift has accumulated as the ice flowed past. (Photo by author.)

known as *stratified drift,* is deposited beyond the glacier as the water flows away, and it is often well sorted and well stratified. Such sediments may develop alternating coarse- and fine-grained, thin-bedded layers due to seasonal changes in the weather, representing summer-winter depositional cycles. These sediments are known as varves. Other outwash sediments, carried outward and deposited by meltwater can be extensive, but are usually dominated by sand-sized particles. Extremely fine, glacially ground-up rock is called *rock flour,* and is often dispersed and suspended in lakes where light refraction produces a very light green color to the water. Streams, on the other hand, have more of a brown cast due to the presence of coarser particles being carried in suspension.

Other types of deposits from glaciers include *dropstones,* rocks dropped by passing glaciers or icebergs. When found in marine sediments these are called *ice rafted debris.* Such materials are also called glacial *erratics* because they are carried from elsewhere and left behind. This material can be useful flow direction indicators, as can *boulder trains.* These are strings of erratics that all originate from a common source and lie along the direction the glacier is traveling, or they may form fan-shaped deposits that open in the direction of flow.

PLEISTOCENE AND PRE-PLEISTOCENE GLACIATION

Over the last 2 million years or so, glacial activity has been extensive in North America (Figure 13.9). The most recent glacial maximum occurred about 22,000 B.P., but good evidence exists to indicate that glaciers have accumulated and advanced periodically back to approximately 1.7 B.P. These cycles of climatic changes, starting with cold times of glacial advances followed by warm, interglacial periods, have been well documented and dated by glaciologists in North America and Europe.

Work by geologists and glaciologists has now documented that as much as 29% of Earth's surface has formerly been glaciated. Evidence also exists that glaciers were forming on Earth as far back as the Proterozoic. The best documented indication of glaciation comes from the end of the Permian Period, when Gondwana, lying at the South Pole (Figure 13.10), was heavily glaciated, leaving tillites, striations, and moraines behind to show that glaciers had been there.

CAUSES OF GLACIATION

The concern of geologists and the question they ask is, "Can Earth rapidly revert to a glacial climate, and if so when, and what effect will it have on civilization?" We know that climate has deteriorated rapidly in the past. For example there was a period called "the Little Ice Age" in the 1600s that caused serious problems for people then living on Earth. What would more severe climate deterioration do? To answer these questions, geologists and climatologists are studying the causes of glaciation and its effects, recorded in the sedimentary record. Recent discoveries from studies of ice cores drilled in Greenland have shown that climate can switch from glacial to interglacial conditions, or from interglacial to glacial conditions in just a few years. Archaeologists are also studying the effect of glaciation on ancient peoples living during the last glacial maximum and before.

There are a number of factors that can cause climate to deteriorate. Most of these affect the atmosphere in some way, resulting in changes in the magnitude of solar

FIGURE 13.9 Extent of past glaciation on the North American continent.

FIGURE 13.10 Reconstruction of Gondwana during the Permian Period showing the areas that, based on glacial evidence, were glaciated.

radiation reaching Earth's surface (filtering), thus regulating climate. Geologic processes producing dust, such as volcanic eruptions or meteor impacts, can seriously reduce solar radiation through the atmosphere. Changes in carbon dioxide content changes the filtering effects. Fluctuations in solar radiation can impact climate, and it has been argued that changes in the orbit or tilt of Earth can also affect climate, a hypothesis suggested by *Milankovitch,* a Yugoslav geophysicist working in the early part of the 20th century. Other important climatic effects include plate tectonics, causing ocean circulation changes and global cooling, or increases in continental elevation, providing areas where ice can more easily form and accumulate.

REFERENCES

BENNETT, R., ed. 1980. *The new America's wonderlands our national parks,* p. 463. Washington, D.C.: National Geographic Society.

BEUS, S. S., ed. 1987. *Decade of North American Geology, DNAG, Centennial Field Guide.* Vol. 2. *Rocky Mountain Section of the Geological Society of America,* p. 475. Boulder, Colorado: Geological Society of America.

CHRONIC, H. 1984. *Pages of stone, geology of western national parks and monuments.* Vol. 1. *Rocky Mountains and Western Great Plains,* p. 168. Seattle: The Mountaineers.

HARRIS, A. G., and TUTTLE, E. 1990. *Geology of national parks.* 4th ed., p. 652. Dubuque, Iowa: Kendall Hunt Publishing Company.

HARRIS, D. V., and KIVER, E. P. 1985. *The geologic story of the national parks and monuments.* 4th ed, p. 464. New York: John Wiley & Sons, Inc.

HILL, M. L., ed. 1987. *Decade of North American geology, DNAG, Centennial field guide.* Vol. 1. *Cordilleran Section of the Geological Society of America,* p. 490. Boulder, Colorado: Geological Society of America.

The Marine System

The marine system affects the continents in a number of ways. Sediments eroded from continents are deposited near-shore, and are often uplifted to become exposed. Ocean circulation is important in controlling climate, and winds cause waves that directly sculpt and mold shorelines. Surface waves have the most direct impact and are driven by the wind. In fact, while in the deep marine system, particle motion in a wave is circular; a cork would describe a perfect circle as a wave passes. However, as the seafloor shallows, near-shore particle motion becomes elliptical. This means that at the base of waves, along the bottom, particles move back and forth.

WAVES AND SHORELINES

As waves move toward shore, *wave refraction,* a change in wave direction, occurs. The wave swings parallel to bottom contours due to drag along the bottom as the water shallows. The result is that wave action against the shoreline is concentrated toward topographic highs that stick out into the water (Figure 14.1). Eventually, these will be smoothed away by the eroding effect of the waves and deposited as beaches along topographic lows, ultimately creating a long, straight shoreline.

As waves break against the shore, the water rushing up the beach and the sediment it is carrying moves onshore in the direction that the wave is moving, usually oblique to the shoreline (Figure 14.2). However, as the water, carrying some sediment, washes back down the beach, it moves perpendicular to the contours of the beach. This has the effect of moving the sediment along the beach, and the net water flow also tends to be in this direction. The sediment moved by this process is called *longshore drift,* and the currents moving in the same direction are called *longshore currents.*

Another important characteristic of waves is that they steepen and break in shallow water. The waves breaking along the shoreline (*surf zone*) release a great deal of energy, causing abrasion, concussion, and particle rounding. This is accompanied by strong erosion or *winnowing,* where very fine particles are selectively carried off,

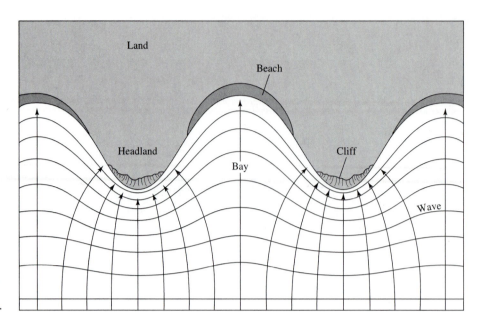

FIGURE 14.1
Schematic illustrating
wave refraction, erosion,
and deposition of
beaches along shorelines.

leaving coarser particles behind. This process sorts the sediment by size and often results in cross-bedded deposits. Loose sediment is transported by waves as a result of
the back-and-forth particle motion at the base of waves, with overall sediment transport being moved slowly seaward. Particle size decreases away from shore due to decreasing energy.

Shoreline Depositional Features

Beaches are the best-known shoreline features, with quartz sand being most commonly
found. The quartz originates as sediments from cliffs eroded by waves, from river input,

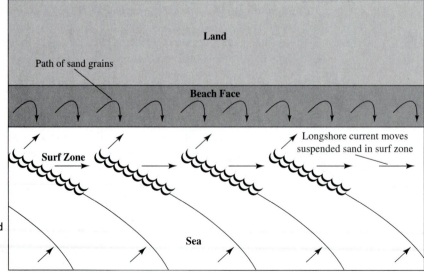

FIGURE 14.2
Schematic illustrating
the generation of
longshore currents and
longshore drift as the
result of waves
breaking on a beach.

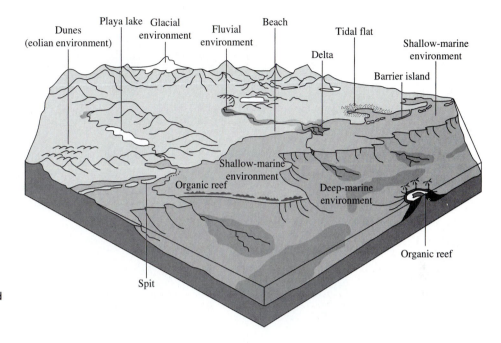

Playa lake Glacial
Dunes   environment
(eolian environment)

Fluvial
environment

Beach

Tidal flat

Shallow-marine
environment

Delta

Barrier island

Shallow-marine
environment
Organic reef

Deep-marine
environment

Organic reef

Spit

FIGURE 14.3
Schematic of a generalized coastline, showing the marine environments in which sediments are deposited and from which sediments are derived.

or from being carried in and away by longshore currents. *Spits* and *bay barriers* result from accumulations of sand and gravel at points where projections from land end in open water (Figure 14.3), at times growing so much that they may completely block bay mouths. These features are found in many areas in the national seashores. The sand deposited at these sites is usually clean and well sorted, built from longshore currents and longshore drift. Most *barrier islands* originated during low sea-level stands (see Chapter 13). Beaches formed along ancient shorelines, then dunes, stabilized by vegetation, built up behind these beaches. Some of these dunes are quite large, and in fact the Wright brothers flew gliders off a high dune in what is now Wright Brothers National Memorial, before their powered flight experiments. Later, as sea level rose, water flooded behind the dunes forming *lagoons*. Periodic storms break through these barriers to form *inlets*. Our national seashores like Capes Hatteras and Canaveral are places where natural beach processes can be observed.

 ## CAPE HATTERAS NATIONAL SEASHORE

Cape Hatteras National Seashore (see Figure 1.22) was the first national seashore to be established. Created in 1937 on the outer banks of North Carolina, it is composed of three barrier islands and the lagoons lying behind them. The seashore is greater than 50 mi (80 km) long and contains the tallest lighthouse, 208 ft (63 m) high, in the U.S. (Figure P14.1). Built in 1870, it was originally located more than a half mile (800 m) from the ocean. Today it is right on the beach, and the National Park Service has been working hard to save it from destruction by ocean erosion. The giant sand bags evident in Figure P14.1 are a temporary measure. Plans exist to try and move the lighthouse by sliding it on skids away from the beach toward the southwest, but the cost will be very high, and as yet the money is not available.

FIGURE P14.1 Cape Hatteras Lighthouse in Cape Hatteras National Seashore, North Carolina. The large sandbags are an attempt to save the lighthouse from erosion by the ocean. (Photo by author.)

Geology. During the last major glaciation, at its maximum approximately 22,000 B.P., sea level was nearly 100 m lower than it is today. At that time dunes built up behind the beach as they do along modern beaches, and vegetation established a foothold in them, stabilizing the sand. When the ice melted and sea level rose, the dunes remained as barrier islands, which are being sculpted and eroded by ocean wave action.

 ## CANAVERAL NATIONAL SEASHORE

Canaveral National Seashore (see Figure 1.22) was established in Florida in 1975. It contains a barrier island/lagoon system that extends for approximately 25 mi (40 km). Along its southern margin sits the Kennedy Space Center, currently the launch site of the Space Shuttle. The barrier islands here formed during times of low sea level (Chapter 13).

Archaeology. Canaveral National Seashore contains two Native American shell mounds: Turtle Mound, a large double mound approximately 50 ft (15 m) high (Figure P14.2), and Castle Windy, a somewhat smaller mound (Figure P14.3). Both were built from discarded seashells, sometime between 800–1400 A.D., by the Timucuan people living in the area at that time. It is interesting that this was also the time that the Anasazi cultures were flourishing in the southwestern part of the United States (see Chapter 16). Most of the shell mounds in Florida are not protected and several have been mined and used as "road metal." One shell mound, a few miles to the north of Canaveral, has a private residence built on top of it.

FIGURE P14.2 The top of Turtle Mound, built from seashells by Native Americans as long ago as 1200 B.P., in Canaveral National Seashore, Florida. These shell mounds were used as observation posts. (Photo by author.)

FIGURE P14.3 Castle Windy, built from seashells by Native Americans as long ago as 1200 B.P., in Canaveral National Seashore, Florida. (Photo by author.)

WRIGHT BROTHERS NATIONAL MEMORIAL

Wright Brothers National Memorial (see Figure 1.22), in eastern North Carolina's outer banks, part of the Cape Hatteras barrier island system (see Cape Hatteras National Seashore for a discussion of the geology), is the site of Wilbur and Orville Wright's first powered flight in 1903. For their early flights they used a glider built in 1900, launching it many times from Big Kill Devil Hill Sand Dune (Figure P14.4), the highest point in the area and an exceptionally large example of the sand dunes found stabilized by vegetation on barrier is-lands along the eastern and southern coastlines of North America. A large granite monument to the 1903 first powered flight is now located there (Figure P14.4). The longest of the first four powered flights lasted only 59 seconds and covered a distance of 852 ft (260 m); the shortest lasted only 12 seconds for 120 ft (37 m). But by 1905 the first practical airplane had been con-structed and flights up to 38 minutes in duration were routine.

FIGURE P14.4 Big Kill Devil Hill sand dune at Wright Brothers National Memorial, North Carolina. Monument represents the point from which the Wright Brothers launched their gliders before they turned to powered flight. Shown also in Figure 1.9. (Photo by author.)

The lagoons behind barrier islands are traps for fine sediment brought into the la-goon by tidal cycles and suspended in rivers. Tidal and river deltas accumulate sedi-ments in lagoons as velocities slow. Furthermore, lagoons have highly variable salini-ties due to restricted ocean water circulation: fresh water input from rivers produces a mix of salt and fresh water called brackish water, and high evaporation rates produce very salty, hypersaline water. Lagoons are located in numerous locations along the At-lantic and Gulf coasts.

Another shoreline area where deposition occurs is on *tidal flats,* flat areas around the rim of lagoons that are alternately dry and flooded during tidal cycles. Here fine sed-iments and organic material accumulate in marshes and may form peat, which eventu-ally changes into coal (*marsh coal*).

Extending beyond the beaches where the ocean is shallow is the Continental Shelf. Here ocean water floods the extensions of continents, and this zone represents 10% of the world's continental area. Many islands, such as those at Channel Islands National Park, are located on the continental shelves that, during low sea-level stands, would be part of the continental mainland. The end of the shelf is marked by the *shelf break*. It is at this point that much of the sediment washed into the ocean collects. Much of Earth's sedimentary strata is formed at the shelf break.

CHANNEL ISLANDS NATIONAL PARK

Channel Islands National Park (see Figures 1.14 and 8.4) originated as a national monument in 1938 and was updated to a national park in 1980. Composed of five main islands off the southern California coast, the park can be visited by boat or small plane with the permission of the National Park Service. The islands include Anacapa, Santa Barbara, Santa Rosa, San Miguel, and Santa Cruz. Along the coastlines of these islands more than 380 sea caves have been identified, with over 9 mi (15 km) of mapped passages.

Three-quarters of Santa Cruz Island is owned by the Nature Conservancy. It is the longest island, being 22 mi (35 km) long and 6 mi (10 km) wide. Most of the island is mountainous and a portion of it was formed from basalt, as was Anacapa Island. The islands were formed during the Miocene as the westward moving North American Plate rode over the now extinct Farallon Plate. All the islands are being uplifted as the North American Plate continues to override the Pacific Plate to the west, while ocean erosion is actively destroying them (Figure P14.5).

Archeology. Some of the oldest sites of human occupation are located in the Channel Islands. These are dated at over 10,000 years B.P. Interesting fossil finds also exist in the islands. For example, extremely unusual miniature woolly mammoth skeletal remains have been found there, the animals standing only 6–8 ft (2 m) at the shoulders.

FIGURE P14.5 Erosion at Channel Islands National Park, California. (Photo courtesy of the National Park Service.)

The Continental Shelf is affected by a number of important geological processes, including progradation (building out by accumulating sediments), compaction, and isostatic depression due to sediment loading. These processes produce sequences of interbedded sediments that dip landward and grow (prograde) over previously deposited offshore sediments. The result is a very complex sedimentary package.

On continents where elevations are very low, semi-isolated *epicontinental seas* may result from broad flooding of these low elevation continental areas. A modern example is Hudson Bay in Canada. In epicontinental seas broad areas of limestone, shale, and evaporites may accumulate when conditions are suitable. In Hudson Bay, for example, conditions are not suitable because temperatures are generally too cold for most limestone-secreting organisms to live and evaporation rates are too low for most evaporites to form. However, at several times in the past, many states were flooded by broad epicontinental seas, and limestones, shales, and evaporites developed in these areas on the North American continent as a result. In fact, many of the limestones and shales we see exposed in the central and eastern parts of North America were formed in this way. Mammoth Cave National Park (Chapter 11) developed in such limestones.

ORGANIC REEFS

In tropical regions with low terrigenous sediment input, coral organisms may secrete skeletons, made of calcium carbonate ($CaCO_3$), which build up into *reefs*. This growth occurs within the *euphotic zone* (the near-surface ocean layer through which sunlight can penetrate), a zone containing many living organisms that is continually collecting other cementing organisms, such as *coralline algae*. The presence of the algae requires sunlight, and, in fact, the algae and coral organisms have a symbiotic relationship, the coral providing a home and the algae providing some food, although much of the coral organism's food comes from filtering the water (*filter feeding*) in which they live.

The result of coral growth is the secretion of a porous reef with a complex internal structure, poorly bedded, with some of the matrix voids filled by $CaCO_3$ fragments. Seaward of the reef is a zone of high energy, where waves break on the leading edge of the system, the energy decreasing leeward, to the rear of the reef. Along the leading edge high energy breaks up the reef, creating a talus deposit, with $CaCO_3$ blocks accumulating as a debris pile seaward of the reef. Wave erosion along the top of the reef, as well as continued growth, produces a reef flat, the horizontal upper reef surface (Figure 14.4). Lithified examples of coral reefs are found in Carlsbad Caverns National Park (Chapter 11) and Guadalupe Mountains National Park (Chapter 10).

There are several types of reefs that develop. For example, *patch* or *pinnacle reefs,* such as those found at Biscayne National Park, form as small, isolated reefs, often found in lagoons behind elongate or barrier reefs. *Fringing reefs* are found along coastlines with no lagoons. And *atolls* form as circular rings, sometimes horseshoe shaped, that surround volcanic islands that have subsided into the sea (Figure 14.5). The high growth rate during subsidence of the island keeps the corals in the euphotic zone.

FIGURE 14.4 Segment of coral reef top at Biscayne National Park, Florida. Rubble is broken coral fragments thrown onto the reef by wave action. (Photo by author.)

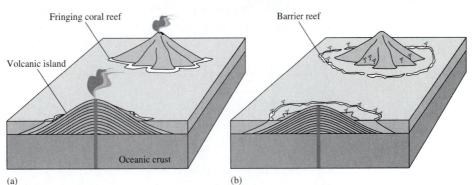

FIGURE 14.5
Schematic of the development of atolls. They begin as fringing reefs around volcanic islands, and, as ocean crust sinks or the island is eroded, the coral continues to grow to maintain its position within the water column and in the euphotic zone. Eventually only the coral reef remains as a ring around what used to be the island in the middle.

BISCAYNE NATIONAL PARK

Biscayne National Park (see Figures 1.22 and 8.17) was established in 1968 as a national monument and upgraded in 1980 to national park status. It includes 43 relatively small islands west of Biscayne Bay in southern Florida and 1 relatively large island, Elliott Key, on which a visitors' center is located. A second visitors' center is located on the mainland. Elliott Key is just north of Key Largo, the first large key leading into the main chain of Florida Keys. Access to the park is by boat. Park attractions include glass-bottom boat trips and reef snorkel/scuba trips. The most interesting part of the park is in its underwater *coral reefs* and its varied wildlife that includes a bird sanctuary and Florida lobster sanctuary. The clear, warm water off the southern tip of Florida has been ideal for the growth of coral reefs.

In 1992 Hurricane Andrew caused major destruction in the park, ripping out many trees on the keys, damaging corals, and completely destroying buildings, including the mainland visitors' center (Figure P14.6). The damage illustrates the serious nature of these storms. Nearby Everglades National Park was also seriously damaged, with walkways and park improvements being destroyed. The visitors' center was also damaged, but not destroyed.

Geology. During the last interglacial time, temperatures were warmer and sea level was higher than today. While sea level was high, a system of small coral patch reefs developed that extends from the tip of Florida into the Gulf of Mexico and makes up a portion of the Florida Keys. These reefs produced the Key Largo Limestone that is today, a time of slightly lower sea level, exposed in Biscayne National Park and elsewhere in the northern part of the keys. Following the last major glaciation, sea level fell significantly, exposing all the continental shelf around the tip of Florida. Erosion during this time sculpted and destroyed portions of the reefs, caused local dissolution of the limestone, and widened the channels between reefs. Then as sea level rose during glacial melting, the area was again flooded and smaller reefs began to grow in protected areas behind the remaining patch reefs.

FIGURE P14.6
Destruction in the area of the mainland visitors' center caused by Hurricane Andrew in 1992 at Biscayne National Park, Florida. (Photo by author.)

NATIONAL PARKS IN U.S. TERRITORIES

There are a number of U.S. Territories where the United States has a strategic interest. Two of these, Hawaii and Alaska, have achieved statehood. Congress has, in two cases recently, established national parks that are not within the boundaries of the 50 United States of America, one in the Virgin Islands in 1956 and the other in American Samoa in 1988. There is precedent for this, however, because Congress established parks in both Alaska and Hawaii before they achieved statehood. These include Hawaii National Park, established in the Territory of Hawaii in 1916, and what was then called Mt. McKinley National Park, established in the Territory of Alaska in 1917.

 ## VIRGIN ISLANDS NATIONAL PARK

Virgin Islands National Park was established in 1956. The Virgin Islands, a U.S. Territory, are situated in the Lesser Antilles (Figure P14.7). The park is located mainly on St. John Island and includes Hassel Island in St. Thomas Harbor. It provides the visitor with excellent examples of coral reefs and their plant and animal communities.

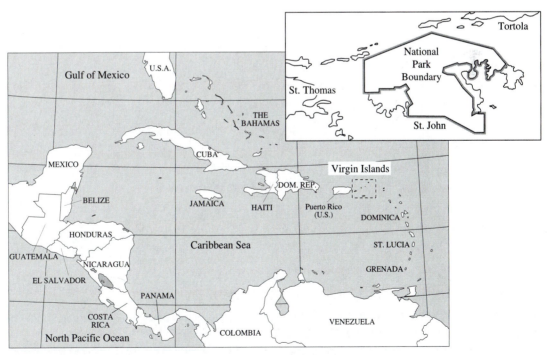

FIGURE P14.7   Map of the Virgin Islands in the Lesser Antilles, the location of Virgin Islands National Park. Insert: Virgin Islands.

NATIONAL PARK OF AMERICAN SAMOA

Authorized by Congress in 1988, the National Park of American Samoa is an interesting outreach by our government. The park is subject to land-use agreements with the local government in American Samoa that go beyond the normal controls placed on other U.S. national parks.

American Samoa, a U.S. Territory, is located in the southwestern Pacific Ocean (Figure P14.8). It is part of a 2,000-mi (3,200 km) long island arc, still relatively active, that had a volcanic eruption on the island of Tutuila in 1906. The territory contains five volcanic islands with fringing reefs, and two coral atolls. American Samoa is the center of the over 3,000-year-old Samoan culture.

The park is located on three islands and includes an area of approximately 9,000 acres. This includes a small segment on Tutuila Island, the largest island, with capital Pago Pago; a 5,000-acre or so segment on Ta'U Island, containing Lata Mountain, the highest peak; and land on Ofu Island, with excellent examples of healthy coral reefs.

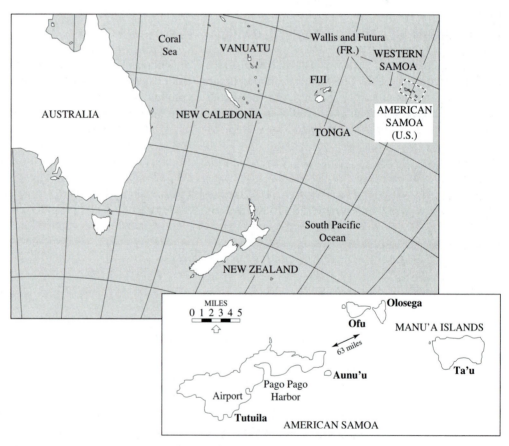

FIGURE P14.8 Map of the South Pacific showing the location of American Samoa, location of the National Park of American Samoa. Insert: American Samoa.

OTHER MARINE CARBONATES

Carbonate platforms, like that found at the tip of Florida where Everglades National Park is located, are large areas of carbonate sediments that form in warm, shallow water over ocean bottom that stands higher than the surrounding seafloor. The sediments are made up, in part, of reef fragments, large quantities of carbonate *fecal pellets* (excreted by marine organisms), and by oolites, sand-sized and larger carbonate, concentrically layered spheres formed by slow carbonate accretion. *Stromatolites* or stromatolite mats are also common on carbonate platforms and contribute to the sedimentation. They are formed by *cyanobacteria,* and have been an important producer of oxygen through photosynthesis since the Archean. Stromatolites incorporate layers of carbonate muds in their structure, which is usually dome, stool, or column shaped.

EVERGLADES NATIONAL PARK

Everglades National Park (see Figures 1.22 and 8.17) was authorized in 1934 to protect the southern tip of Florida (Figure P14.9). Because of its unique character and wildlife it has been designated a world heritage site. Much of the park is composed of swamp, having fresh water to the north that turns salty (brackish) toward the tip of Florida, as seawater mixes with the freshwater flow. The very near-surface, Miocene Tamiami Formation limestones (Figure P14.10) seriously retard soil formation, and the result is that only grasses, dwarf cypress, and mangrove trees grow in much of the park. There are many small keys (islands) within the park's southern edge, with the rocks here mainly lime muds and limestone (Figure P14.11).

The park is a sanctuary for saltwater crocodiles, an animal unique to the United States. It is also sanctuary for rare birds, including the southern bald eagle, great white heron, wood stork, anhinga, and roseate spoonbill. In some areas *hammocks* develop, dry mounds with trees and one of these contains the largest mahogany tree in the United States. There are many walking trails, canoeing areas, and the Seminole Reservation lies just outside the park to the north.

The park's alligators are very important to the ecological system of the area. When alligators were heavily hunted earlier in this century their numbers dropped drastically and the ecology of the park suffered. This is because the "gators" create for themselves a small lake by rooting around and pushing the mud created up and out of the holes they are creating. Eventually some of these "bowls" become quite large and in these small lakes (Figure P14.12) the female gators raise and guard their young. During times of drought, when the rest of the everglades dries up, fresh

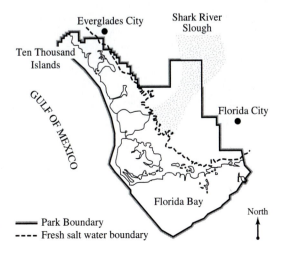

FIGURE P14.9 Map of Everglades National Park, Florida.

—— Park Boundary
- - - - Fresh salt water boundary

Everglades City
Shark River Slough
Ten Thousand Islands
GULF OF MEXICO
Florida City
Florida Bay
North

FIGURE P14.10
Limestone is only a few inches below the surface in Everglades National Park, allowing only dwarf trees and grasses to grow. Note that the walkway is disrupted, damaged by Hurricane Andrew in 1992. (Photo by author.)

FIGURE P14.11
Southern tip of Florida in Everglades National Park, where lime muds are exposed. This area is saltwater and includes a number of small islands (keys) where trees adapted to the environment cling to life and stabilize the keys with their root systems. (Photo by author.)

FIGURE P14.12 "Gator bowls" in the "glades," small lakes cleaned out by alligators to raise their young, in Everglades National Park. (Photo courtesy of the National Park Service.)

water for animals still can be found in the "gator bowls." This semisymbiotic relationship also provides a source of food for the alligator!

Geology. Florida is underlain by a thick sequence of relatively undeformed Cretaceous and Tertiary marine sedimentary rocks, mainly limestones. During the Miocene, the Everglades area was covered by very fine, mainly carbonate sediment that was compacted and cemented to produce the Tamiami Formation that crops out over large areas, including the Tamiami Limestone to the northwest and the Miami Limestone in central portions of the park. During the last interglacial time, sea level was higher than it is today, flooding the area that now includes the park and causing erosion and dissolution of Tamiami units. During this time the marls that were accumulated and only partially consolidated now form some of the sediment outcropping in the southern portion of the park. Later, the last major glacial advance over parts of the North American continent lowered sea level and caused erosion and sculpting of the reefs located in the southern and southwestern portions of the park. As the ice sheets melted, the area was flooded to present-day levels.

TURBIDITY CURRENTS

Turbidity currents are very important in transporting sediment accumulations on the continental shelves out into the deep marine environment. *Turbidity currents* are an underwater, rapid flow of dense, sediment-rich water that flows downslope, sometimes very fast (40–55 kilometers per hour), eroding as it goes. These currents are driven by gravity and are generated by the buildup of sediment due to overloading and then slumping, sometimes during earthquakes; they are also generated spontaneously.

The sediment deposited from turbidity currents is called a turbidite, and these have been extensively studied. Turbidites accumulate on submarine slopes and on abyssal plains, and are also found in lakes. Their deposition results in graded sediments with fine grains at the top and coarser grains deeper in individual turbidites. This grading results from high energy in the basal part of the flow carrying coarse particles, the tail end of the flow coming along later with much finer particles being deposited from much lower energy currents, the last gasp, so to speak. It is interesting that grain alignments in the bottom of the flow are at right angles to the flow direction, particles aligning much like pencils rolling on a table. The finer particles, however, align in the direction parallel with the flowing current.

PELAGIC SEDIMENTS (DEEP OCEAN)

In deep ocean basins sediment from many sources is constantly accumulating. Terrigenous sediments get into the deep ocean after being blown off land as windblown dust, usually quartz, or as turbidites, deposited from far-reaching turbidity currents (folded examples can be seen along the coast in Olympic National Park), or as ice rafted debris falling from passing icebergs, or as clays produced from weathering of submarine volcanoes.

Biogenic sediments of three primary types make up marine sediments. These are calcium carbonate, usually produced in mid- to low-latitude surface waters and found as thick, *carbonate ooze* deposits at water depths above 4,000 m. These are lithified into marine limestones. Below approximately 4,000 m depth, carbonates begin to dissolve and eventually may disappear from the sediment. *Siliceous sediments* are also important as marine sediments. They are produced in mid- to high-latitude surface waters where temperatures are relatively cold or produced in nutrient rich upwelling zones from the shells of silica-secreting organisms, radiolarians and diatoms being the most abundant. Siliceous sediments are lithified into chert layers (also sometimes called flint), composed of SiO_2. A third commonly found component of marine sediments is *sponge spicules,* long, thin needlelike spines secreted by deep-sea sponges and left on the bottom after the sponges die and decay, creating a very unusual texture in the sediments. Marine sediments deposited in near-shore environments may later be uplifted and exposed on land. Excellent examples are found in Olympic National Park and at Redwood National Park.

 ## OLYMPIC NATIONAL PARK

Olympic National Park (see Figures 1.12 and 8.4) was established in 1909 as a national monument and in 1938 upgraded to park status. It is located in northwestern Washington and has recently been designated a world heritage site because of its spectacular mountains and beaches. The park is divided into two sections, one section including the Olympic Range inland to the east, with Mt. Olympus at 7,965 ft (2,428 m) the highest peak. The Olympic Range has relatively low rainfall, but the section of the park, to the west, along the Pacific coast, has the highest rainfall in the continental United States. This portion of the park has high, wave-cut cliffs (Figure P14.13) and intricate sea arches.

Geology.   The rocks exposed in the park are mainly marine that are Eocene in age. They have been deposited on the ocean floor, then pushed up as North America moved to the west, subducting the oceanic plate in the west under the continent. The marine rocks were folded, faulted, and metamorphosed (Figure P14.14), and some volcanism was added; then the system was pushed up and exposed in what is now the Olympic Range. In places in these mountains we can see pillow basalts (Figure 5.8), once extruded onto the oceans floor, but now exposed at over 5,000 ft (1,500 m) elevation. The mountains are highly glaciated and today contain 60 active glaciers.

The coastal part of the park is experiencing strong erosion as the Pacific Ocean drives against the shore. As the result of plate motion to the west, the coastal areas are being uplifted, so wave action has extensively sculpted the coastline forming sea arches, spires, and cliffs (shown in Figure 1.13).

FIGURE P14.13 Wave-cut cliffs in the western segment of Olympic National Park, Washington. (Photo by author.)

FIGURE P14.14 Tilted and folded Tertiary marine sediments now exposed high in the Olympic Range, Olympic National Park. (Photo by author.)

REDWOOD NATIONAL PARK

Redwood National Park (see Figures 1.14 and 8.4) was established in 1968. Located on the coast in northwestern California, it was designed to save 58,000 coastal redwoods from logging activities. The coastal redwoods are the world's tallest trees, the tallest one being the Harry Cole Tree. Within Redwood National Park there are three California state parks that predate the national park and contain the best stands of trees. Because logging operations were endangering these three areas, Congress agreed to place the intervening land under federal control to protect all the trees. Due to the unique character of the trees, the area has been designated a world heritage site. It is an area of high rainfall, and the park extends along the Pacific Coast, with beautiful beaches and some sea caves located in the park.

Geology. The Jurassic-Cretaceous Franciscan Formation, initially deposited in a trench to the west of the North American continent, is exposed here. These rocks were exposed during late Cretaceous time, have been hydrothermally altered, badly broken, and chemically weathered, yielding a yellowish, clay-rich, friable sediment. In early Tertiary time the coast was further uplifted during the Coastal Range Orogeny, with Tertiary marine sediments being thrust over Franciscan rocks. The Coastal Range Orogeny is still active all along the coast of California. As is the case in Olympic National Park, wave action has extensively sculpted the coastline forming sea arches, spires, and cliffs (Figure P14.15).

FIGURE P14.15
Coastal erosion of the Franciscan Formation exposed in Redwood National Park, California. (Photo by author.)

REFERENCES

BENNET, R., ed. 1980. *The new America's wonderlands, our national parks,* p. 463. Washington, D.C.: National Geographic Society.

CHRONIC, H. 1986. *Pages of stone, geology of western national parks and monuments.* Vol. 2. *Sierra Nevada, Cascades & Pacific Coast,* p. 170. Seattle: The Mountaineers.

HARRIS, A. G., and TUTTLE, E. 1990. *Geology of national parks.* 4th ed., p. 652. Dubuque, Iowa: Kendall Hunt Publishing Company.

HARRIS, D. V., and KIVER, E. P. 1985. *The geologic story of the national parks and monuments.* 4th ed., p. 464. New York: John Wiley & Sons, Inc.

HILL, M. L., ed. 1987. *Decade of North American geology, DNAG, Centennial field guide.* Vol. 1. *Cordilleran Section of the Geological Society of America,* p. 490. Boulder, Colorado: Geological Society of America.

KENNETT, J. P. 1982. *Marine geology,* p. 813. Englewood Cliffs, New Jersey: Prentice-Hall, Inc.

NEATHERY, T. L., ed. 1986. *Decade of North America geology, DNAG, Centennial field guide.* Vol. 6. *Southeastern Section of the Geological Society of America,* p. 457. Boulder, Colorado: Geological Society of America.

15

Stratigraphic Concepts and Fossils

One of the most spectacular national parks in the park system is the Grand Canyon. It has been called the "crown jewel of the national park system." It is impossible to describe its impact to those who have not seen it, even though they may have seen pictures and movies. The canyon is awesome when first seen, and remains impressive every time it is visited. This impact is due, in part, to its size, but also because of the many colors and widespread, relatively flat-lying strata exposed there.

Strata are distinct, bedded layers caused by process changes that result in different sediment being deposited or a distinct break in sedimentation. Lava flows and ash flow tuffs are also often bedded. The study of strata and the sedimentary processes responsible for forming these beds is called stratigraphy. In the Grand Canyon there are distinct cliffs formed by strata that are resistant to weathering and gentle slopes where the strata are easily weathered. In many cases these layers may contain fossils that can be used to date the deposition of the enclosing sediments. It was the study of strata and the fossils they contain that led to the development of the Geologic Time Scale (Chapter 3). Thus at the Grand Canyon, and other parks where extensive strata are exposed, segments of the Geologic Time Scale are represented—slices of time.

GRAND CANYON NATIONAL PARK

Grand Canyon National Park (see Figures 1.14 and 8.7), cut by the Colorado River, is located in northwestern Arizona. It has had a varied history under the jurisdiction of the federal government, starting as a forest reserve in 1893, then becoming a game preserve in 1906, a national monument in 1908, and finally a national park in 1919. It has been designated a world heritage site. When the park is first seen from the south rim, the flat, Permian Kaibab Limestone capping the canyon rim in the distance (Figure P15.1) is striking. The Kaibab is the youngest of the mainly Paleozoic sediments exposed in the canyon (Figure P15.2). Very deep in the canyon Precambrian sediments and metamorphic and igneous rocks are well exposed. The

FIGURE P15.1   View from the South Rim of the Grand Canyon National Park, Arizona. (Photo by author.)

Grand Canyon exists primarily due to Pliocene uplift, occurring between 5 to 10 million years B.P., causing the Colorado River to cut its way through this emerging obstacle. On the north side of the park there has been some late volcanism, occurring about 1.2 Ma.

Geology.   The geologic story of the Grand Canyon is complex. During early Precambrian time, sediments were deposited that were later metamorphosed during Proterozoic orogenesis and accompanied by igneous intrusions to form the Vishnu Schist, exposed deep in

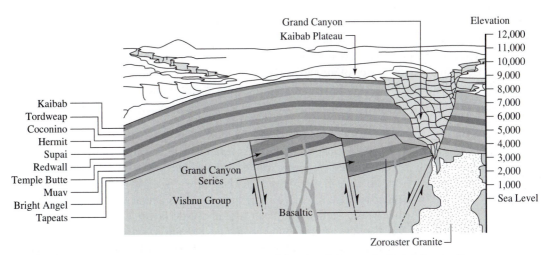

FIGURE P15.2   Very general geologic cross section of the Grand Canyon in Grand Canyon National Park. Rocks exposed in the canyon are mainly Paleozoic, with the Tapeats Sandstone being the lowest Paleozoic unit. Precambrian rocks include the Vishnu Group, metamorphic rocks; the Zoroaster Granite, some basaltic intrusions; and the Grand Canyon series of tilted sedimentary rocks. Elevations at the south rim of the canyon are greater than 7,500 ft. (Modified from a diagram by Peter Coney and Dick Beasley.)

FIGURE P15.3   Cross-bedded eolian sandstone in Grand Canyon National Park. (Photo by author.)

the canyon. The younger Zoroaster Granite, in turn, has been emplaced into the Vishnu Schist (Figure P15.2). This was followed by Precambrian volcanic activity and cycles of faulting, deposition of sediments, tilting, folding, uplift, and extensive erosion. Beginning in the Cambrian with deposition of the Tapeats Sandstone, periodic Paleozoic deposition, uplift, and erosion then occurred, followed by Mesozoic deposition, folding, faulting, and erosion. Uplift of the Colorado Plateau during the Miocene resulted in the erosion of all Mesozoic rocks from the immediate area of the canyon, leaving the Permian Kaibab Limestone exposed along the rim. And finally, further uplift and tilting during the Pliocene changed the course of the Colorado River, increasing the stream gradient and thus cutting the canyon.

A broad range of sediments are exposed throughout the canyon, and upon close examination, the layers show their often unusual character. For example, we can see cross-beds, indicative of eolian deposition (Figure P15.3), a broad range of fossils characteristic of marine limestones, and many other features. Geologists have studied (and continue to study) the canyon for many years to learn more about Earth's history.

Very early stratigraphers established that sediments are usually deposited in horizontal layers, the *principle of original horizontality,* and that younger beds were then deposited later, on top of older beds, the *principle of superposition.* If we were to find older beds on top of younger beds in a sedimentary sequence, this would indicate that the whole section had been turned upside down. Early geologists also examined the breaks between flat-lying sequences of similar strata, a layering typical of many sediments, and they named these bedding planes. Their studies showed that bedding planes represent process changes parallel to bedding, mainly the result of composition and grain-size changes.

Much more distinctive, usually major, breaks between sedimentary beds are called unconformities and represent significant changes in deposition, nondeposition, or erosion. Figure 15.1 represents the deposition of sedimentary strata and the development of unconformities. If unconformities separate flat-lying strata, the breaks are called *disconformities* and are found in many national parks such as Capitol Reef Na-

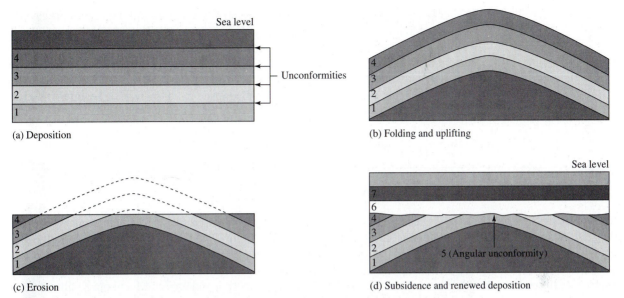

FIGURE 15.1 Deposition of sedimentary beds and the development of disconformities and angular unconformities. (a) Horizontal beds accumulate with breaks representing changes in the environment. (b) After lithification these beds are uplifted and folded. (c) Erosion then cuts the beds and (d) allows further horizontal deposition of sediments above, producing an angular unconformity.

tional Park and Colorado National Monument, known for their outstanding stratigraphy. In some cases the strata have been tilted, the beds then eroded and flat-lying sediments deposited on top. This represents a great amount of time, because the originally flat-lying sediments were first tilted, usually deep within the Earth, then uplifted and finally eroded. The boundary between the flat-lying beds above and the angular/tilted beds below is called an *angular unconformity* (Figure 15.1). The last type of unconformity, a nonconformity, represents breaks between flat-lying sediments deposited over eroded igneous or metamorphic rocks, such as those observed at the top of Mt. Moran in Grand Teton National Park (Figure 15.2; see also Chapter 7).

FIGURE 15.2 Well-known nonconformity at the top of Mt. Moran in Grand Teton National Park, where Cambrian sediments were deposited on eroded metamorphic and igneous rocks. (Photo by author.)

Changes in color, often seen in sedimentary sequences and sometimes divided by single bedding planes, can be misleading. For example, the Entrada Sandstone is dark brown, the result of hematite staining, while the Navajo Sandstone, which lies immediately below at Arches National Park (Chapter 9), is essentially identical in composition, but is light tan (see Figure P9.1). In this case the disconformity is enhanced by the color change, but clearly the color is not diagnostic as to the type of sediment making up the bed.

CAPITOL REEF NATIONAL PARK

Capitol Reef National Park (see Figures 1.18 and 8.7) was established as a national monument in 1937, then upgraded to a national park in 1971. Located in central Utah, it contains some superb stratigraphy, mainly Mesozoic rocks, but some Paleozoic rocks are also exposed in the park (Figure P15.4). Also, there has been Tertiary igneous activity in the area, with some examples of dikes and sills exposed.

The origins of the unusual name, Capitol Reef, are interesting. The word "reef," as used in this context, refers to a barrier to travel and is a term used by the first travelers in the area, who were sailors. It is the scarps (cliffs) in Mesozoic rocks that made a travel barrier (Figure P15.5). The word *capitol* derives from the many domed-shaped formations (shaped like the Capitol Building dome) that are actually due to erosion of the Navajo Sandstone exposed in the park (Fig-

ure P15.6). Many sedimentary structures can be seen in the park. Figure P15.7 shows an example from the Triassic Moenkopi Formation, where ancient, fossilized ripple marks lie next to modern ripple marks made from weathering and eroded Moenkopi sediment.

Geology. The oldest rocks exposed in the park are Permian in age, the Cutler Sandstone, seen also at Natural Bridges National Monument (Chapter 11), and the Kaibab Limestone, the capping unit along the Grand Canyon rim. These, as well as the overlying Mesozoic rocks in the park, were folded into a large, asymmetrical anticline during the late Cretaceous (Figure P15.4). In the Tertiary, igneous intrusions from the east extended into the area, and uplift, faulting and erosion have created the exposures seen in the park today.

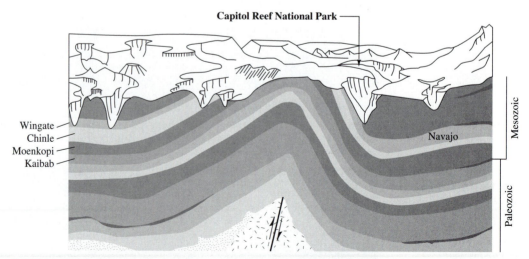

FIGURE P15.4 Very general geologic cross section at Capitol Reef National Park in Utah. Broadly folded Mesozoic and Paleozoic sediments have been eroded, and some units are exposed at the surface in the park. (Modified from a diagram by George Billingsly and William Breed.)

FIGURE P15.5
Mesozoic rocks at Capitol Reef National Park, Utah, that make up the barrier (reef) to travel. These include the Wingate Sandstone (top), Chinle Formation (center), and Moenkopi Formation (dark, bottom; see also Figure P15.4). (Photo by author.)

FIGURE P15.6 Domed exposures of the Navajo Sandstone (like the Capitol Dome in Washington, D.C.) in Capitol Reef National Park, from which the park, in part, gets its name. (Photo by author.)

FIGURE P15.7
Moenkopi Formation ripple marks next to modern ripples formed from eroded and redeposited Moenkopi sediment in Capitol Reef National Park. (Photo by author.)

 # COLORADO NATIONAL MONUMENT

Colorado National Monument (see Figure 1.18) was established in 1911 in western Colorado. It includes part of the Uncompahgre Plateau, which was uplifted in the late Tertiary, thus causing bending and faulting of the initially flat-lying stratigraphy exposed there. One result is a broad *monocline* (Figure P15.8) formed along the eastern margin of the monument. Another result is the striking erosional features, including resis-

tant mesas, deep box canyons, high scarps, and some interesting pinnacles (Figure P15.9).

Geology. Exposed at the base of the geologic section throughout the park is a nonconformity, where sedimentary rocks of the Triassic Chinle Formation are lying on eroded Precambrian metamorphics, gneisses, and schists (Figure P15.10). Most of the rocks exposed

FIGURE P15.8
Monocline on the eastern margin of Colorado National Monument, Colorado. (Photo by author.)

FIGURE P15.9 Unusual exposures of the eolian, cross-bedded Wingate Sandstone in Colorado National Monument, Colorado. (Photo by author.)

in the park are flat-lying Mesozoic sediments that are capped by the Mancos Shale, late Cretaceous in age. At the base of the Mesozoic sequence is the Chinle Formation, with most cliffs in the park being made of the Wingate Sandstone, a Triassic eolian sandstone with carbonate cement, very much like the Navajo and Entrada Sandstones. Also included in the monument are some granitic intrusions.

FIGURE P15.10 A major nonconformity exists in Colorado National Monument, Colorado, where an erosional unconformity lies between Triassic sediments above and eroded Precambrian metamorphic rocks below. (Photo by author.)

SEDIMENTARY STRUCTURES

Besides the very broad-scale features in stratigraphic sequences, such as unconformities, there are many fine-scale structures that give an indication of how and in which environment the sediment formed. Such features as *graded bedding,* with uniform changes in grain size, fine grains at the top, coarse at the bottom, may indicate that deposition occurred from turbidity currents. Geologists can then look for a source of the emplacement of the beds as well as understand its mechanism. Cross beds often indicate dunes, and so forth.

Ripple marks, produced by wave action, are often found on the tops of sedimentary layers (Figure P15.7), as are *mud cracks,* produced by drying. Both examples usually indicate very shallow water depths. In the first, the water was moving fast enough to form the ripples. In the second, the system dried out, allowing mud cracks to form. Surface impressions made by trails, tracks, rain drops, burrows, etc. are also found on the tops of beds. Sometimes the bottom of sedimentary beds also has marks where an object has pushed its way through the bed and its impression remains. These are called *sole marks* and are formed as casts filling these surface impressions.

Another very important sedimentary structure or texture is that of *mottled bedding,* usually resulting from disruption or burrowing by biological organisms, called *bioturbation.* Such disruption indicates active biological invasion into the sediment during feeding by organisms and is recognized mainly in oxygenated environments. Where bioturbation has occurred, primary sedimentary features may have been destroyed.

Unusual Sedimentary Features

There are some very unusual features found in strata that can be very beautiful and often valuable. Because of their appearance, some of these features are cut and sold as home decorations or included in museum exhibits, and gem and mineral collectors spend a lot of time looking for them. Sedimentary nodules, for example, are unusual mineral concentrations. These can be of some value as display items if they are large and/or interesting in appearance. *Concretions* are localized concentrations of cemented sediment that may have additional mineral concentrations or unusual cementation patterns within. Some of these are remarkable when cut and polished. *Geodes* are hollow, semi-spherical rocks that have internal crystal concentrations, often of the mineral quartz or calcite, but sometimes of other, more exotic minerals.

FACIES AND SEA LEVEL CHANGES

As in metamorphic rock, the term *facies* is also applied to sedimentary rocks to indicate unique appearance and characteristics. Thus, sedimentary facies and *facies changes* characterize a rock that represents a local environment and/or environmental changes within a rock, for example, shallow versus deep water facies. One critical aspect of depositional environments is water depth. Along shorelines or in shallow marine environments, apparent sea-level changes may be caused by tectonic factors that are independent of global sea-level variations. At any one locality, sea level may appear to fall, a change called *regression,* where the shoreline migrates seaward with the corresponding facies changes also prograding or growing seaward, the newly deposited sediments covering those already accumulated. The opposite effect, *transgression,* is used to characterize apparent sea-level rises and migration of shorelines landward, with the sedimentary facies, in turn, shifting landward.

NOMENCLATURE

As in every area of endeavor, in order to adequately communicate in the most effective manner, it is necessary to understand the specific nomenclature used. In stratigraphy geologists are dealing with the concept of geologic time, as well as the rocks that contain the geologic story. Sometimes the rocks can be dated, sometimes they cannot, and sometimes geologists are only interested in geologic time independent of the rock. Therefore, several sets of names have evolved to deal with these different concepts. Three of these are used in geology most often today. *Time units* are based on *biostratigraphic* names, that is, they are based on the fossils identified as living at a given time and are independent of rocks. *Time-rock units* are given as *chronostratigraphic names,* where the rocks are correlated to ages. These are shown in Table 15.1. There are also simply *rock units,* where *rock-stratigraphic names* are used that are independent of time estimates. Of the rock units, geologic *formations* are the most useful nomenclature. Formations are mapable rock units that have definite boundaries (contacts) and include some unique, identifying characteristics.

TABLE 15.1   STRATIGRAPHIC NOMENCLATURE

Time Units	Chronostratigraphic Units	Rock Units
Era	Erathem	Supergroup
Period	System	Group
Epoch	Series	Formation
Age	Stage	Member
—	Zone	Bed

Boundary stratotypes, "ideal" stratigraphic sections that are globally unique and characterize system or other chronostratigraphic boundaries (terms used most often by working biostratigraphers), are very useful when dealing with rock ages. These stratotypes have been established all over the world by the international geological community of researchers so that anyone can examine those rocks that, theoretically, best represent the time of interest.

FOSSILS, RECORDS OF ANCIENT LIFE

Fossils are records or traces of ancient life that are found mainly in sedimentary rocks. The most spectacular examples are the large dinosaurs such as the *Tyrannosaurus rex* pictured in Figure 15.3, but there are also delicate fossils, like jellyfish, found in some shales. In very special circumstances fossils may also be found in some lavas and ashflows. For example, when a lava flows around an ancient tree and cools and the tree leaves an impression in the lava or a hole in the lava where the tree once stood, this is a fossil. Usually only the hard parts of organisms are preserved, and that happens only in special situations, such as in areas where fast burial occurs in relatively low-energy (e.g., no waves) environments.

The word "ancient" in this case means older than 10,000 years B.P., the beginning of the Holocene Epoch, a date arbitrarily used to denote the beginning of the historical record of human history. Many geologists look at modern processes, but usually any evidence of anything living earlier than 10,000 years B.P. is considered a fossil.

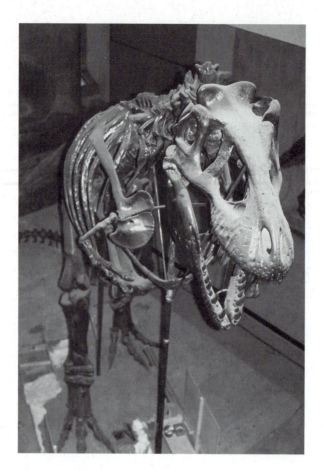

FIGURE 15.3
Tyrannosaurus rex at the Carnegie Museum, Pittsburgh, Pennsylvania. (Photo by author.)

In the National Park System there are eight park areas that are important fossil localities. These are:

- Petrified Forest National Park, famous for abundant and beautiful petrified logs;
- Badlands National Park (Chapter 9), famous for important fossil mammals, including giant pigs that lived from 37 to 23 million years B.P.;
- Dinosaur National Park, famous for a large number of Mesozoic dinosaur fossil discoveries;
- Fossil Butte National Monument, famous for 50 million-year-old fossil fish from the Green River Formation, which are mined outside the monument and sold all over the world;
- Florissant Fossil Beds National Monument (Chapter 10), famous for 1,200 species of Eocene/Oligocene insects;
- Agate Fossil Beds National Monument, famous for 20 million-year-old mammal fossils, including extinct species of rhinoceros;
- John Day Fossil Beds National Monument, famous for a diverse and changing fossil fauna, animals living from 50 to 5 million years B.P. as the climate changed from forests to grasslands and deteriorated during the Tertiary; and
- Hagerman Fossil Beds National Monument, famous for 3 million-year-old mammal fossils, including 150 specimens of the extinct mastodons as well as the Hagerman horse.

These parks have yielded fossil plants, insects, reptiles, dinosaurs, birds, and mammals that cover the development and evolution of life throughout the Mesozoic and Cenozoic Eras, and are places where active research results in new discoveries. Our understanding of many fossil forms has come from these national park areas.

Fossils are defined as natural remains or evidence of ancient organisms. Evidence of an organism includes *casts* and *molds*; *trace fossils*—traces of the animals passing, such as burrows, tracks (footprints), and trails; *stomach stones* used by animals such as chickens and dinosaurs to grind up food in their gizzards; *coprolites,* fossilized fecal material; and other evidence. *Fossilization* usually takes place only in an organism's hard parts. Bones and especially teeth are are easily preserved, although this requires immediate burial in low-energy environments. Helping the preservation process is petrifaction, the result of circulating mineral-bearing waters, and the more general process, *mineralization,* a term denoting the replacement of organic material by some mineral or combination of minerals.

Petrifaction means to turn to stone. The process involves mineralization, where replacement, *recrystallization,* or *permineralization* (material added to pore spaces within the original) of materials, organic or inorganic, can involve many different minerals. In most instances, however, organic materials are replaced either by SiO_2, and therefore *silicified,* or replaced by $CaCO_3$, and therefore *calcified.* Another interesting process that is responsible for fossilization is called *carbonization.* In this case, distillation of organic carbon, under the high pressures generated by deep burial, causes the organic material to be absorbed by rocks and leaves a carbon print of the organism, usually plants, on the rock. These carbon impressions or outlines are found on the bedding planes in shale beds (see Florissant Fossil Beds National Monument, Chapter 10 and Figure P10.5, and Fossil Butte National Monument, this chapter). In other, relatively rare instances, preservation and fossilization have resulted from freezing or the mummifying of organisms. Woolly mammoths and ancient humans have been found frozen in glaciers, and ancient humans and other organisms have been found mummified, including the remains of one human found in Mammoth Cave National Park (Chapter 11).

PETRIFIED FOREST NATIONAL PARK

Petrified Forest National Park (see Figures 1.14 and 8.7) in northeastern Arizona was proclaimed as a national monument in 1906 and became a national park in 1962. Its northern end contains the Painted Desert, a multicolored area of badlands. The park is dominated by the Triassic Chinle Formation, with its multicolored shales and petrified logs (Figure P15.11). The southern end of the park contains many petrified logs, as well as some ruins left by Native American communities. In fact, many areas in the Southwest contain Native American artifacts. Some of these areas will be discussed in Chapter 16.

Geology. The logs preserved in the park are of conifers now extinct. During the Tertiary, pyroclastic volcanism covered the area, killing, burning, and quickly burying trees. Fine sand, silt, and clay was washed over the area, adding their volume to rapidly bury more trees. These sediments were further covered by volcanic ash-flow tuffs. Fluids circulating in the sediments covered by the tuffs replaced the wood fragments with silica, silicifying them. This involved permineralization and some replacement, thus petrifying (turning into stone) the logs. The beautiful colors exhibited by the petrified logs are the result of the presence of slight amounts of iron and manganese. In the Cretaceous, the area was covered by seas that deposited thick sequences of marine shales and sandstones. Then during the Cenozoic, a complex period of uplift, volcanism, erosion, lacustrine sedimentation, and more erosion has produced the features seen in the park today.

FIGURE P15.11
Petrified logs in Petrified
Forest National Park,
Arizona. (Photo by
author.)

 FOSSIL BUTTE NATIONAL MONUMENT

Located in the desert country of southwestern Wyoming, Fossil Butte National Monument, established in 1972, contains sediments deposited in a freshwater lake, Fossil Lake, that once existed in the region 50 million years B.P. Preserved in the lake were millions of fish fossils (Figure P15.12), including gars and stingrays, as well as plants, insects, turtles, and some mammals. Fossil Butte (Figure P15.13) stands above the surrounding valley and is the result of erosion after Pliocene-Holocene uplift in the area. Sediments outside the monument are being quarried for the well-known fish fossils that are sold all over the world.

Geology. Thrust faulting and uplift in western Wyoming during the Cretaceous and Tertiary Periods represent the strong tectonic activity along the Idaho-Wyoming Thrust Belt, which built the topography in the area. During this period a number of isolated, freshwater lakes developed in the region, in which abundant flora and fauna flourished. The monument contains an abandoned quarry that exposes the Green River Formation, the important stratigraphic unit containing the fossils from Fossil Lake. Included in this section are finely laminated shales and thin beds of ash-flow tuff. Also exposed in the park are finely laminated limestones containing kerogen. These beds are called oil

shales, and someday, when the price of oil is much higher than it is today, they will be an important source of oil. The laminations and fish kills, represented by the abundant well-preserved fossils, were caused by changes in the environment of the lake. Geologists have speculated that similar to the Great Salt Lake today, the ancient lake at Fossil Butte National Monument may have fluctuated from fresh water at the margins, due to seasonal rain input, to salt water in the interior. The massive fish kills may have been the result of changes in salinity. It is also possible that, like Lake Ontario in New York, where summer high temperatures cause massive fish kills, heating of the shallow Fossil Lake during the summer months may have been responsible for some of the fish kills.

Also exposed in the monument are Wasatch Formation rocks (Figure P15.13), the Eocene unit from which the Fairyland Canyon hoodoos in Bryce Canyon National Park, Utah (Chapter 9), are formed. At Fossil Butte, the fluvial Wasatch rocks are often red or purple, representing abundant iron with oxidation to hematite and are usually found as deltas, interlayered with the lake sediments. Wasatch rocks contain unique mammal fossils, including the remains of primitive horses and reptiles.

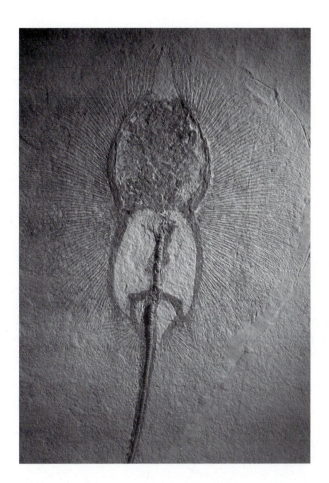

FIGURE P15.12 A fossilized stingray in the Fossil Butte National Monument Visitors' Center. (Photo by author.)

FIGURE P15.13 Fossil Butte at Fossil Butte National Monument, Wyoming. The low eroded hills in the foreground are rocks of the clastic Wasatch Formation. (Photo by author.)

Fossils have many uses in geology. They are important in reconstructing ancient environments that existed where and when the animals or plants lived and were deposited. For example, if the rocks under study are limestones, containing fish skeletons or shells, these fossils usually indicate a marine environment, although in two important examples in the national park system freshwater fossil fish have been identified, such as those found at Fossil Butte National Monument and Florissant Fossil Beds National Monument (Chapter 10). By studying living relatives of these fossil organisms it is sometimes possible to predict the water depth at which they lived, as well as the water temperature, salinity, and other factors in their environment. If the fossils happen to be land dinosaurs, land mammals, trees, etc., then they were fossilized in terrestrial environments, such as those fossils found at Dinosaur National Monument and Agate Fossil Beds National Monument. Coal found in swamps usually indicates mild climates, and there are many other environmental factors that can be determined from fossils. Fossils are also useful in correlating rock sequences from place to place. Such analyses rely on guide or index fossils, unique organisms that live for only a short time, but that are abundant and easily recognized in the rock record.

It is the distinctive groups of organisms that have allowed us to develop the Geologic Time Scale. The concept of *faunal succession* or organic evolution, where fossil groups (assemblages) succeed one another, allows us to recognize characteristic groups that existed for a specific period of time and then disappeared. Groups, rather than individuals, allow better time-period identification because there is a better chance that at least some individuals in the group will be preserved in the fossil record, thus allowing the group and therefore the time to be identified.

DINOSAUR NATIONAL MONUMENT

Dinosaur National Monument (see Figure 1.18) was established in 1915 by President Woodrow Wilson. Located in Colorado and Utah in the east-west trending Uinta Mountains, it contains the site of some of the most famous dinosaur fossil discoveries in the world.

The visitors' center at the monument is actually a working laboratory where bones are studied and excavated (Figure P15.14). In some cases, whole skeletons are preserved in the rock, and a few of these are on display (Figure P15.15). One of the most famous dinosaur fossils to be discovered at Dinosaur National Monument was the *"Brontosaurus,"* illustrated in Figure P15.16. The image we had of this dinosaur was based on the *Camarasaurus*-like (Figure P15.15) head in Figure P15.17, but late in the 1970s it was discovered that the wrong head was hanging on the specimen on display in the Carnegie Museum! The head (Figure P15.17) was replaced with the "correct" head (Figure P15.18), changing the dinosaur. In fact, there is no *"Brontosaurus!"* These problems confront paleontologists when they try to reconstruct fossil organisms from the limited fossils available (Figure P15.14). Most rock exposures and the fossils in them are poor, and many bones, pieces in the puzzle, are missing, so the problem of reconstruction and identification can be very difficult. Remember, no living organism now exists on Earth to tell scientists how these animals were put together!

Box canyons are abundant in the monument, and in some, ranchers corralled their cattle (Figure P15.19). Beautiful vistas abound, where modern ranchers still graze their cattle (Figure P15.20). Areas containing early Native American petroglyphs, or rock paintings, and surrounding ranch lands have been added to the monument since it was founded, to make it more of a recreation area. Also added have been many miles of hiking trails, camping areas, and boating accessibility.

Geology. Forming a large asymmetrical anticline, the Uinta Mountains at this locality contain Morrison Formation rocks, from which many unique, some one of a kind Jurassic dinosaurs have been recovered. During the Jurassic this was an area of low-lying swamp that was highly vegetated and that provided food for

FIGURE P15.14
Researcher excavating dinosaur bones inside the visitors' center at Dinosaur National Monument. (Photo courtesy of the National Park Service.)

FIGURE P15.15
Complete dinosaur (*Camarasaurus*) skeleton on display inside the visitors' center at Dinosaur National Monument. (Photo by author.)

FIGURE P15.16 Artist's conception of a "*Brontosaurus,*" based on fossils recovered at Dinosaur National Monument.

the dinosaurs. Also accumulated were very fine muds and sands, deposited in a low-energy backswamp environment. Dinosaurs roamed, fed on the plants and on each other, died, and were preserved here. Preservation often occurred in point bars where the bones were rapidly buried by streams. Because many of the bones are very large, dinosaurs are more readily preserved than are more delicate organisms.

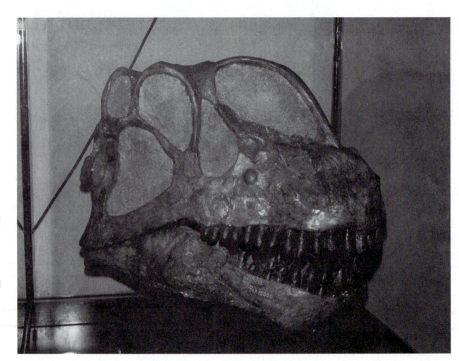

FIGURE P15.17 Wrong head placed on the body of the "*Brontosaurus*" (now removed) giving the incorrect impression that the organism looked like Figure P15.16. These fossils are located at the Carnegie Museum in Pittsburgh, Pennsylvania. (Photo by author.)

FIGURE P15.18 Correct head now hanging on the *Apatosaurus,* the correct name for the dinosaur often incorrectly called *"Brontosaurus."* The old head (Figure P15.17) is in the case at the upper left. Fossils at the Carnegie Museum in Pittsburgh, Pennsylvania. (Photo by author.)

FIGURE P15.19 Box canyon, part of old historic farm at Dinosaur National Monument. (Photo by author.)

FIGURE P15.20 Modern ranch and eroded mountains within Dinosaur National Monument. (Photo by author.)

AGATE FOSSIL BEDS NATIONAL MONUMENT

Named for the agate exposures in the area, Agate Fossil Beds National Monument is another important fossil locality, producing a rich range of mammal fossils, including rhinoceros, early dogs, giant pigs, and several other unusual Miocene mammals. Authorized as a national monument in 1965, it is located in northwestern Nebraska and contains fossils that are about 19 million years old, the fossils resulting from animals living and dying in an ancient marshy grassland or savanna. Quick burial, perhaps as the result of being trapped in bogs or quicksand, protected the skeletons from being destroyed. Today, erosion by the Niobrara River has formed the small hills that contain the fossils. Quarrying since the 1800s has resulted in many fossils being removed from a thin, approximately 1-meter-thick, horizontal sandstone bed, but in some areas rhinoceros bones have been left in place and protected so that visitors can see how abundant the fossil-containing layers are (Figure P15.21). Research by paleontologists, currently working at Agate Fossil Beds National Monument, provides important information concerning Tertiary life in North America.

FIGURE P15.21 Extinct rhinoceros bones exposed in place and on display at Agate Fossil Beds National Monument. (Photo by author.)

PALEONTOLOGY

Paleontology is the study of fossils. It is a study area that is broken down into many specialties, because the record of life on Earth is so vast and complex. In very general terms there are paleontologists who specialize only in the study of *invertebrate organisms,* organisms with no backbone, such as scallops or squids. Others study only *vertebrate organisms,* those containing a backbone, such as squirrels or dinosaurs. There are many different subdisciplines in these areas. For instance, a person studying squirrels probably wouldn't be studying dinosaurs or fish. One of the important subdisciplines is *micropaleontology,* the study of organisms only identifiable using a microscope. Some geologists study fossil plant remains, the area of *paleobotany,* or one of its subdisciplines, such as *palynology,* the study of spores and pollen.

The modern emphasis in paleontology is to look at fossils from an ecological and environmental viewpoint, incorporating disciplines such as stratigraphy and statistics. As a result, some paleontologists now prefer to be called *paleoecologists* or *biostratigraphers*. In the past, paleontology was more descriptive than quantitative, but today some disciplines are very quantitative, for example, *biometrics,* the area of paleobiological statistics.

CLASSIFICATION OF FOSSIL ORGANISMS

Classification of fossils is a complicated process because we must rely on evidence in the rock, and the rock record is not complete. Biologists are able to classify organisms as *species,* based on studying living organisms; paleontologists also define individual species, but they work at a distinct disadvantage. The species concept relies on the fact that organisms of different species cannot interbreed to produce viable offspring (an offspring that can in turn produce offspring). But the species concept becomes an enigma if the organism is extinct. Therefore, paleontologists are forced to use *morphological traits,* shape, size, and proportions of organisms to classify them by species. On the face of it this doesn't seem too difficult, especially if a scientist is comparing very different organisms, such as a squirrel, a horse, or a dinosaur. But what happens if a paleontologist studies something that looks like a squirrel, but it stands 6 feet tall at the shoulders. Some might say, "Now that's a squirrel!" But others might say, "No it's not, it's a different species!" How do scientists determine the species, since the breeding potential of two fossil organisms cannot be tested. This brings up one of the many questions that paleontologists face. It has been called the debate between the "lumpers" (those who lump everything together as one, regardless of size) versus the "splitters" (those who see a new species in every adult population with only a small size difference). The lumper versus splitter viewpoint can be found in all areas of study, but in paleontology the size issue is a knotty one to solve.

Take the example of modern-day humans versus those living in England at the time of William the Conqueror, around 1066 A.D. Knights, the elite soldiers of the period, were significantly smaller than people living today. Their suits of armor are smaller than are many modern 12-year-old children. If skeletons of knights were found in one stratigraphic level and modern human skeletons in another, a splitter might tend to classify them as two different species. But if we found fossil human skeletons that were 1 ft (0.5 m) tall, would they represent a different species? The lumpers would argue that they do not. It has been clear to paleontologists that additional morphological information is needed, structural changes in bones for example, to help us finely differentiate one species from another.

Other examples of problem areas in identifying species differences are abundant. For example, there are some modern organisms that create shells that coil to the right as they grow at one temperature, and when it's colder they coil to the left, but in every other respect they appear to be the same organism. If they were found as fossils, would they be considered different species? Shape, as well as size and differing proportions, are also factors. One of the major difficulties in trying to define species is that many of them were never fossilized, or at least their remains have never been found. This makes comparisons very difficult, and points to the "missing links" problem. When only a few organisms in a fossil lineage are preserved and comparisons are made, evolutionary changes from one species to the next appear to be major. However, when many fossils in a lineage are preserved the changes appear to be minor because the very slight changes from one species to the next become available for study. As we find more and more fossils, we begin to fill in the gaps in our knowledge concerning ancient life on Earth, and the unusual often can be explained by the next discovery.

THE CONCEPT OF EVOLUTION

One of *Charles Darwin*'s major contributions to earth sciences was his book, *On the Origin of Species,* published in 1859. He argued that slow, constant change was occurring in organisms that originated by evolving from less complex forms and that this change was continuous, with no past discontinuities. He further argued that survival was based on advantage, the principle of *natural selection,* driven by the instinct for self- or race preservation. The challenge facing organisms was that of obtaining food; one important response was not to become food. Those animals that were able to meet these challenges and responses in the most effective manner survived, and any useful mutation that helped was incorporated as a change in the gene pool.

The production of a new species is the result of spontaneous mutation of chromosomes that dictate what an organism will be like. It has been argued that there are two ways that this can happen, one as the result of constant but very slow change (as proposed by Darwin), or the second, as the result of rapid, unusual mutations, a process called *punctuated change,* proposed by Stephen Gould at Harvard University. Both processes are driven by competition for food, resulting in an organism's physical adaptation to the environment, and requires an organism's isolation from the gene pool, through migration, *tectonic isolation,* or *biological isolation.* Biological isolation results when an organism is so different that it becomes socially isolated from the main breeding gene pool, perhaps by that individual's choice or by the choice of the rest of the population. Modern scientists believe that punctuated evolution probably better explains the changes we see, but "survival of the fittest" (natural selection) still drives evolution.

The concept of evolution is based on *anatomical* evidence and includes *homology,* the study of structures that have the same ancestral origin but that serve different functions, such as a flying squirrel's "wings." Other evidence includes *vestigial organs* or appendages that are relatively small and lacking in complexity. Another example is a part of an organism that has no function, such as a whale's pelvis, which is similar to a functioning pelvis in other organisms. Our understanding of evolutionary changes in organisms comes from important fossil discoveries from localities around the world, including those localities supervised by the National Park Service.

Generally, when we look at organisms there are two evolutionary trends, one that involves changes in the gene pool resulting in *phylogeny,* where new species arise from older species, and one that involves changes in single individuals during their lifetimes, or *ontogeny.* Phylogenetic changes throughout Earth's history have resulted in greater complexity in organisms and generally increasing body size. It has now become clear that most changes are due to rapid speciation and that evolutionary changes are not reversible.

Organisms are constantly migrating into environments that they find favorable, and this slow spread is facilitated by land bridges between continents, such as the Isthmus of Panama and its closure approximately 3 million years B.P. that connected North and South America. When the isthmus closed, extremely hardy North American animals crossed to South America and drove many of the less hardy organisms living there to extinction. The removal of such a land bridge will have the effect of creating two populations that can no longer interbreed. A species on one continent may go to extinction, while the same species on another continent may continue to exist.

Besides migration, organisms are distributed by *dispersion.* This can result from atmospheric processes, such as wind dispersal, or from other factors, being carried by organisms such as birds, for instance. Oceanic dispersal may be accomplished by floating organisms, such as the larvae stages in marine life cycles, or as the result of attachments to floats.

The factors driving species formation are mainly environmental. Climate is the most important of these, the primary factors being temperature and moisture availability. *Paleoclimatology* examines past changes in climate. An interesting result of climate and climatic change is the isolation of organisms from environments. For example, deserts and jungles act as barriers to migration. Another factor controlling speciation involves sea-level changes that affect both marine and terrestrial organisms. Transgression of marine waters over land can create barriers to migration, while regression opens pathways. Such a pathway across the Bering Straits during the last major glaciation between Alaska and Siberia may be how ancient humans arrived in North America. Kobuk Valley National Park (Chapter 16) was established to protect sites that may yield information concerning such a migration. Glaciation affects speciation by cooling the climate, and glaciers create barriers to migration and cause isolation. Orogenic events also create barriers against migration, as do lakes and rivers. Such barriers can become factors in speciation through isolation of groups and their gene pools, thus allowing speciation to occur.

As species adjust to new environments, they tend to become specialized. This trend is called *adaptive radiation* and can cause a species to diverge from a common ancestor, a trend known as *divergence,* or the species may show that similar forms are produced in different species, a process known as *convergence.* Usually these trends are controlled by the available environment, a good example being the many varieties of plants that have taken on the tree form to better compete for light. Organisms tend to adapt to better use the resources available to them. For example, animals that live primarily in water, such as fish, whales, and penguins, are very different species, but they all have developed the ability to swim much more efficiently than have land-based organisms. Some varieties have adapted to mimic an already successful organism. There are many examples, but one from the southeastern United States is interesting, that of the hognose snake. When threatened, it flattens its head into a hood, like that of a cobra, and spits and threatens to strike. It puts on quite a frightening display, but the snake is harmless. Locally the snake is called a puff adder because of the display; this snake should not be confused with the very deadly poisonous puff adder found in Africa.

EXTINCTION

Extinction involves species disappearance, and *mass extinctions,* identified in the rock record, should be of great concern to all of us because they represent events where large groups of organisms became extinct all at once. Such mass extinctions are the basis for defining the major time boundaries in the Geologic Time Scale, and they are global in scale. So what happened to kill off all these organisms? To answer that question, we need to look at how organisms manage to survive.

There are certain reactions that promote survival. These reactions rely on the fact that physical changes that impact organisms are usually gradual, and it is some physical change or changes that ultimately drives organisms to extinction. So far the Earth has been organically dynamic, that is, it never kills everything all at once, and those slow physical changes that are severe drive relatively standard survival reactions in species. For example, during times of extreme stress, species tend to produce excess offspring and to develop a means for dispersing those offspring.

Large-scale, but slowly changing factors contributing to mass extinctions are many. Plate movements cause changes in climate, isolation of populations, and formation of land bridges that bring populations together. *Food chain* (food webs) disruptions can occur, where one critical species is eliminated through excess predation, climatic changes, or other factors. A population's size can become too small to withstand problems of disease or predation, such as the population of the California condor. Low variability (low *genetic diversity*) does not allow organisms to readily adapt to changes in the physical environment. Examples can be found in some of the crops that are grown by humans, which have become too genetically specialized. Narrow adaptation, the result of overspecialization, can result in extinction if the physical environment changes. Isolation can, in turn, add to these problems. Competition for food and unrestrained predation can kill off species. The passenger pigeon (Figure 15.4), in the late 1800s numbering many millions of birds, was killed off by humans before 1925 as the result of unrestrained predation. The American buffalo nearly went the same way. Disease can also kill off a species. Dutch elm disease has almost completely eliminated one of the most beautiful trees growing in North America.

Extreme, rapid changes in physical environment can quickly cause mass extinctions. Primary among these are large atmospheric effects that rapidly change climate, such as major volcanic eruptions or bolide impacts that pump large amounts of dust into the atmosphere, reflecting solar radiation back into space or limiting solar radiation penetration through the atmosphere. It has been hypothesized that the dinosaurs

FIGURE 15.4 In the early 20th century there were millions of passenger pigeons. By 1925 they were extinct.

became extinct as the result of a bolide impact at the end of the Cretaceous Period. The location of the impact is still unknown, but one locality suggested as a possible impact site is southern Mexico (Figure 15.5). What will happen to us if a large meteorite hits the Earth in the near future? Geologists are studying impact events to try to answer this question. It has been argued by some geologists that the feature in Canyonlands National Park known as an upheaval dome (Figure P11.3), was created by one of a series of bolides that hit Earth at the end of the Cretaceous and killed off the dinosaurs, as well as many other animals living on Earth at that time.

Changes in atmospheric gases, such as carbon dioxide and other gases, can heat the atmosphere by trapping solar radiation. Changes in solar radiation (the Sun heating or cooling) can also greatly affect climate, and can, in turn, lead to the buildup of ice on continents and further alter climate. Another rapid change that has an effect on organisms is a lethal increase in chemicals being brought into the environment, the result of geological or manufacturing processes. Nutrient depletion in the oceans can cause reduced *phytoplankton* (microscopic surface-water plants, important in producing oxygen and food) production, the result of a lack of upwelling due to water mass stability or tectonic stability and reduced runoff. The end result is that a critical part of the food chain, the main building block on which other organisms feed and on which their survival depends, would be gone.

Another question that has been asked is, "When the Earth's magnetic field polarity reverses, does it cause extinctions?" The question is asked because the magnetic field of the Earth supports the Van Allen radiation belts that nearly surround the Earth (Figure 3.10). These belts are made up of highly charged protons and electrons, constituents of the Solar Wind, that are streaming toward the Earth from the Sun; the belts protect the Earth from radiation. It has been argued by some that when the magnetic field reverses, the Van Allen radiation belts would collapse or partially collapse, thus allowing highly charged particles to reach the Earth's surface and destroy life. There is some evidence in favor of such a hypothesis, but most evidence is against it. When the next reversal occurs, scientists will have an opportunity to directly test the hypothesis.

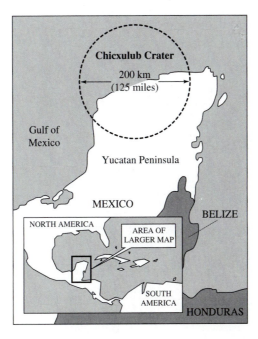

FIGURE 15.5 Location of a bolide impact in Mexico that may have been responsible for the extinction of the dinosaurs and many other organisms at the end of the Cretaceous Period.

REFERENCES

BENNET, R., ed. 1980. *The new America's wonderlands, our national parks,* p. 463. Washington, D.C.: National Geographic Society.

BEUS, S. S., ed. 1987. *Decade of North American geology, DNAG, Centennial field guide.* Vol. 2. *Rocky Mountain Section of the Geological Society of America,* p. 475. Boulder, Colorado: Geological Society of America.

BOGGS, S. 1994. *Principles of sedimentology and stratigraphy.* 2nd ed, p. 800. New York: Macmillan College Publishing Company.

CHRONIC, H. 1984. *Pages of stone, geology of western national parks and monuments.* Vol. 1. *Rocky Mountains and Western Great Plains,* p. 168. Seattle: The Mountaineers.

CHRONIC, H. 1988. *Pages of stone, geology of western national parks and monuments.* Vol. 4. *Grand Canyon and the Plateau Country,* p. 158. Seattle: The Mountaineers.

HARRIS, A. G., and TUTTLE, E. 1990. *Geology of national parks.* 4th ed., p. 652. Dubuque, Iowa: Kendall Hunt Publishing Company.

HARRIS, D. V., and KIVER, E. P. 1985. *The geologic story of the national parks and monuments.* 4th ed., p. 464. New York: John Wiley & Sons, Inc.

ROBISON, R. A., and TEICHERT, C., eds. 1979. *Treatise on invertebrate paleontology.* Part A. *Introduction,* p. 569. Boulder, Colorado: The Geological Society of America; and Lawrence, Kansas: The University of Kansas.

16

Cultural Areas

A large number of national park areas are devoted to history or prehistory, which, in this book are called "cultural areas." Included in this chapter are a number of Native American sites of cultural interest within the park system, situated primarily in the southwestern United States. Also included in this chapter are forts and military parks, located primarily along the eastern seaboard. These range in age from battlefields of the earliest colonial days to forts active through World War II. We have already encountered a few cultural area examples, including the Native American shell mounds built by the Timucuan people in Canaveral National Seashore (Chapter 14), effigy burial mounds built by the Woodland peoples at Effigy Mounds National Monument (Figure 11.3), and Fort Jefferson on Garden Key in Dry Tortugas National Park (Chapter 10). The geologic setting was important in determining where each of these cultures settled.

ANASAZI

The Anasazi lived in the southwestern United States, developing a flourishing culture until their disappearance around 1600 A.D. The word *Anasazi* comes from the Navajo language and has a variety of meanings, "the ancient enemies," "foreigners," and "those who came before." They were prehistoric agricultural people who raised maize, beans, and squash, and who domesticated dogs and turkeys.

The Anasazi primarily lived in southern Colorado, Utah, much of New Mexico, and eastern Arizona (Figure 16.1). They lived from A.D. 1–1600 and developed into five separate cultures that are classified based upon locality. The Chaco Anasazi in central New Mexico, identified from A.D. 500–1300, seem to have built the largest interconnected community (see Chaco Culture National Historic Park and Aztec Ruins National Monument). The Mesa Verde Anasazi culture in southwestern Colorado and southeastern Utah, also identified from A.D. 500–1300, are more well known to most Americans because of Mesa Verde National Park (see Mesa Verde National Park and Hovenweep National Monument). A third group, the Little Colorado Anasazi, located along

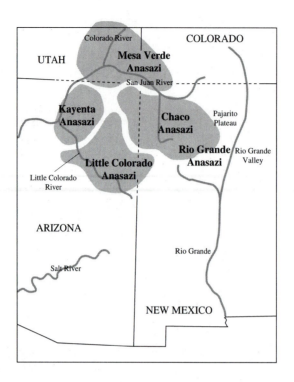

FIGURE 16.1 Map of Anasazi cultural areas.

the Little Colorado River, developed later and lasted longer, from A.D. 850 to 1500. The fourth group, the Kayenta Anasazi in eastern Arizona, lived at the same time as the Chaco and Mesa Verde groups, A.D. 500–1300, but their cultural influence was not as extensive. Last, the Rio Grande Anasazi in northern New Mexico lasted the longest as a distinct culture, from A.D. 600–1600. They are best represented at Bandelier National Monument.

Besides locality classifications, the Anasazi have been further divided by cultural classifications. These begin with the Basketmaker II time period, A.D. 1–450, whose people were primarily hunter-gatherers who commonly constructed pit houses. By Basketmaker III time, A.D. 450–750, the Anasazi had added agriculture to hunting and gathering and began to build ridge-top or mesa-top houses. During Pueblo I time, A.D. 750–900, Anasazi houses were bigger and more extensive, and they had given up hunting and gathering for agriculture. Then in Pueblo II time, A.D. 900–1100, a few large villages were built (there are excellent examples in Chaco Culture National Historic Park), and they began living in surface rather than subsurface rooms. Finally, during Pueblo III time, A.D. 1100–1300, the Anasazi began to move off mesa tops into canyons, possibly due to defense considerations, suggesting that they were being attacked by other groups.

Because the Anasazi are so well studied, it is possible to classify their culture in detail. Archaeologists also classify Native American cultures in the Southwest, based on cultural changes, using a *time* classification system. The first of these, the Paleo-Indian Period (?14,000–5500 B.C.), includes the earliest Native Americans in North America, the oldest being the Clovis Culture, a people who were big-game hunters who lived when the climate was cold and probably relatively dry. The following Archaic Period (5500 B.C.–A.D. 600) saw warmer temperatures, small-game hunters and gatherers, and the beginnings of the Anasazi culture. During the Developmental Period

(A.D. 600–1175) agriculture became well established, as did the community. But during the Coalition Period (A.D. 1175–1325) climate deteriorated and communities began to break up. During this time there was a late influx of people who brought new pottery styles. The Classical Period (A.D. 1325–1600) saw communities being re-established, but they were not as elaborate as those of the Developmental Period. The Classical Period ended in 1598 when Spanish colonists led by Don Juan de Onate arrived in New Mexico. This was the beginning of the Historic Period (A.D. 1600–present), so called because historical records exist. But more importantly, the Historic Period marks a time of confrontation between Native Americans and Europeans.

CHACO CULTURE NATIONAL HISTORIC PARK

Chaco Culture National Historic Park (see Figure 1.17) was proclaimed as Chaco Canyon National Monument in 1907. Located in central New Mexico, it was renamed in 1980 and designated as a world heritage site in 1987. Chaco Canyon was the main center of the Chaco Anasazi culture (A.D. 500–1300), which flowered between A.D. 900–1150. Cultural decay and slow social disintegration followed climate deterioration because survival conditions became too difficult.

At its zenith, the Chaco Anasazi had a network of large towns, containing a total population of 2,000–5,000 people. Excavation of the largest town, Pueblo Bonito (Figure P16.1 a, b), began in 1896, and it is now well exposed. Containing 34 kivas (circular ceremonial areas), more than 600 rooms, and walls up to four stories high, the city was built from sandstone blocks (Figure P16.2 a, b). Two of the kivas at Bonito are known as Great Kivas, due to their large size.

Some archaeologists believe that Great Kivas originated in other cultures and later were adopted by the Anasazi around A.D. 700. Traditionally kivas were thought to be places of religious or ceremonial functions, but Great Kivas may also have been administrative centers.

Connecting Pueblo Bonito with other cities was a network of over 400 mi (650 km) of beautifully designed roads, some 30 ft (10 m) wide and straight as an arrow. Because of extensive excavation efforts, we know a great deal about the people living in Pueblo Bonito. For example, we know that they were an agricultural community that used irrigation, and that Pueblo Bonito was a major trade center in southwestern North America.

The Anasazi are famous for their stone calendars. These calendars kept track of the year by using sunlight shining through a crack in an overhanging rock onto a design. Using these calendars, they controlled their agricultural activities. Chaco Canyon is the location of the best surviving example of these stone calendars.

FIGURE P16.1
(Top) Pueblo Bonito today in Chaco Culture National Historic Park, New Mexico.
(Photo by author.)
(Bottom) Drawing of the pueblo located at the site showing how it might have once looked.

FIGURE P16.2
(Top) Four-story-high wall in Pueblo Bonito, in Chaco Culture National Historic Park. Note that the Park Service has braced the wall to keep it from falling.
(Bottom) The rate at which such walls at the site are moving is monitored with a strain gauge, illustrated here.
(Photos by author.)

AZTEC RUINS NATIONAL MONUMENT

Aztec Ruins National Monument (see Figure 1.17) was established in 1923, but it has nothing to do with the Aztec cultures. Located in northwestern New Mexico, the area was occupied from approximately A.D. 1100–1200 by Chacoan Anasazi. It was then abandoned, but later occupied from A.D. 1225 to approximately 1300 by Mesa Verdean Anasazi. The reasons why either culture left the area are unknown; perhaps a severe drought forced them to leave.

Aztec contains the largest reconstructed Great Kiva in existence (Figure P16.3). The site at one time contained approximately 500 rooms and from 400 to 700 people. Some smaller kivas were also constructed (Figure P16.4). As with most Chacoan sites, walls were built from sandstone blocks, some incorporating very intricate designs.

FIGURE P16.3
Reconstructed Great Kiva (circular ceremonial area) at Aztec National Monument, New Mexico. (Photo by author.)

FIGURE P16.4
Excavated small kiva at Aztec National Monument. (Photo by author.)

MESA VERDE NATIONAL PARK

Mesa Verde National Park (see Figures 1.18 and 8.7) is one of two national parks exclusively designed to protect Native American archaeological sites. The other is Kobuk Valley National Park in Alaska. Mesa Verde was established in 1906, in southwestern Colorado. Because of its unique archaeological sites and beautiful location it was designated a world heritage site in 1978. It is the center of the Mesa Verde Anasazi culture, A.D. 500–1299.

The geological setting was important to the Mesa Verde Anasazi because it provided them with both shelter and protection. In late Cretaceous time the Mancos Shale was deposited. This was followed by shallow-water sand deposition and lithification that formed the Point Lookout Sandstone. Then some coal and softer rocks were deposited, and the entire section was capped by the massive Cliff House Sandstone. During the Laramide Orogeny, uplift brought the plateau to an elevation of approximately 8,000 ft (2,400 m). Erosion, then Pliocene isostatic rebound and uplift, gave the park its present-day elevation.

Weathering and mass wasting processes shaped the park into something useful to the Anasazi. For example, the Cliff House Sandstone forms a resistant overhang, protecting the homes built into spaces left by the less resistant rocks that had been eroded away (Figure P16.5). Some of the structures they built were large and unique (Figure P16.6).

FIGURE P16.5 Cliff Palace, one of many sites built under the protection of the Cliff House Sandstone at Mesa Verde National Park, Colorado. (Photo by author.)

FIGURE P16.6 Some of the structures built at Mesa Verde National Park are unique, including Square Tower House, the tallest building pictured here. (Photo by author.)

HOVENWEEP NATIONAL MONUMENT

Hovenweep National Monument (see Figure 1.18) was established in 1923 in southeastern Utah. It was occupied by Mesa Verde Anasazi from approximately A.D. 1100–1300. Built in and along canyon rims (Figure P16.7), the buildings reflected the changes in Pueblo III culture; people were beginning to worry about defense, but they still needed access to their crops for farming.

FIGURE P16.7 House built on a canyon rim by Anasazi people at Hovenweep National Monument. (Photo by author.)

 ## BANDELIER NATIONAL MONUMENT

Bandelier National Monument (see Figure 1.17) was established in 1916. Located in the Jemez Mountains of northeastern New Mexico, Bandelier was established to protect this center of Rio Grande Anasazi culture. The park is divided into two areas, both include homes that were carved from the Bandelier Tuff, Quaternary volcanics erupted from the Valles Caldera located to the northwest of the park.

The central part or core of the park is located in Frijoles Canyon, carved by stream erosion through the Bandelier Tuff (Figure P16.8). Unique Anasazi ruins are preserved here, where homes were carved into the tuff along the canyon walls (Figure P16.9). The area was settled by the Rio Grande Anasazi around A.D. 1100 and then abandoned around A.D. 1600. A second portion of the park is located a few miles away and occupies a hilltop area that has not been excavated. But it has obviously been occupied, and here, also, the homes are cut into the Bandelier Tuff. In places deep grooves are worn in the tuff where the Anasazi people walked (Figure P16.10).

FIGURE P16.8 Frijoles Canyon cut through the Bandelier Tuff and containing most of the excavated Anasazi sites at Bandelier National Monument, New Mexico. (Photo by author.)

FIGURE P16.9 Homes carved and built into the Bandelier Tuff at Bandelier National Monument, New Mexico. (Photo by author.)

FIGURE P16.10 Path worn into the Bandelier Tuff at Bandelier National Monument by the Anasazi, as they went about their daily business. (Photo by Sue Ellwood.)

OTHER NATIVE AMERICANS IN THE SOUTHWEST

Although the Anasazi remained the dominant culture in the Southwest, there were other Native American peoples living at the time of the Anasazi, and in some cases, trading with them. These included the Sinagua (with cultural sites found at Walnut Canyon National Monument and Wupatki National Monument), Hohokam (see Casa Grande Ruins National Monument), and the Panhandle Pueblo Cultures (with cultural sites at Alibates Flint Quarries National Monument).

 ## WALNUT CANYON NATIONAL MONUMENT

Walnut Canyon National Monument (see Figure 1.14) was established in 1915 in northern Arizona, just east of Flagstaff. Initially controlled by the U.S. Forest Service, it was transferred in 1933 to the National Park Service. The site contains approximately 900-year-old cliff dwellings, built by the Sinagua people who arrived in the area at about A.D. 700, and left before A.D. 1400. The ruins at Walnut Canyon represent part of the culture also found at Wupatki National Monument.

The area is capped by the Kaibab Limestone (the Paleozoic capping limestone along the Grand Canyon rim [see Chapter 15]) that forms an overhang for the dwellings that occupy space formed when the underlying Toroweap and Coconino Sandstones were eroded away (Figure P16.11). These units are calcite cemented, cross-bedded eolian sands like the Navajo and Entrada Sandstones encountered in a number of other parks and monuments. In places the overhanging Kaibab Limestone has collapsed onto the underlying dwellings; we can only wonder what may be buried in the rubble (Figure P16.12)!

FIGURE P16.11 Kaibab Limestone layers (upper beds) forming overhangs for dwellings in Walnut Canyon National Monument, Arizona. Note the distinct cross-beds in the Coconino Sandstone at the base of the section. (Photo by author.)

FIGURE P16.12 Kaibab Limestone layers forming overhangs for dwellings in Walnut Canyon National Monument. Note that in the middle of the photograph, a large piece of the limestone has fallen on the living area below. (Photo by author.)

WUPATKI NATIONAL MONUMENT

Wupatki National Monument (see Figure 1.17) was established in 1924 and is in northern Arizona. The monument contains an interesting mix of lava flows and Sinagua culture ruins. At one time some Anasazi also occupied the site, which was abandoned approximately A.D. 1225, due to drought. The dwellings here are built from Quaternary lava, Permian Kaibab Limestone, and Triassic Moenkopi Sandstone fragments along the tops of canyon walls (Figure P16.13).

FIGURE P16.13
Buildings on the canyon rim at Wupatki National Monument. (Photo by author.)

CASA GRANDE RUINS NATIONAL MONUMENT

Casa Grande Ruins National Monument (see Figure 1.14) was set aside for protection in 1892 by President Benjamin Harrison, even before President Theodore Roosevelt established the first national monument in 1906. Case Grande was designated a national monument in 1918. Located between Phoenix and Tucson in southwestern Arizona, it was built in the early 1300s by the Hohokam people, whose culture lasted from approximately A.D. 100–1450. These people were the first farmers in the American Southwest. Hohokam means "those who are gone" or "all used up."

Casa Grande Ruins has a large standing building, called Casa Grande, that was constructed from slabs of caliche (calcareous cemented sands) cemented together using mud (Figure P16.14). This four-story building includes 11 rooms and is a one-of-a-kind building, believed to have been an observatory. It is located in the center of a village that includes other still-standing buildings. The village was abandoned in the late 1300s, perhaps due to extreme climate variations, probably drought. Because the type of building material used has a tendency to dissolve (Figure P16.15), a large roof has been constructed over the structure (seen in Figure P16.14) to preserve it.

FIGURE P16.14   Casa Grande at Casa Grande National Monument, Arizona. (Photo by author.)

FIGURE P16.15   Rain tends to erode the caliche and mud mix used for building at Casa Grande National Monument. (Photo by author.)

ALIBATES FLINT QUARRIES NATIONAL MONUMENT

Alibates Flint Quarries National Monument (see Figure 1.17) is the only National Monument in Texas. Located near Amarillo in the Texas Panhandle, it was established in 1965. Initially it was named Alibates Flint Quarries and Texas Panhandle Pueblo Culture National Monument, but, thankfully, it was renamed in 1978. The site had been quarried for approximately 10,000 years, but only permanently occupied from approximately A.D. 1200 to 1450 by the people of the Panhandle Pueblo Culture who quarried for flint (chert) [SiO$_2$] that was then traded with other cultures. Ruins of a village area within the monument have not been excavated and are not accessible to the public.

The flint was formed by replacement in the Permian Alibates Dolomite [Ca(Mg)CO$_3$] during late stage diagenesis, and is found in distinct bands (Figure P16.16), sometimes greater than 2 m thick. The color is spectacular, containing most of the colors of the rainbow, and is distinctive; thus it can be recognized when found associated with other cultural sites.

FIGURE P16.16 Chert (known by archaeologists as "flint") exposure in dolomite at Alibates Flint Quarries National Monument. (Photo by author.)

OLDEST CULTURES IN NORTH AMERICA

Exactly when people first arrived in North America is a hotly debated issue, as is the question concerning how they got here. One argument is that people arrived via a Bering land bridge into Alaska from Siberia during the last major glaciation, sometime around 22,000 B.P. They could cross at that time because sea level would have been low enough that the Bering Strait would have been dry land. Others think these people arrived earlier than 22,000 B.P., perhaps as much as 40,000 B.P. or more. A third argument suggests early peoples arrived by boat, moving along the Pacific coastline, and therefore were not constrained by the glaciers that may or may not have existed at that time on the North American continent. To resolve these issues, it is necessary to find archaeological sites that represent evidence of early occupations, and to establish their ages. One such site is located in Alaska, at Kobuk Valley National Park. Other examples may be found along the Pacific coastline. Channel Islands National Park (Chapter 14), for example, has some very old archaeological sites that are being evaluated, but so far they correlate with Clovis Culture's Paleo-Indian time. The Coastal Range Orogeny, accompanied by uplift, rejuvenation, and erosion, may have destroyed sites along the Pacific Coast that could have helped determine whether early peoples arrived by land or sea.

KOBUK VALLEY NATIONAL PARK

Kobuk Valley National Park (see Figure 1.11) was created as a National Monument in 1978 and upgraded to national park status in 1980. Located in northern Alaska, north of the Arctic Circle, it is one of only two parks designed to preserve archaeological remains, the other being Mesa Verde National Park. Preserved here are more than 14,000 years of human occupation. The question being asked is "Are the artifacts located here more than 14,000 years old?" It has been argued that early humans came through Kobuk Valley to America even before the last glacial maximum 22,000 B.P. The materials preserved here may allow us to discover if this is true.

The park consists of a segment of the Baird Mountains to the north and the Kobuk River valley flatlands to the south. It is an area of permafrost and glacial deposits and has two active sand-dune fields, covering approximately 30 square miles, in the southeastern corner of the park. There are also extensive inactive Pleistocene dune fields (Figure P16.17), mainly south of the Kobuk River. These extensive dunes found north of the Arctic Circle illustrate the point that sand dunes are not deposited exclusively in hot, dry, desert climates. The permafrost in the park has the advantage of protecting archaeological sites from erosion and from degradation by plants, animals, and bacteria.

FIGURE P16.17   Sand dunes in Kobuk Valley National Park in Alaska. (Photo courtesy of National Park Service.)

MILITARY PARKS

There are a large number of military and historical parks in the United States, covering a period in American history of more than 400 years. Many of these parks are situated along the eastern Coastal Plain (Figure 8.17) and are military sites that were defensive in nature. But a few, like Gettysburg National Military Park, are the sites of encounters between armies. A number of the military parks are battlefields, while others never saw a shot fired.

VALLEY FORGE NATIONAL HISTORIC PARK

Valley Forge National Historic Park (see Figure 1.23) was authorized on July 4, 1976, in eastern Pennsylvania, 200 years after the United States of America declared its independence from Britain. The park was the location of General Washington's headquarters (Figure P16.18) and army during the harsh winter of 1776–1777, where the troops suffered severely.

Located in the foothills of the Appalachian Mountains of eastern Pennsylvania, the rocks at Valley Forge have been metamorphosed and are chemically weathered. The area is now rolling farmland.

FIGURE P16.18
General George Washington's headquarters at Valley Forge National Historic Park, Pennsylvania. (Photo by author.)

GETTYSBURG NATIONAL MILITARY PARK

The battle of Gettysburg was fought for three days in early July 1863. It also served as the turning point of the Civil War. After Gettysburg the Confederacy was fighting exclusively a defensive war. Gettysburg National Military Park (see Figure 1.23) in southern Penn-

sylvania was established in 1895 and is the site of a national cemetery. Because national cemeteries are "federal," no Confederate soldiers are buried in them, even in those in the South, unless individual soldiers distinguished themselves in other ways or in later wars. It is is also the location of President Lincoln's famous Gettysburg Address (Figure P16.19).

The terrain around Gettysburg dictated how the battle was fought. At one end of the battlefield are igneous intrusive rocks that are relatively resistant to erosion, forming Little Roundtop and Big Roundtop, strategic high ground. Devils Den, in the same area, is also composed of igneous rocks that are fractured and exfoliated (Figure 9.1). The fractures provided cover from which Confederate soldiers fired at Union troops on the Roundtops, and in these fractures many of the Confederate soldiers died. Pickett's Charge was directed against the high ground along Seminary Ridge, which is also composed of igneous rocks and more resistant to erosion, thus forming the ridge that ended for all time the advance of Confederate troops into the United States (Figure P16.20).

FIGURE P16.19 Site of President Lincoln's Gettysburg Address in Gettysburg National Military Park, Pennsylvania. This is the Soldier's Memorial in the Gettysburg National Cemetery. (Photo by author.)

FIGURE P16.20
Farthest point of advance of Confederate soldiers during the Civil War, known as the "High Water Mark," in Gettysburg National Military Park. (Photo by author.)

 VICKSBURG NATIONAL MILITARY PARK

A second major battle, the siege of Vicksburg, Mississippi, ended at about the same time that the battle of Gettysburg was fought, July 1863. Vicksburg National Military Park (see Figure 1.22) was established in 1899, and is the site of a national cemetery. The siege of Vicksburg turned into a very bloody battle. Because Vicksburg is located on the Mississippi River, Union forces were able to bring gunboats and mortar boats in to bombard the city. The Confederates were able to sink some of these with mines. One boat has been raised and is being restored at the site. Confederate artillery batteries are fired twice a day during the summer to demonstrate the noise and smoke associated with a Civil War cannon (Figure P16.21).

Close hand-to-hand fighting was the rule during the siege of Vicksburg, due to the geological setting. Limestones and loess in the area have been actively weathered and eroded as groundwater drained into the nearby Mississippi River. The result is a karst (hummocky) topography that provided soldiers on both sides with natural defensive positions. On many occasions, opposing troops were dug in just a few feet from each other, but the smoke from firing guns and cannon made it almost impossible to see each other over those short distances.

FIGURE P16.21
Cannon firing at Vicksburg National Military Park, Mississippi, demonstrating the large amount of smoke generated on the battlefield. (Photo by author.)

FORT SUMTER NATIONAL MONUMENT

The Civil War officially began at Fort Sumter in Charleston Harbor, South Carolina, on April 12, 1861. Confederates initiated the conflict and Union troops, isolated in the fort and not able to get reinforcements or supplies, surrendered after two days on April 14.

Fort Sumter National Monument (see Figure 1.22) was established in 1948 and includes Fort Moultrie across the harbor on Sullivan Island to the north. Fort Moultrie was originally constructed as a Revolutionary War fort and defended Charleston Harbor against the British on June 28, 1776 (Figure P16.22). It was maintained as a fort until 1947 (as a World War II coastal battery), after which it was closed and incorporated within Fort Sumter National Monument. Chief Osceola, who led the Seminoles against the invading U.S. Army in Florida, was held captive here for a time. He was captured under a flag of truce, when asked to partake in peace negotiations. Osceola died in 1838 and is buried in the courtyard at the entrance to Fort Moultrie (Figure P16.23).

Because they are located on the Coastal Plain, Fort Moultrie and Fort Sumter were ideally positioned to protect the major commerce center of Charleston.

Fort Sumter is built on a sandbar in the harbor, where sediment had accumulated as the result of tidal effects. The sandbar was strengthened by adding ballast, before the fort was built. Erosion has been a problem for the forts, as has been rust from saltwater (weathering of iron to hematite). Sand has also been a problem, being blown by coastal winds throughout the forts and building dunes along the shorelines.

History.   Construction began on the fort in 1829, resulting in a five-sided fort that was almost complete in 1860. Construction on this fort, as with many other forts along the Atlantic Coastal Plain, was originally started in the aftermath of the War of 1812, due to the desire to protect the nation's shores from further invasion. After surrendering the fort to the Confederates, Union troops were able to move into position to initiate their own bombardment that continued throughout most of the Civil War. The result was that the fort was badly damaged and the upper stories were completely shot away. In 1899 Sumter was converted into a large coastal battery (Figure P16.24).

FIGURE P16.22 Fort Moultrie in Fort Sumter National Monument, South Carolina. Fort Sumter can be seen in the middle left distance of the photograph. (Photo by author.)

FIGURE P16.23 The grave of Chief Osceola, Chief of the Seminoles in the early 1800s, is located at the entrance to Fort Moultrie in Fort Sumter National Monument. The stone is marble, metamorphosed limestone. (Photo by author.)

FIGURE P16.24 Fort Sumter as it looks today, in Fort Sumter National Monument. Most of the fort was destroyed during the Civil War, and what remains is the first story and harbor batteries built after the Civil War. (Photo by author.)

 ## CASTILLO DE SAN MARCOS NATIONAL MONUMENT

Castillo de San Marcos National Monument (see Figure 1.22) was established as Fort Marion National Monument in 1924, with the name change coming in 1942. Located in St. Augustine on the east coast of Florida, it was the site of the first permanent European settlement, in 1565. The fort is the oldest masonry fort in the United States, and was built in 1672 by the Spanish. It was later modified in the 1700s. The fort (Figure P16.25) was designed to protect St. Augustine from attack.

Located on the Coastal Plain, Castillo de San Marcos is situated along the lagoon behind the barrier islands in that part of Florida. Initially it was probably much nearer to the shore, but constant sedimentation in the area has resulted in its further isolation from the sea. The fort is built of coquina limestone blocks (Figure 10.3) that were quarried locally by the Spanish. These blocks contain the many loosely lithified shells distinctive of this type of limestone.

FIGURE P16.25 Castillo de San Marcos Fort at Castillo de San Marcos National Monument, St. Augustine, Florida. (Photo by author.)

FORT MATANZAS NATIONAL MONUMENT

Fort Matanzas National Monument (see Figure 1.22) was established in 1924, also on the east coast of Florida near St. Augustine. It was built on a back barrier island, Rattlesnake Island, to guard Matanzas Inlet, a breach in the main barrier island and a tidal inlet. Its purpose was to prevent attack from the Atlantic and to protect St. Augustine from approach from the rear. Built in 1569, the fort was a wood structure that was later rebuilt of coquina limestone, sometime between 1740–42 (Figure P16.26). The fort's shape is a unique form used by the Spanish and Portuguese (Figure P16.27), with a low gun platform backed by a tower

FIGURE P16.26 Fort Matanzas, the oldest fort in the United States, is located on Rattlesnake Island at Fort Matanzas National Monument, Florida. (Photo by author.)

FIGURE P16.27 The Tower of Belem in Lisbon, Portugal. This is a typical shape for forts built by the Spanish and Portuguese in the 17th and 18th centuries. Fort Matanzas was patterned after this style. (Photo by author.)

where troops were housed. The coquina used in construction of the fort was quarried from nearby Anastasia Island.

Besides Rattlesnake Island, the monument also includes part of Anastasia Island, the main barrier island that fronts on the Atlantic Ocean. Along the ocean are developing dunes stabilized by plants such as sea oats, an endangered species shown in Figure P16.28. The name for the park, *Matanzas,* is Spanish for massacre; it was chosen as the name for the inlet at the southern end of Anastasia Island after the Spanish, in 1565, massacred 245 French prisoners here.

FIGURE P16.28 The Atlantic Ocean from barrier island dunes. This is typical of areas now preserved in the National Park System. (Photo by author.)

FORT McHENRY NATIONAL MONUMENT AND HISTORIC SHRINE

Fort McHenry National Monument and Historic Shrine (see Figure 1.23) is located in Baltimore Harbor, Maryland. During the War of 1812 the fort was attacked by the British and successfully defended. As a result of these heroic efforts in defense of the United States flag, Francis Scott Key wrote his famous song, "The Star Spangled Banner." The shrine was authorized as a national park under the War Department in 1925 and later was transferred to the Park Service. The current name was designated in 1939 and is the only historic shrine in the National Park System. Fort McHenry (Figure P16.29) was built in 1799, following the Revolutionary War, to protect the city of Baltimore from attack.

FIGURE P16.29 Fort McHenry in Baltimore Harbor, in Fort McHenry National Monument and Historic Shrine, Maryland. A War of 1812 battle at this fort in which the American defenders beat off attacks by the British, inspired writing of the "Star Spangled Banner." (Photo by author.)

FORT RALEIGH NATIONAL HISTORIC SITE

Fort Raleigh National Historic Site (see Figure 1.22) was set up in 1941 in eastern North Carolina and reconstructed in 1950 (Figure P16.30). The fort is part of the "Lost Colony" established by Sir Walter Raleigh in what was then called Virginia, in 1585. The colonists living there disappeared sometime before 1590, and their fate has yet to be discovered. This site is that of the second oldest fort in the United States, but the actual fort is gone. As with many of the forts built along the eastern seacoast, this site is situated on the Atlantic Coastal Plain. It is located to the west of the lagoon that lies just behind North Carolina's Outer Banks.

FIGURE P16.30 Fort Raleigh, the second oldest fort in the United States, was reconstructed in 1950 in Fort Raleigh National Historic Site, North Carolina. (Photo by author.)

FORT FREDERICA NATIONAL MONUMENT

Fort Frederica National Monument (see Figure 1.22) was established in 1936 in eastern Georgia, 200 years after the fort was built. Fort Frederica was a military town on the colonial Georgia frontier and was established by Governor James Oglethorpe to protect the colony. Extensively fortified, the town was built in 1736 and was occupied until 1749. In a pivotal battle against the Spanish in 1742, the Spanish were turned back, keeping the colony British. The fort (Figure P16.31) is located on a Coastal Plain barrier island, fronting on the lagoon between St. Simons Island and the mainland in southeast Georgia.

FIGURE P16.31 Fort Frederica, in Fort Frederica National Monument, Georgia. Here, in 1742, the British defeated the Spanish in their attempt to settle the rest of the New World. (Photo by author.)

FORT PULASKI NATIONAL MONUMENT

Fort Pulaski National Monument (see Figure 1.22) was established in 1924 on the eastern Georgia Coastal Plain, as one of a series of forts built to protect the United States after the War of 1812. Construction began in 1829, but construction was not complete by 1860. Confederate troops assumed control after Georgia seceded on January 19, 1861. Beginning on April 10, 1862, Union troops shelled Fort Pulaski with rifled cannon. The rifled cannon had such a devastating effect on the fort that it surrendered just 30 hours later. The quick surrender of Fort Pulaski and the demonstrated effectiveness of rifled cannon (Figure P16.32), forced military thinkers to completely change how forts were built.

FIGURE P16.32 Pits left by strikes of rifled cannon shells on Fort Pulaski in Fort Pulaski National Monument, Georgia. The shells can still be seen in some of the pits. (Photo by author.)

 ## FORT LARAMIE NATIONAL HISTORIC SITE

Located in the eastern part of Wyoming, Fort Laramie became an important stopping point for settlers, as the East Coast population began to move westward along the Oregon Trail. The fort was proclaimed a national monument in 1938, and then redesignated as a national historic site in 1978. The site of an early 19th century trading post, the fort preserves military buildings con-structed during the mid to late 19th century. Nestled in a meander bend of the Laramie River, Fort Laramie was well protected from attack. A number of buildings are preserved here and have been restored on the site (Figure P16.33), including officers quarters and enlisted soldiers' barracks.

FIGURE P16.33 Reconstructed buildings at Fort Jefferson National Historic Site. (Photo by author.)

REFERENCES

Beus, S. S., ed. 1987. *Decade of North American geology, DNAG, Centennial field guide.* Vol. 2. *Rocky Mountain Section of the Geological Society of America,* p. 475. Boulder Colorado: Geological Society of America.

Chronic, H. 1986. *Pages of stone, geology of western national parks and monuments.* Vol. 3. *The desert southwest,* p. 168. Seattle: The Mountaineers.

Chronic, H. 1988. *Pages of stone, geology of western national parks and monuments.* Vol. 4. *Grand Canyon and the Plateau Country,* p. 158. Seattle: The Mountaineers.

Lasca, N. P., and Donahue, J., eds. 1990. *Decade of North American geology, DNAG, Centennial special.* Vol. 4. *Archaeological Geology of North America,* p. 633. Boulder, Colorado: Geological Society of America.

Neather, T. L., ed., 1986. *Decade of North American geology, DNAG, Centennial field guide.* Vol. 6. *Southeastern Section of the Geological Society of America,* p. 481. Boulder, Colorado: Geological Society of America.

Roy, D. C., ed. 1987. *Decade of North American geology, DNAG, Centennial field guide.* Vol. 5. *Northeastern Section of the Geological Society of America,* p. 481. Boulder, Colorado: Geological Society of America.

Glossary

A

A-horizon   The uppermost soil layer that includes decaying plant material, where iron and calcium minerals have been dissolved and precipitated deeper in the B-horizon below.

Aa   Hawaiian term for lava flows that form in blocky, rough, jagged fragments.

Ablation   The processes by which snow and ice are lost from a glacier, including melting, calving, sublimation, and wind erosion.

Abrasion   Mechanical removal of rock by friction or impact.

Absolute ages   The geologic age of a material, usually in years.

Acadian Orogeny   A middle Paleozoic mountain-building event identified in the Appalachian Mountain region.

Accreted terranes   Allochthonous continental or oceanic terranes added to the margin of a continent by collision.

Adaptive radiation   Division of a group of organisms into groups of organisms, usually occurring within short geologic time periods.

Aerobic photosynthesis   Photosynthesis carried on in the presence of free oxygen.

Age   The formal geochronologic unit of lowest rank, below epoch.

Aleutian Range   Island arc extending to the west from Alaska, where under-thrusting of the Pacific Plate has resulted in volcanism at the surface, producing an arcuate-shaped chain of volcanic islands.

Alkali flats   Level area in an arid region where evaporation from ponded water has concentrated salts, usually potassium or sodium.

Alleghenian Orogeny   Late Paleozoic mountain-building event in eastern North America.

Allochthonous   Formed elsewhere from where found, as opposed to autochthonous.

Alluvial fans   A fan-shaped accumulation of loose rock, usually in semiarid regions, deposited by streams where they issue from narrow mountain valleys onto broad plains or valleys beyond.

Amphibole   A family of dark ferromagnesian minerals, including hornblende and others.

Amygdules   A vesicle in igneous rock filled with a secondary mineral, such as calcite, copper, or other minerals.

Anatomical   The physical structure of animals and plants.

Andesite   A medium- to dark-colored extrusive igneous rock.

Angular unconformity   A break between two sets of sedimentary rocks whose bedding planes are not parallel, where the older beds below are tilted relative to the younger beds above.

Anhydrite   A mineral with the chemical formula [$CaSO_4$].

Anion   A negatively charged ion.

Anthracite   Coal of the highest metamorphic grade, generating the highest heat output for a coal during burning.

Anticline   A fold with the beds producing a convex upward shape.

Aphanitic   Igneous rock texture where the crystals can not be seen with the naked eye.

Appalachian Mountains   Mountains that extend along most of the eastern seaboard of North America.

Appalachian Plateau   A high-elevation, relatively horizontally layered plateau region to the west of the central Appalachian Mountains.

Apparent polar wander (APW)   Paleomagnetic poles calculated for rocks of different ages, indicating the drift of continents through geologic time.

Aquifers   A permeable rock body that allows water to flow through it.

Aragonite   A mineral with the chemical formula [$CaCO_3$], the same formula as calcite, but with a different crystal structure.

Arbuckle Mountains   Mountains located in southern Oklahoma.

Archimedes   A 2d century B.C. scientist who said that the supporting force of objects placed in a fluid is equal to the weight of the displaced fluid (*Archimedes' principle*).

Archipelago   Numerous islands in an island group.

Arctic Rocky Mountains   Extension of the Rocky Mountains extending east-west along the north coast of Alaska.

Arêtes   A narrow ridge produced by backward growth from cirque glacial erosion into two sides of a mountain.

Arkose   A feldspar rich sandstone.

Artesian flow   Natural flow of groundwater to the surface.

Ash   Fine pyroclastic material erupted from a volcano.

Asthenosphere   Part of the upper mantle, it is a shell of Earth below the lithosphere that is a zone of weakness, allowing plate movement to take place.

Asymmetric fold   A fold with axial limbs at different angles, resulting in an offset of the fold axis.

Atolls   Coral reefs that form as circular rings, sometimes horseshoe shaped, surrounding volcanic islands that have subsided into the sea.

Atom   The smallest part of a chemical element that can take part in a chemical reaction without being permanently changed. Atoms have balanced charges.

Atomic number   Defined by the number of protons in the nucleus of the atom.

Atomic weight   Defined by the number of protons and neutrons in the nucleus of an atom.

Aureole   A contact metamorphic zone surrounding igneous intrusions.

Authigenic   Formed in place.

Autochthonous   Geologic sequences formed or produced where found, as opposed to allochthonous.

Autosuspensions   A fluid flow supporting particles by turbulence and driven by gravity.

Axial plane   A planar axial surface.

Azoic Period   A term, no longer used, referring to a time of no life in the geologic record. Used by 19th century geologists.

B

B-horizon   Lies immediately below the A-horizon in soils. The zone of accumulation and precipitation of ions, especially calcium and iron.

Back swamp   The swampy area behind natural levees developed along the banks of rivers.

Badlands topography   Finely eroded topography developed in areas of little or no vegetation.

Bajada   Broad surface of sediment at the base of arid mountains, where alluvial fans have coalesced into a broad apron.

Barchan dune   Wind-deposited sand in a distinctive horseshoe shape, with the points of the horseshoe pointing in the direction the wind is blowing.

Barrier island   A long narrow island, roughly parallel to the coastline and separated from it by a lagoon, which developed due to sand build-up above the high-tide line during times of low sea-level stand.

Barrier reef   A long, narrow reef roughly parallel to the coastline and separated from it by a lagoon.

Basal slip   The sliding of a glacier on its bed; also known as basal sliding.

Basalt (basaltic)   A dark-colored igneous rock, usually extrusive, but sometimes intrusive, having relatively low silica content.

Base level   The lowest level to which a stream can erode.

Basin and Range province   Topography in the western United States characterized by block-faulted valleys, separated by mountains.

Basin   A depressed valley with only internal drainage, often a block-faulted valley

Batholith   A large igneous intrusion, usually discordant with the intruded rocks, with more than 100 km^2 surface exposure.

Bay barrier   A sand pit that has grown across the mouth of a bay, so that the bay is no longer connected to the main body of water.

Beach   Unconsolidated sediment covering a slope from the low-water mark to a change in slope, such as a cliff, or to the point where permanent land plants are located.

Bed   The smallest formal lithostratigraphic unit of sedimentary rocks.

Bed load   Grains rolled along the bottom as a carpet of material.

Bedding   The arrangement of a sedimentary rock in layers of varying thickness and character.

Bedding plane   The planar surface or boundary that separates each sedimentary layer.

Benioff zone   A planar zone of seismic activity dipping at an angle of 45° or so beneath Pacific Ocean trenches.

Biogenic sediment   A sedimentary rock made up of biologically produced components.

Biological isolation   An organism is so different that it becomes socially isolated from the main breeding gene pool, perhaps by that individual's choice or by the choice of the rest of the population.

Biometrics   The area of paleobiological statistics.

Biostratigrapher   A person who studies stratigraphy based on the paleontologic aspects of rocks.

Biostratigraphic name   A name given to rock units derived based on the fossil content of those rocks.

Bioturbated (bioturbation)   Reworked or disturbed by organisms.

Bituminous   Medium coal grade, usually broken down into: A, best; B, intermediate; and C, lowest.

Black Hills of South Dakota   Unusual area of local uplift in western South Dakota, containing mineralization that includes gold.

Blue Ridge Mountains   High Appalachian Mountains in the central and southern part of the eastern United States that get their name from the haze and humidity that lend a blue cast to the mountains much of the year.

Blueschist   High grade metamorphic rocks resulting from the low temperatures, but high pressures, produced in subduction zones.

Body waves   P and S seismic waves that travel through the interior of Earth.

Bolide A meteor fireball that strikes Earth.

Bolson A desert basin with interior drainage from the surrounding mountains, floored with alluvial fan sedimentation.

Bond The chemical forces holding ions together in a mineral or particle aggregate.

Bottomset beds See prodelta.

Boulder train Strings of glacial erratics all originating from a common source and lying along the direction of travel of the glacier, or as fan-shaped deposits that open in the direction of flow.

Boundary stratotypes "Ideal" stratigraphic sections that are globally unique and that characterize system or other chronostratigraphic boundaries.

Bowen N. L. Bowen, the originator of the igneous magma cooling concept explaining how minerals form from the melt, known as the Bowen reaction series.

Bowen reaction series See Bowen.

Boxwork structure A network of intersecting mineral blades or plates deposited in cavities along fracture planes, from which the original host rock has been dissolved.

Braided stream deposits An overloaded stream that divides into an interlacing network of small branching and reuniting channels.

Breccia A coarse-grained rock consisting of cemented, angular, broken rock fragments.

Brecciation Formation of a breccia by breaking the rock into coarse angular fragments.

Brevard Fault Zone A major fault zone in the southern Appalachians identified as a thrust fault; however, it disappears in the middle Appalachians.

C

C-horizon The lower portion of a soil zone consisting of mainly unconsolidated rock material that has been little affected by soil-forming processes.

Calcified Replacement of organic material by calcium carbonate during fossilization.

Calcite A mineral with the chemical formula [$CaCO_3$].

Calcrete Near-surface sand and gravel layer cemented with calcium carbonate precipitated from solution.

Caldera A large, basin-shaped, generally circular volcanic depression caused by collapse after the eruption of lava.

Caledonian Orogeny European term for orogeny that occurred in western Europe during the Acadian Orogeny.

Caliche A term used in the Southwest to characterize the accumulation of salts or calcium carbonate as secondary stony layers in near-surface sediments.

Calve When blocks of ice break off from the leading edge of a mass of ice into water, forming icebergs.

Canadian Shield The ancient central core of the North American continent, located mainly in Canada.

Capillary action The action by which a fluid, such as water, is drawn up (or down) in small pores or interstices by surface tension.

Capillary fringe The zone immediately above the water table moistened by capillary action.

Carbonate ooze The term used by marine geologists to describe high sediment concentrations of microfossils made almost exclusively of calcium carbonate shells.

Carbonate platforms Shallow marine flats, like those associated with the Bahamas, usually found at low latitudes where warm, shallow water allows the production of calcium carbonate sediment.

Carbonates A group of minerals whose anion is [CO_3]$^{-2}$; for example, the mineral calcite [$CaCO_3$].

Carbonation A process of chemical weathering controlled by weak acids produced by carbon dioxide dissolution in water.

Carbonization Replacement by carbonates, as during fossilization.

Cascade Range A range of mountains extending northward from north-central California through Oregon, Washington, and into Canada.

Cast The secondary rock or mineral that fills a natural mold, such as a shell, leaving an impression or cast of the original.

Cataclastic The term used to describe structures produced in a rock as the result of severe mechanical stress during dynamic metamorphism, including bending, breaking, and granulation of minerals.

Cation An ion with a positive charge.

Caver A cave explorer.

Cementation A diagenetic process producing lithified hard sediments from loose material by precipitation of minerals in sediment pore spaces.

Channel roughness Turbulence caused by material lying within a stream channel.

Chert A hard, sedimentary, very fine-grained, usually nodular rock composed mainly of extremely small (less than 30 microns) crystals of [SiO_2] as an organic or inorganic precipitate or as a replacement product. Dark varieties of chert are often called flint.

Chronostratigraphic names Rock units classified on the basis of their age or time of origin.

Cinder cone A conical hill formed by accumulating volcanic pyroclastic debris, normally of basaltic or andesitic composition.

Cirque Semicircular, steep-walled hollow, produced by the eroding activity of a mountain glacier.

Cirque glacier A small glacier occupying a cirque.

Clastic dike A sedimentary dike consisting of a variety of sedimentary particles derived from sedimentary beds below or above; for example, sandstone dike.

Clastic sedimentary rocks Detrital sedimentary rocks formed from the broken pieces of other rocks (sedimentary, igneous, or metamorphic).

Cleavage The breaking of a mineral along its crystallographic planes.

Cleavage planes One of the surfaces along which a rock tends to split due to its cleavage.

Coal A readily combustible rock containing more than 70% carbonaceous material by volume and formed by heating and compaction during burial of accumulated plant material. It is found in many grades, which are based on the impurities and heat output when burned.

Coastal plain A low, usually broad plain bordering the ocean shoreline, with sediments that are usually horizontal or gently sloping toward the water; it is flooded or emerges as sea level rises or falls.

Colluvium   Loose, heterogeneous soil and/or rock fragments deposited by rainwash, sheetwash, or slow continuous downslope creep, usually collected at the base of a slope.

Column   A columnar deposit formed by the union of a stalactite with its corresponding stalagmite.

Columnar jointing   Parallel prismatic columns, polygonal in cross section, sometimes found in intrusive and extrusive igneous rocks.

Composite sequence   A single sequence of beds that includes all lithologic types in the order in which they occur.

Composite volcano   See stratovolcano.

Compression   Forces that tend to decrease the volume of, or shorten, a substance.

Conchoidal   A rock fracture that produces a curved, smooth surface.

Concordant   Rocks of a differing type displaying parallelism with rocks of another type with which they are in contact.

Concretions   Localized concentrations of cemented sediment that may have additional mineral concentrations or unusual cementation patterns within.

Cone of depression   A local depression in the surface of the water table due to pumping and removal by a well at a rate that does not allow the entire water table to adjust to the removal.

Conglomerate   A coarse-grained clastic sedimentary rock composed of subrounded or angular fragments that are greater than 2 mm in diameter.

Contact marble   Marble produced by contact metamorphism of a limestone.

Contact metamorphism   Thermal metamorphism of country rock due to heating by igneous emplacement.

Continental drift   Term no longer used, associated with the theory proposed by Wegener in 1912 that there was relative displacement between continents on Earth.

Continental shelf   That part of the continental margin between the shoreline and the continental slope, where offshore slopes steepen to angles of 3–6°.

Continuous reaction series   In igneous magmas, where early-formed crystals react with the changing magma without abrupt crystal phase changes.

Convection   Mass movement of subcrustal or mantle material either laterally or as upward- or downward-moving convection cells, driven by thermal processes within Earth.

Convergence   An evolutionary process where similar forms are produced in different biological species.

Convergent boundary metamorphism   Metamorphic effects at the boundary between two converging plates, resulting in low temperature, high pressure metamorphism, producing blue schist facies metamorphic rocks.

Coprolites   Fossilized fecal material.

Coquina   Detrital seashells and seashell fragments loosely cemented together into poorly consolidated limestone.

Coral reef   A coral-dominated organic mound or ridge of in-place coral colonies and accumulated coral fragments and carbonate sand.

Coralline algae   Calcareous algae that form encrustations resembling coral.

Country rock   As used here, the host rock that encloses igneous intrusives.

Crater   Basinlike rimmed structure, usually at the summit of a volcanic cone.

Craton   An ancient part of Earth's crust that has attained relative tectonic stability.

Creep   The extremely slow, but continuous, downslope movement of surface and near-surface geological materials under the influence of gravity.

Crust   The outermost layer or shell of Earth located above the mantle and having lower density, more silica- and aluminum-rich rocks than the mantle.

Crystal habit   The general shape of crystals.

Crystal structure   The regular, orderly, repeated arrangement of atoms in mineral crystals, distinctive to each mineral.

Cutoff   A short channel that forms when a stream cuts through the short strip of land between the meander bend of a stream.

Cyanobacteria   The group of bacteria that produces chlorophyll and oxygen through photosynthesis and is most responsible for the formation of stromatolites.

D

Charles Darwin   The author of the book *On the Origin of Species.*

Debris fall   The free collapse from a cliff or steep slope of relatively unconsolidated rock debris.

Debris flow   A moving mass of material, rock, soil, and mud, with more than half the particles being larger than sand size.

Debris slide   A slide involving observable downslope movement of relatively dry and unconsolidated geologic debris in which the debris slides or rolls forward forming irregular hummocky deposits.

Decade volcano   Volcano chosen for a decade of intensive study.

Declination   The horizontal angle at any location between magnetic and true north.

Deep focus earthquake   An earthquake occurring between 300–700 km depth.

Deep-sea drilling   Rotary drilling into the deep ocean bottom; technologically difficult.

Deflation   The sorting out and removal of loose, fine-grained particles through wind erosion, leaving coarse particles behind.

Delta   Sediment accumulating at the mouths of rivers or tidal inlets.

Delta fronts   A narrow zone where delta deposition is most active, forming relatively coarse, horizontal forset beds.

Deltaic sediments   Sediments deposited in deltas.

Dendritic   Forming in a treelike or branching pattern.

Dendrochronology   The study of annual growth rings in trees for the purpose of dating.

Density   Mass per unit volume.

Desert pavement   Residual, relatively coarse, and polished rock particles left behind by wind erosion.

Detrital sedimentary rocks Clastic sedimentary rocks formed from the pieces of other rocks (sedimentary, igneous, or metamorphic).

Diagenesis All the chemical, physical, and biological changes, except weathering and metamorphism, experienced by sediments after deposition and that result in the formation of a sedimentary rock.

Diatom A microscopic single-celled marine and freshwater phytoplankton that secretes an SiO_2 shell.

Differential erosion Erosion that occurs at varying rates caused by differences in the resistance and hardness of minerals being eroded.

Dike Intrusive, tabular igneous rock that cross-cuts intruded bedding or foliation.

Dike swarm A group of dikes that are usually parallel or radiate outward from a common central point.

Diorite Rocks that are intrusive, plutonic, and intermediate in composition between basaltic and granitic magmas.

Discharge The rate of flow at a given moment in volume per unit of time.

Disconformity Special form of unconformity, where the bedding planes above and below the break are parallel, indicating a significant break in the deposition in a sequence of sedimentary rocks.

Discordant Lack of parallelism between adjacent strata.

Dispersing system Distributary system of a delta dispersing water and sediment throughout the delta as the stream dumps its load.

Dispersion The range of scatter of values about a central tendency; statistical variability.

Displaced terrane Continental or island arc fragments that collect along a continental margin located on the leading edge of a moving plate.

Distributary A divergent stream that flows away from the main stream and does not return to it, such as in a delta.

Divergence An evolutionary trend in which an organism's lineage becomes less similar through time and may evolve into two distinct lineages.

Divide The high terrain marking the boundary between two drainage basins.

Dolostone Rock composed mainly of the mineral dolomite $[Ca,Mg(CO_3)_2]$.

Dome An uplift or anticline structure, either circular or elliptical in shape, in which the strata dip gently away in all directions.

Drainage basin An area, bounded by divides, that is drained by tributaries into a main stream.

Drift A general term applied to all rock material transported and deposited by glacial or melt-water processes associated with glaciers.

Dropstone A stone usually dropped by an iceberg into sediments lying on the bottom of the water in which the iceberg is floating.

Ductile deformation The deformation of a rock under stress without fracture or faulting, said to be 5–10% deformation before failure.

Dune The wind deposit of sand grains, usually quartz, controlled by obstacles, such as vegetation, that cause wind velocities to slow and sand grains to collect.

Dunite A mafic mineral composed mainly of olivine.

E

East Pacific Rise Fast spreading ocean ridge located in the eastern Pacific Ocean off the coast of Central and South America.

Elastic deformation Deformation that disappears when the deforming forces are removed.

Electron spin resonance The resonance occurring when electrons undergoing energy level transitions in an element are irradiated at an electromagnetic frequency that produces a maximum absorption.

Electrons Negatively charged particles orbiting the nuclei of atoms.

Element The simplest form of a substance that cannot be divided further by ordinary chemical means; identified by the number of protons in the nucleus.

Emplacement mode A term usually referring to how igneous rocks formed; for example, intrusive versus extrusive.

End moraines A ridgelike, glacial margin accumulations of drift accumulating from an actively flowing glacier.

Eolian A term pertaining to wind processes, especially erosion and deposition.

Eon The geochronologic unit of highest rank; for example, Phanerozoic.

Epicenter The point on the surface immediately above the focus of an earthquake.

Epicontinental sea A sea on a continental shelf or within a continent.

Epoch The geochronologic unit longer than an age and shorter than a period.

Era The geochronologic unit next in order from an eon, longer than a period, but shorter than an eon.

Erode The general processes whereby Earth crustal materials are loosened, dissolved, and moved from one place to another.

Erodibility How readily a substance yields to erosion.

Erratics Rock fragments carried by glacial ice and deposited some distance from their source.

Euphotic zone That part of the ocean in which light penetrates to sufficient depth to allow photosynthesis; average depth about 80 m.

Euramerica An ancient continent consisting of Europe and North America combined.

Evaporites A nonclastic sedimentary rock composed mainly of minerals precipitated from saline solutions, as the result of partial or total evaporation of the solution.

Evolutionary stage How advanced an organism is along an evolutionary lineage.

Exfoliation The process by which concentric plates or shells of rock are stripped from the bare surface of a rock mass, rounding the rock, much like peeling an onion.

Exfoliation dome A dome-shaped form produced by large-scale exfoliation processes of exposed massive plutonic rock.

Extrusive An igneous rock that has been erupted onto Earth's surface.

F

Fabric The complete spatial and geometric configuration of identifiable components that make up a rock.

Facies   A term that indicates unique appearance and characteristics of sedimentary and metamorphic rocks and usually reflects their origin.

Facies changes   Characteristics within a rock that represent a local environmental change.

Failed rift   A continental area along which spreading begins, then for some reason ceases.

Farallon Plate   A plate hypothesized by Tanya Atwater that has been completely subducted under North America.

Fault block   A crustal unit formed and completely or partially bounded by faults.

Fault breccia   Jagged rock fragments that remain where the rock has been crushed, and often with fine-grained material filling in between blocks.

Fault drag   Where beds adjacent to faults have been bent by motion along the fault.

Faults   Fractures in the crust along which displacement has occurred.

Faunal succession   The observed chronological succession of life forms through geologic time.

Fecal pellets   Material excreted by marine organisms.

Feldspars   A group of abundant and important rock-forming minerals with the general formula $[MAl(Al,Si)_3O_8]$, where M = potassium, sodium, calcium, barium, rubidium, strontium, or iron.

Ferromagnesian minerals   Dark, mafic igneous minerals that contain iron or magnesium.

Filter feeding   An animal that obtains its food by straining out organic matter from water as it passes through some part of its body.

Fins   Large, vertical, long thin bands of rock.

Fissile rock   Rock that is easily broken along layers.

Fission track   The path of radiation damage made by nuclear particles in a mineral or glass by the spontaneous fission of radioactive elements contained within the mineral.

Fissure eruptions   An eruption that takes place from an elongate fissure rather than from a central vent.

Fjords   Glaciated valleys or troughs that are now partly submerged by seawater.

Flexural slip   Bending of beds.

Flint   See chert.

Flood basalts   A term applied to those basaltic lavas that occur as vast accumulations of horizontal or subhorizontal lava flows erupted in rapid succession over a large area. Also called plateau basalts.

Floodplain   The relatively smooth plane adjacent to a river channel, created by the present river and covered with water when the river overflows its banks during floods.

Fluorescence   A type of luminescence in which the emission of light ceases when the external stimulus ceases.

Fluted   A differentially eroded rock with a surface shaped into flutes, small scale ridges, and depressions.

Fluvial sediments   Sediments deposited in streams or by stream action.

Flysch   A marine sedimentary facies composed of thick, thinly bedded, graded deposits, mainly of calcareous sandy shales and muds, interbedded with conglomerates, coarse sandstones, and graywackes.

Focus   The point within Earth where the earthquake actually occurs.

Fold axis   The surface intersection with the axial plane.

Foliation   A general term for the planar arrangement of textural or structural features within rocks.

Food chain   The passage of energy and materials from primary producers, plants, to progressively larger and larger consumers, animals.

Foredeep depressions   An elongate depression bordering an island arc or other orogenic belt, such as a trench.

Formation   A lithostratigraphic name for a usually mappable rock unit recognized based on lithic characteristics and stratigraphic position; next in rank below group.

Forset beds   See delta fronts.

Fossil assemblages   A group of contemporary fossils that occur at the same stratigraphic level.

Fossilization   All processes involving the eventual preservation of all, a part, a trace, or evidence of ancient plants and animals.

Fossil   Any remains, trace, or imprint of a plant or animal that has been preserved in rocks from some past geologic or prehistoric time.

Fractional crystallization   Crystallization in which early formed crystals are prevented from equilibrating with the magma from which they grew, producing residual liquids different from the original.

Fracture   The breaking of rocks, other than along cleavage, due to mechanical failure.

Fringing reefs   Reefs found along coastlines with no lagoons.

Frost heaving   Uneven lifting and distortion of soils due to subsurface freezing of water and growth of ice crystals.

Frost wedging   Where jointed rock is forced apart by expanding water and ice during freezing.

Frosted grains   An effect seen in sand dunes produced by eolian processes as the result of the impact of sand grains blowing against one another and scratching their surfaces.

G

Gabbro   A mafic plutonic rock, generally the intrusive equivalent of basalt.

Genetic diversity   Organisms that are not too closely related are said to have genetic diversity, the result of differences in the gene pool.

Geodes   Hollow, semispherical rocks that have internal crystal concentrations, often of the minerals quartz or calcite, but sometimes of other more exotic minerals.

Geologic Time Scale   Chronological sequence of geologic events used as a measure of a relative or absolute geologic time.

Geomorphology   The study of landforms.

Geothermal gradient   The rate of temperature increase with depth within Earth.

Geysers   Hot springs that intermittently erupt hot water and steam.

Glacial till   Jumbled, unsorted, and unstratified sediment deposited directly from glaciers.

Glacial valleys   U-shaped, steep-sided valley carved by a glacier.

Glacier   A large body of slowly flowing ice formed by compaction and recrystallization of snow.

Glaciologists   Scientists who study glaciers, their effects and deposits.

Glassy   Texture of some rapidly cooled and crystallized igneous rocks, similar to broken glass.

Gneiss   A foliated, light- and dark-banded metamorphic rock.

Gneissic layering   Alternating light and dark layers found in gneisses.

Goethite   A mineral with the chemical formula [FeO(OH)].

Gondwana   Southern supercontinent proposed by Wegener, composed mainly of South America, Africa, Australia, Antarctica, and India, which existed in the late Paleozoic and early-middle Mesozoic.

Graben   An elongate, depressed crustal unit bounded by normal faults on its sides.

Graded bedding   Uniform changes in grain size, with fine grains at top, coarse grains at bottom.

Granite (granitic)   Silica-rich, plutonic igneous rock made mainly of quartz and feldspar, with some mica.

Granite gneiss   A metamorphic rock with the composition of granite derived from a granite or sedimentary rock and with gneissic layering.

Granodiorite   A plutonic rock intermediate in composition between granitic and andesitic rocks.

Granulite   A metamorphic rock consisting of even-sized interlocking mineral grains with little or no preferred orientation.

Gravity   The force of attraction between all bodies with mass.

Graywacke   Dark gray, indurated coarse-grained sandstone consisting of poorly sorted angular fragments of quartz and feldspar in a compact clayey matrix.

Great Plains   The relatively flat region of North America immediately to the east of the Rocky Mountains.

Greenschist   A metamorphic facies, generally green in color due to the presence of the minerals chlorite, epidote, or actinolite.

Grenville Orogeny   Orogenic event that is identified along the eastern seaboard of North America, dating to the late Precambrian.

Ground moraine   Widespread but relatively thin accumulations of glacial debris.

Groundmass   The very fine-grained material between the phenocrysts in igneous rocks.

Groundwater   The subsurface water contained within the zone of saturation; including underground streams.

Group   The lithostratigraphic unit next in rank above formation, and consisting of named formations.

Guide fossils   A fossil with supposed value as an indicator of the age of the strata in which it is found; usually it has unique, easily identifiable characteristics.

Gypsum   Mineral with the chemical formula [$CaSO_4 \cdot 2H_2O$].

H

Halite   Mineral with the chemical formula [NaCl].

Hammocks   Dry, humus-rich mounds with trees, which develop in Everglades National Park and elsewhere.

Hanging valleys   A glacial valley whose mouth is at a high level on the steep side of a larger glacial valley.

Heat flow   Flow of heat out of Earth, generated by radioactive decay within Earth.

Hematite   Mineral with the chemical formula [Fe_2O_3].

Hercynian Orogeny   Late Paleozoic orogenic event involving Europe; synonymous with the Variscan Orogeny.

High grade   Metamorphic rocks usually generated by high temperature and pressure; for example, high grade coal (also high rank).

Hominid   A primate of the family Hominidae of which Homo sapiens is the only living species.

Homology   The study of structures with the same ancestral origin, but that serve different functions.

Horn   A high, rocky, sharp-pointed mountain peak resulting from the intersecting walls of three or more cirques cut back into the mountain by headward erosion by glaciers.

Hornfels   Usually a contact metamorphic rock composed of fine, equidimensional grains with no preferred orientation.

Horst   An elongate, uplifted crustal unit bounded on its long sides by normal faults.

Hot spot   An area of rising magma originating in the mantle and persisting at one location for millions of years.

Humus   The well-composed, usually dark, relatively stable organic matter in the soil.

Hydrated   A mineral compound in which water is part of the chemical composition.

Hydraulic gradient   The change in flow rate along the length of an aquifer.

Hydrologic system   The water-containing units and the interrelated physical and chemical relationships controlling water on Earth.

Hydrolysis   A decomposition reaction involving water.

Hydrothermal alteration   The alteration of rocks and minerals by the reaction of hydrothermal water with preexisting minerals.

I

Ice rafted debris   Glacial drift and other material carried by icebergs and dropped as icebergs melt.

Igneous rock   Rock formed by cooling from a hot liquid magma.

Ignimbrite   Rock formed by the eruption and consolidation of ash flows and nuées ardentes.

Impact metamorphism   Alteration effects resulting from the impact of a bolide or meteor.

Inclination   The angle at which the magnetic field lines dip.

Index fossils   A fossil that allows the containing strata to be identified and dated; usually broadly distributed, easily identified, and fairly common.

Infiltration   The flowing of a fluid through pores or small openings that fills the strata.

Inlets   Small, narrow breaks in barrier islands that allow penetration of the ocean into the lagoon behind.

Inner core   The solid, central part of Earth's core extending from approximately 5,100 km to the center of Earth (6,371 km).

Interbedded   Beds alternating with or lying between beds with different characteristics.

Intermittent streams   Streams that seasonally dry up because the groundwater seepage is absent or too low to maintain the stream flow.

Intrusive   When igneous rock cools within Earth before it reaches the surface.

Invertebrate organisms   Organisms with no backbone.

Ion   A positively (cation) or negatively (anion) charged atom caused by an imbalance in the number of protons in the nucleus versus the number of electrons surrounding the nucleus.

Island arc   A curved belt of volcanoes produced as the result of ocean crust subducting under ocean crust.

Isostasy   Equilibrium, essentially floating, of lithospheric units above the asthenosphere. Loading or unloading of the crust causes adjustments.

Isostatic adjustment   Readjustment of segments of the lithosphere in response to changes in loading due to factors such as erosion or melting of ice sheets (see isostatic compensation).

Isostatic compensation   Adjustment to maintain equilibrium of the lithospheric units of varying mass and density (see isostatic adjustment).

Isotope   One of two or more species of the same chemical element, with the same number of protons in the nucleus, but with different numbers of neutrons.

J

Joints   Fractures without displacement.

K

Kaolinite   Clay mineral with the general composition $[Al_2O_3 \cdot 2SiO_2 \cdot 2H_2O]$.

Karst topography   Topography mainly formed by dissolution of limestone to produce sinkholes, caves, and irregular hilly topography.

Kenoran Orogeny   A mountain-building event, now eroded away, that occurred during the Archean Eon in the central part of Canada.

Key beds   A well-defined and easily identifiable sedimentary strata used in geologic mapping.

Klippe   An isolated rock unit that is the erosional remnant of a thrust sheet or nappe.

L

L wave   See Love wave.

Lacustrine sediments   Sediments that were deposited in lakes.

Lagoon   A shallow, elongated stretch of water lying between the mainland and a barrier island.

Lahars   Volcanic, debris-laden mudflows.

Laminae   Thin sedimentary layers that result from compaction of fine, clay-sized particles to form sheets.

Laminar flow   Water flow in which the stream lines remain distinct and the direction of these lines remains unchanged with time.

Landslide   A general term covering a wide range of downslope mass movements of material driven by gravity.

Laramide Orogeny   A mountain-building event identified in the eastern Rocky Mountain area from the late Cretaceous to the end of the Paleocene.

Lateral moraine   The glacial drift deposited as elongated ridges at the sides of moving glaciers.

Laterite   The highly weathered red subsoil (B-horizon) that develops in tropical or warm forested to temperate climates.

Laurasia   The protocontinent of the northern hemisphere consisting of most of North America, Europe, Asia (excluding India), and Greenland.

Lava   A molten extrusive rock and its solidified equivalent.

Lava tube   The hollow space beneath the surface of a lava flow formed by the withdrawal of magma after the surface has solidified.

Leaching   The dissolution and removal of selected mineral constituents from a soil or rock.

Leeward   The downwind side of an area toward which the wind is blowing.

Limbs   The area of a fold between adjacent fold hinges.

Limestone   A sedimentary rock composed primarily of $CaCO_3$.

Limonite   A general term for a group of brown, hydrous iron oxide minerals.

Lithic sandstone   A sandstone containing abundant rock fragments.

Lithification   The formation into rock of deposited loose detrital material, involving processes such as cementation, desiccation, crystallization, and compaction.

Lithosphere   A zone approximately 100-km thick, including the crust and part of the upper mantle. It is the region that is broken into plates that move on Earth's surface.

Lithostatic pressures   The vertical pressure at a point in Earth's crust equal to the weight of a column of overlying rock.

Load   The material that is carried by a natural transporting agent, such as a stream, a glacier, the wind, the tides, or the ocean currents.

Loess   A fine-grained, light brown glacial drift carried by wind and deposited in a nonstratified, porous, slightly coherent layer usually less than 30 m thick. It consists mainly of silt particles, with some clay and fine sand also included, and is usually bioturbated.

Longitudinal profile   The profile of a stream valley drawn from its head (source) to its mouth.

Longshore current   The ocean current produced as waves approach a shore at an angle. Constrained by the shoreline, the water swings parallel to the shore.

Longshore drift   The material that is carried parallel to the shore by longshore currents.

Love wave   A surface seismic wave with particle motion that is transverse to the direction of travel of the wave.

Low grade   A metamorphic rock produced by relatively low temperature and pressure; for example, lignite coal.

Luster   The quality and intensity of light reflection from the surface of a mineral.

M

Magma   Naturally occurring liquid and mobile rock within Earth from which igneous rocks are formed.

Magma pulse   The intrusion of magma at midocean ridges contributing to seafloor spreading.

Magmatic differentiation   A process where homogeneous magmas differentiate, allowing a variety of minerals to crystallize at different temperatures.

Magnetite   The mineral with the chemical formula [Fe_3O_4].

Main trunk   In a stream, the dominant channel in a network of stream tributaries.

Mantle   The zone within Earth between the crust and the core, extending to a depth of approximately 2,900 km.

Mantle plume   Magma rising from the mantle, through the lithosphere, and to the surface, forming what are called hot spots.

Marble   Metamorphosed limestone.

Marker bed   A distinctive bed that is characterized by some easily identifiable property, such as crossbedding, and serving as a geologic marker.

Marsh coal   Coal that is derived from plant material, formed under marshy conditions, and from sapropel, a sludge composed of plant remains under strong anaerobic (low oxygen) conditions.

Mass extinction   The geologically instantaneous extinction (death) of many species of organisms.

Mass wasting   The gravity-driven downslope movement of loose rocks and sediment.

Meander scars   A crescent-shaped trace of a meandering stream that has since changed its course, leaving behind a channel bend that becomes isolated from the rest of the stream.

Mechanical weathering   The physical processes of weathering, such as frost wedging, involving no chemical change.

Medial moraine   The ridgelike accumulations of drift formed between two merging valley glaciers.

Member   A lithostratigraphic term for one of perhaps several distinct units that make up a formation.

Mercalli intensity scale   An arbitrary earthquake intensity scale based on observed phenomena, such as people running from buildings as opposed to simply being upset; named after Giuseppi Mercalli, an Italian geologist who devised it in 1902.

Mesosphere   Earth's mantle below the asthenosphere.

Metaconglomerate   Metamorphosed conglomerate.

Metamorphic facies   A set of metamorphic mineral assemblages providing a predictable relationship between mineral composition and chemical composition.

Metamorphic grade   The intensity of metamorphism that a rock undergoes.

Metamorphic rock   Any rock derived from pre-existing rocks as the result of chemical, mineralogical and/or structural changes that occur in the solid state as the result of temperature, pressure, and/or chemical alteration.

Metasomatism   Simultaneous solution and deposition of a new mineral in the host body of an old mineral.

Micrite   Precipitated carbonate mud with crystals less than 4 microns in diameter, interpreted as a lithified ooze.

Microcrystalline   Textures containing very fine crystals.

Micropaleontology   A branch of paleontology that deals with the study of organisms only identifiable using a microscope.

Mid-Atlantic Ridge   A continuous range of ocean bottom mountains running through the central portion of the Atlantic Ocean; the Atlantic Ocean spreading center.

Midocean ridges   A continuous range of ocean bottom mountains running for over 50,000 km, generally through the median part of all the ocean basins; where the ocean crust is formed.

Milankovitch   A Yugoslav geophysicist (mathematician) working in the early part of the 20th century who proposed an astronomical theory of glaciation controlled by variations in Earth's rotational precession, tilt of the rotational axis, and orbital eccentricity.

Mineral   A naturally occurring, inorganic, crystalline solid.

Mineralization   A term denoting the replacement of organic material by some mineral or combination of minerals.

Moho   The sharp seismic discontinuity between the crust and the mantle, first identified by Andrija Mohorovicic, a Croatian seismologist; the term is based on an abbreviation of his name.

Mojave Desert   A desert region in southern California.

Molasse deposit   A partly marine, partly terrigenous/deltaic sedimentary facies.

Mold   A mark or primary depression made on a sedimentary surface; filling such depressions produces a cast.

Monadnock   An upstanding hill or mountain rising conspicuously above the general level of a peneplain in an eroded landscape.

Monocline   A single flexure fold.

Moraine   Local accumulation of unsorted or unstratified glacial drift dumped from a moving glacier.

Morgan   Jason Morgan of Princeton University coined the expression plate tectonics in the early 1970s, based on the global distribution of earthquakes.

Morphological traits   The shape, size, and proportions of organisms.

Mottled bedding   Discoloration of sediments that look like the material has been partially stirred or mixed, rather than having a homogeneous appearance; often associated with bioturbation.

Mud crack   An irregular crack in a general polygonal form resulting from shrinking during drying of thin-bedded, fine-grained sediments deposited on the ground surface.

Mudflow   General mass movement of mainly fine-grained, liquefied material that may contain a wide range of matter, including large rocks, and thus may be very destructive.

Mudstone   Rock composed of silt and finer, flat clay grains.

Mylonite   A fine powdered rock "flour" crushed by the movement of rocks within a fault zone.

N

Nappes   Large stacked and overturned folds and thrusts.

Natural levees   Low ridges of fine sand that accumulate along the edge of streams during flooding, ultimately containing the stream within its banks during time of normal stream discharge.

Natural selection   Individuals with characteristics advantageous for survival constitute an increasing proportion of the population.

Neutron   A particle found in the nucleus of atoms and having mass but no charge.

New England province   The geologic province that includes only the states of Vermont, New Hampshire, Maine, Massachusetts, Connecticut, and Rhode Island.

Nonconformity   An unconformity separating horizontal sedimentary beds above that were deposited on eroded igneous or metamorphic rocks below.

Normally magnetized   A direction of Earth's magnetic field similar to that seen today, where a compass needle will point generally toward the north pole of Earth.

Novaculite   A dense, hard, even-textured, light-colored, fine-grained metamorphosed siliceous rock used in the manufacture of whetstones.

Nucleus   The central part of an atom containing protons and neutrons.

Nuée ardente   A fast, turbulent cloud of hot gas and pyroclastic material explosively erupted from a volcano.

O

Obsidian   A dark volcanic glass usually of rhyolite composition and exhibiting conchoidal fracturing.

Obsidian hydration   The weathering rim on obsidian due to water vapor slowly diffusing into the glass and producing a hydrated layer or rind.

Ocean Drilling Program (ODP)   A multinationally funded program designed to drill into the deep ocean crust and analyze samples thus obtained. Headquarters are located at Texas A&M University.

Olivine   Mineral with the chemical formula $[(Mg,Fe)_2SiO_4]$.

Ontogeny   The changes in single individuals during their lifetimes.

Oolitic   Spherical particles, similar to small pearls, making up some limestones.

Opal   Hydrated mineral with the composition $[SiO_2 \cdot nH_2O]$.

Ophiolite   Igneous elements of oceanic crust that have been uplifted and are now exposed on land.

Orogenesis   The term that characterizes the totality of the mountain-building process.

Orogeny   A particular mountain-building event.

Orthoclase   A mineral of the alkali feldspar group with the composition $[KAlSi_3O_8]$.

Ouachita Orogeny   The extension of the Appalachian orogenic event into the Central Mountains province.

Outer core   The outer, liquid (based on seismic evidence) portion of the core of Earth, lying below the mantle, from 2,900 to 5,100 km depth.

Overbank deposits   Fine-grained sediments deposited on a floodplain during flooding events.

Overland flow   Rain flowing over the ground surface as surface runoff before it infiltrates or reaches streams.

Overturned folds   A fold whose limbs have tilted beyond the perpendicular.

Oxbow lakes   Lakes that result from a narrow neck, meander-bend cutoff, isolating the cutoff as a small crescent-shaped lake.

Oxidation   A chemical reaction where the oxygen anion combines with cations to produce oxidized minerals.

Oxides   A group of minerals whose anion is $[O]^{-2}$; for example, the mineral hematite $[Fe_2O_3]$.

P

P wave   Seismic body wave, also called primary or primus, that is the first to arrive at a seismograph after an earthquake; particle motion is push-pull in the direction of wave motion.

Pahoehoe   A Hawaiian term for the ropy structure formed on the surface of relatively nonviscous lava flows.

Paleobotany   The study of fossil plants.

Paleoclimatology   Study of past changes in climate.

Paleoecologist   One who studies ancient fossils and their environments.

Paleomagnetism   The study of the magnetic properties of rocks.

Paleontology   The study of fossils.

Paleosol   Ancient soils often preserved in sediments.

Palynology   The study of spores and pollen.

Pangaea   A supercontinent named by Wegener that existed during the late Paleozoic and very early Mesozoic, including most of the continents on Earth.

Parabolic dune   Sand dune with a crescent shape, convex in the downwind direction, with its horns pointing upwind.

Patch reef   A flat-topped or moundlike organic reef, generally less than a kilometer across.

Peat   An unconsolidated deposit of semicarbonized plant remains, the precurser to coal.

Pediment   A broad, gently sloping surface that is an alluvial veneer found associated with eroding mountains in arid or semiarid climates.

Pediplain   A relatively flat erosion surface usually formed in deserts as the result of coalescing pediments.

Pedogenesis   The formation of soils.

Pedology   The study of soils.

Pegmatite   Very coarse-grained igneous rock with interlocking crystals found as veins, irregular dikes, or lenses, usually at the margins of batholiths.

Pelagic sediments   Sediments deposited in the deep ocean, composed of biological tests of calcium carbonate and silicon dioxide, some terrigenous material, some weathering products from submarine basalts, and other material.

Peneplain   Erosional surface that is "almost a plane."

Peridotite   Mantle igneous rock composed primarily of olivine.

Period   Geochronologic unit lower in rank than an era and higher than an epoch.

Permanent stream   A stream that flows year around, but whose volume may vary dramatically.

Permeability   The property or capacity of a material to transmit a fluid.

Permineralization   Material added to pore spaces within the original.

Petrification (petrifaction)   The process of turning into stone.

Petrology   The study of the history, origin, occurrence, and structure of rocks.

Phaneritic   Igneous rock texture in which the crystals can be seen with the naked eye.

Phenocryst   Distinct crystals in igneous rocks.

Phyllite   A metamorphic rock that is intermediate in grade between slate and mica schist.

Phylogeny   Where new species arise from older species.

Phytoplankton   Microscopic surface-water plants, important in producing oxygen and food.

Piedmont   Lying or formed at the base of a mountain or mountain range.

Piedmont glaciers   Large glaciers lying at the base of a mountain range, formed by spreading out and coalescing valley glaciers.

Pillow lava   General term for lava extruded into water and displaying rounded or elongated pillow structures.

Pinnacle reefs   An isolated reef mound.

Plastic deformation   Permanent deformation of the shape or volume of a substance without rupture.

Plate tectonics   A theory of global tectonics in which the lithosphere is subdivided into a series of interacting ridged plates that interact at their boundaries.

Plateau basalts   See flood basalts.

Playa deposit   Salts evaporated and accumulated from drying playa lakes.

Playa lake   A shallow intermittent lake in arid areas that dries in the summer.

Plucked   Quarried from the bedrock as the glacier passes.

Plug dome   A volcanic dome characterized by pushed up consolidated conduit fill.

Pluton   A large igneous intrusion.

Point bars   Sediment dropped inside a bend that results in cross-bedded, sand-sized sedimentary deposits.

Polarity   Magnetic zones recorded in rocks; normal versus reversed polarities.

Polarity time scale   A dated stratigraphic sequence of measured polarities, used in relative dating of rocks.

Pollen   The several-celled microgametophyte of seed plants enclosed in the microspore wall.

Polymorph (polymorphism)   Two minerals that have the same chemical composition, but different crystal structures.

Porosity   The volume of open spaces within a rock.

Porphyritic   Igneous rock texture where coarse crystals are contained within a fine-grained matrix called a groundmass.

Potassium feldspar   Mineral with the composition $[KAlSi_3O_8]$.

Precambrian shield   Ancient core of a continent.

Pressure head   Height of a column of liquid whose weight is equal to the hydrostatic pressure.

Principle of crosscutting relationships   An important early principle in geology that recognized that a geologic unit cut by another is older than the cutting unit.

Principle of original horizontality   An important early principle in geology that recognized that sediments being deposited in water initially develop horizontal strata.

Principle of superposition   An important early principle in geology that recognized that strata deposited first are older than those deposited on top of them; that is, the oldest sediments are located on the bottom unless the sequence is badly deformed.

Prodelta   Part of a delta lying beyond the delta front that slopes down to the floor of the basin in which the delta is being deposited. The beds deposited are cross-bedded, known as forset beds.

Prograde   Growth or migration seaward of a delta or other sediments.

Protolith   The parent rock of a metamorphic or sedimentary product.

Protons   Positively charged particle in the nucleus of an atom.

Pumice   A light-colored vesicular glassy igneous rock with the composition of a rhyolite.

Punctuated change   Sudden, rapid evolutionary change.

Pyrite   A mineral with the chemical formula $[FeS_2]$.

Pyroclastic   Clastic rock material formed by volcanic eruption.

Q

Quartz   Mineral with the chemical formula $[SiO_2]$.

Quartz arenite   Sandstone containing mainly quartz grains with a clay matrix.

Quartzite   Metamorphosed sandstone where the sand grains are partly fused together.

Quench   Cool very rapidly.

R

R wave   See Rayleigh wave.

Racemization of amino acids   A method of geochronology based on analyzing amino acids.

Radial stream pattern   A pattern developed by erosion outward from a common central high point in elevation.

Radiogenic isotopes   An isotope that was produced by the decay of a radionuclide, but which itself may not be radioactive.

Radiolarian   A single-celled marine protozoan that lives in the ocean's surface waters and secretes an SiO_2 shell.

Rayleigh wave   A surface seismic wave producing elliptical particle, wavelike motion.

Recrystallization   The solid state formation of new crystalline mineral grains in a rock, at the expense of the primary minerals.

Rectangular stream pattern   Patterns that exist where fractures in underlying rock control the erosion pattern with near right-angle bends following bedding planes of steeply inclined strata. Erosion follows joints, faults, or foliation planes.

Recumbent folds   An overturned fold with the axial fold nearly horizontal.

Reefs   A ridgelike or moundlike structure, massive or layered, built by calcareous secreting organisms, especially corals, and mainly consisting of their remains.

Reflected wave   A seismic wave that bounces off layers in the subsurface and is then recorded to give a picture of the subsurface layers.

Refracted wave   A seismic wave that is bent in the subsurface due to density changes with depth.

Regional metamorphism   Large scale metamorphism that is due to the high temperature and pressure effects occurring at the base of mountains.

Regression   The retreat of the sea from land areas; sea level lowering.

Rejuvenate   Stimulate renewed erosion as the result of structural uplift or a drop in sea level, thus lowering the erosion base level and creating a youthful landscape appearance.

Relative ages   Ages for geologic features determined relative to other geologic features instead of in terms of years.

Replacement   A process of fossilization involving substitution of organic matter of an organism with an inorganic mineral substitute.

Resurgent dome   Rising again of a magma dome into the crater formed by an earlier eruption of a volcano.

Reversals   Changes in polarity of Earth's magnetic field; reversals change the polarity from normal to reversed or reversed to normal.

Rheology   The study of fluid flow.

Rhyolite   Extrusive igneous rocks, the extrusive equivalent of granite.

Richter   Carl Richter was a seismologist at Cal Tech who devised a quantitative scale to measure earthquake magnitude.

Ripple marks   An undulatory surface of alternating subparallel small-scale ridges and hollows that developed from wind or wave action.

Rock avalanche   The rapid downslope flow of rock fragments, during which the rocks are further broken.

Rock flour   Very fine rock debris pulverized as rocks move or are crushed by overburden and grinding, such as under glaciers.

Rock glacier   A mass of poorly sorted angular rock fragments with interstitial ice; the whole flows as a glacier.

Rock units   Rock layers that are identified using rock-stratigraphic names that are independent of time estimates.

Rock-stratigraphic names   Rock unit designations that are independent of time estimates.

Rockslide   The sudden downward flow of loose rock fragments.

S

S wave   Seismic body wave that arrives second in the chain of waves arriving from an earthquake; particle motion is transverse to the propagation direction of the wave.

Salt diapir   The upward intrusion of salt into the subsurface.

Saltation   Sediment transport in which particles are moved progressively forward in steps as the result of other particles slamming into them, causing them to jump up and forward, later slamming into other particles and continuing the process.

Salt pan   Small natural depression in which water has accumulated, followed by evaporation and the precipitation of salt that remains.

Sandstones   Medium-grained clastic sedimentary rock containing quartz sand-sized grains in a clay or silt matrix, or firmly cemented together.

Savanna   An open, grassy, essentially treeless plain developed in tropical or subtropical region.

Scarp   A line of cliffs produced by faulting or erosion.

Schist   Strongly foliated metamorphic rock that can be easily split into thin plates, where 50% or more of the minerals are platy minerals such as mica.

Schistosity   The foliation in a schist or other coarse-grained crystalline rock due to parallel, planar arrangement of platy, prismatic, or elongated mineral grains.

Scoria   A pyroclastic rock that is irregular in form and is usually vesicular.

Seafloor magnetic anomaly patterns   Magnetic polarity sequence acquired by basalts during the production of new ocean floor, and recorded by magnetometers at the surface of the ocean.

Seafloor spreading   The hypothesis that the ocean's crust is being created at midocean ridges due to upwelling basaltic magma, then moving away from the ridge axis and ultimately being destroyed at ocean trenches.

Seamount   An elevation on the seafloor 1,000 m high or higher and, in some instances, flat topped (called a guyot).

Sedimentary rock   Rock generally resulting from the lithification of unconsolidated sediment that has accumulated in layers.

Seismic fling   Earthquakes that result from motion in small segments along a fault, like a mole crawling very fast through Earth, the ground opening before it and closing behind; technically called a displacement pulse.

Seismic waves   General term for elastic waves produced by earthquakes or generated artificially.

Selenite   The clear crystalline form of gypsum with a composition [$CaSO_4 \cdot H_2O$].

Series   Stratigraphic unit that ranks below a system and that is always a subdivision of a system; a major chronostratigraphic unit.

Shale   Fine-grained detrital, finely laminated sedimentary rock formed by consolidation of clay, silt or mud; the rock easily breaks into thin layers.

Shear   Deformation resulting from stresses that cause parts of a body to slide relative to each other in a direction parallel to their plane of contact.

Shear wave   (See S wave).

Sheeting or unloading   Results from fracturing parallel to land surface due to erosion, reduction of confining pressure, thus unloading—removing the rock load from above.

Sheet wash   An overland sheetflood occurring in a humid region.

Shelf break   An abrupt change in slope marking the boundary between the continental shelf and the continental slope.

Shield   Stable, ancient continental core or platform.

Shield volcano   A volcano with a flattened shape built mainly by flows of basaltic lavas.

Sialic   Upper crustal rocks rich in silica and aluminum.

Siderite   Mineral with the chemical formula [$FeCO_3$].

Siliceous sediments   Sediments that are rich in SiO_2.

Silicic lava   High silica content lavas; for example, rhyolite.

Silicified   A process of fossilization where the original organic material is replaced by SiO_2.

Silicon tetrahedron   The [SiO_4]$^{-4}$ complex anion, the basic building block in igneous minerals.

Sills   Concordant, tabular igneous intrusives.

Siltstone   Indurated silt with the texture and composition of shale but without the fine laminations.

Sinkholes   Solution cavity often accompanied by collapse in a limestone, typically found in karst regions.

Skeletal   Textures that represent limestones containing microscopic plant and animal remains.

Slate   Very fine-grained metamorphosed shales that can readily split into fine plates.

Slaty cleavage   Parallel foliation of fine-grained, platy minerals developed in slate and other low-grade metamorphic rocks.

Slickensides The place where the fault has abraded, striated, and polished surfaces along which the fault motion occurred.

Slump A sudden landslide characterized by shearing and rotational particle movement, usually producing a series of arcuate, down-stepped slip faces.

Sodium chloride Common table salt, having the mineral name halite and the composition [NaCl].

Solar wind Highly charged particles, mainly protons and electrons, that stream from the sun, collecting in the Van Allen radiation belts, then streaming in at the poles to produce the northern and southern lights.

Sole marks Impressions on the underside of a sandstone or siltstone bed that form as casts filling surface impressions.

Solifluction Slow downslope flow of waterlogged soil and other material, in areas underlain by frozen material.

Sorting Physical processes resulting when sedimentary particles with similar characteristics, such as shape, grain size, or composition, accumulate together.

Sparite Made up of calcite cement with little or no micrite.

Spatter cone A low, steep-sided cone of accumulated volcanic spatter material, usually basaltic in composition.

Specific gravity (Weight of an amount of mineral)/(weight of an equal amount of water).

Species A class of individuals that can breed and produce viable offspring; conversely, unlike species cannot interbreed **and** produce viable offspring (an offspring that can in turn produce offspring).

Speleothem Any secondary mineral formed in caves by fluid action.

Sphericity The quality that demonstrates how closely a particle approximates a sphere.

Spheroidal Generally spherical in shape.

Spit A small point of land extending from the mainland into the ocean, and generally composed of sand, and formed and shaped by longshore currents.

Sponge spicules Long, thin, needlelike spines secreted by deep-sea sponges and left after the sponge dies and decays.

Stage A chronostratigraphic unit that is smaller in rank than a series.

Stalactite A dripstone speleothem, usually cylindrical, that hangs from the ceiling in a cave.

Stalagmite A dripstone speleothem that grows upward from the floor of a cave and is usually cylindrical in shape.

Stock An igneous intrusion resembling a batholith, except that it has less than 100 square km in surface exposure.

Stomach stones The distinctively shaped, smooth rocks used by animals such as chickens and dinosaurs to grind up food in their gizzards.

Stoping Blocks of country rock falling into an intruding igneous magma, and, in part, creating a pathway for the intrusion.

Strain The change in the shape of a body as the result of applied stress.

Strata Layers of sedimentary rock that are visually distinctive from other such layers above and below.

Stratified drift Glacial drift that accumulates in layers due to sorting of grains during deposition from a meltwater stream or from suspension in calm water.

Stratigraphy The study of rock strata, concerned with the interrelationships and age of sedimentary layers.

Stratovolcano A large, steep-sided volcano formed generally from andesitic lavas and tuffs.

Streak plate A white porcelain plate used to scratch or powder minerals for identification of characteristic color.

Stream gradient Angle between the stream channel and the horizontal, measured in the direction of water flow.

Stream head The source of a stream.

Stream mouth The end of a stream.

Stream terraces Level surfaces within a stream valley that represent ancient floodplain surfaces remaining after continued stream erosion has removed part of the ancient floodplain.

Stress The force per unit area acting on any surface within a solid.

Striations Grooves scraped into the bedrock, usually by glaciers.

Stromatolite Ancient small, generally mound-shaped structures produced by cyanobacteria, plantlike organisms that secreted a carbonate structure and carried on photosynthesis.

Subduction The process whereby the edge of one lithospheric plate descends beneath the edge of another.

Subduction zone A long narrow belt along which subduction takes place.

Sublimation The process whereby ice goes directly to the gaseous state (water vapor).

Sulfates A mineral composed of the sulfate anion, with a $[SO_4]^{-2}$ composition.

Sulfides A mineral composed of sulfur $[S]^{-1}$ as the anion.

Supercontinent A composite continent composed of a combination of several continental areas.

Superposed streams A stream superimposed onto buried structures that are exposed during erosion.

Surf zone The beach zone where waves break.

Surface waves Seismic waves (R and L waves) that travel along the interface of layers or along the surface of Earth.

Suspended load Fine grains that are supported by the turbulence of the stream.

Sutured A zone where two continents have been welded together, usually indicated by the remnants of a mountain range created when the continents collided.

Symmetrical fold A fold that is uniform on both sides.

Syncline A concave upward fold.

T

Taconic Orogeny A mountain building event that occurred during the Ordovician period, identified in the northern Appalachian Mountains.

Talus Cone-shaped rock debris found at the base of cliffs.

Tectonic isolation The separation of organisms of the same species by the tectonic building of a mountain range and isolation of the separated organisms, thus allowing speciation to occur in the isolated gene pools.

Tektite A rounded, pitted body of silicate glass of extraterrestrial origin, a few centimeters or less in size.

Tension A state of stress in which tensile stress, stress that causes separation, predominates; stress that pulls materials apart.

Tephra   A general term for the pyroclastic materials that are blown out of volcanoes.

Terminal moraine   A mound of glacial drift that marks the forward extent that a glacier moved before it melted back.

Terrigenous   Continental as opposed to marine environments.

Terrigenous sediments   The sediments derived from continental materials.

Texture   The general physical appearance of a rock.

Thermal maximum   The elevated temperatures of approximately 11,000 years B.P.

Thermoluminescence dating   A method of dating applicable to materials that have once been heated, which uses the release of energy stored during heating as electron displacements in the crystal lattice.

Thrust sheet   The body of nearly horizontally lying rock pushed over a large-scale thrust fault.

Tidal channels   Channels cutting through sediment accumulated in tidal deltas behind inlets in barrier islands, in which tidal flow is highest.

Tidal delta   The fan-shaped sediments deposited behind and in front of tidal inlets through barrier islands.

Tidal flat   A coastal, nearly horizontal, muddy, marshy, and barren tract of land that floods and drains during each tidal cycle.

Till   Jumbled, unsorted sediment dropped by glaciers.

Tillite   A lithified sedimentary rock formed from glacial till.

Time-transgressive bed   A layer of similar material deposited over a long period of time in response to sea-level rise and fall.

Time units   Based on biostratigraphic names, the fossils identified as living at a given time; these are independent of rocks.

Time-rock units   Given as chronostratigraphic names where the rocks are correlated to biostratigraphic units.

Topset beds   The horizontal, sandy sediment deposited on the upper surface of deltas.

Trace fossil   Sedimentary structure consisting of the fossilized evidence of the presence of an ancient organism, such as footprints or burrows left by worms.

Traction carpet   The loose sediment, too heavy to be moved in suspension, that is moved along the bottom of streams or in the ocean by stream flow or ocean currents.

Traction transport   The transport of materials along the bottom as a traction carpet.

Transform fault   A special type of strike-slip fault that allows offset of midocean ridge segments, trench-to-ridge segments, or trench-to-trench segments.

Transgression   The spread of the ocean over land areas, resulting in continental flooding.

Transverse dune   An asymmetrical sand dune that is elongated perpendicular to the prevailing wind direction. The leeward slope is steep, while the windward slope is gentle.

Travertine   A finely crystalline, calcium carbonate precipitate, usually forming a layered or fibrous structure.

Trellis stream pattern   A pattern developed when erosion is controlled by underlying bedrock along zones of differential weakness.

Trench   A narrow, elongate depression in the seafloor that results from subduction of ocean crust.

Tributary   A stream feeding into the main trunk of a stream.

Tsunami   A potentially very destructive seismic sea wave produced by offset of the seafloor during an earthquake.

Tuff   A general term for all consolidated pyroclastic rocks.

Tundra   A treeless plain, characteristic of Arctic or Subarctic regions and underlain by permafrost.

Turbidite   A sediment or rock deposited from a turbidity current, characterized by graded bedding, with fine grains at the top and coarse grains at the bottom.

Turbidity currents   An underwater, gravity-driven, rapid flow of dense, sediment-rich water that flows downslope, sometimes very fast (40–55 km/hr), eroding as it goes.

Turbulent flow   Fluid flow in which flow lines are mixed and chaotic.

Twinning   The development of a twin crystal in a mineral.

Type section   The originally described stratigraphic sequence that constitutes a stratigraphic unit.

U

Unconformity   A substantial break in the rock record between one unit overlain by another, represented by a physical difference in the beds.

Unpaired terrace   A stream terrace on one side of a river valley, where the stream has eroded the other side of the valley, thus destroying the other half of the pair.

V

Variscan Orogeny   See Hercynian Orogeny.

Valley glaciers   A glacier lying in and forming a U-shaped mountain valley.

Varves   A sequence of finely, light- and dark-banded layered sediments deposited yearly, resulting from seasonal variations in weather.

Vent   The opening at Earth's surface through which volcanic materials are extruded onto the surface.

Ventifact   Rocks with carved, fluted, and polished surfaces that result from wind-blown desert erosion.

Vertebrate organisms   Those organisms with a backbone.

Vesicles   Gas pockets produced as the gas expanded and then was trapped as the lava cooled.

Vesicular   A lava containing many gas pockets.

Vestigial organs   Parts of an organism that have no function.

Volcanic ash   See ash.

Volcanic block   A large angular fragment with a diameter greater than 64 mm, ejected from a volcano.

Volcanic bomb   A large rounded fragment with a diameter greater than 64 mm, ejected from a volcano.

Volcanic dome   A steep-sided rounded extrusion of highly viscous lava squeezed out of a volcano, which forms a dome-shaped mass.

Volcanic lapilli   Rock fragments 2 to 64 mm in size that are erupted from a volcano.

Volcanic tuff   See tuff.

W

Water gap   A deep pass in a mountain ridge through which a stream flows.

Water table   A surface below Earth's surface between the water saturated zone below and the zone of aeration above.

Wave refraction   The bending of ocean waves as they approach the shore due to wave-base drag on the bottom as depth shallows.

Weathering   The mechanical and chemical processes by which rocks at Earth's surface are destroyed.

Wegener   Alfred Wegener was a proponent of continental drift and in 1912 proposed the existence of a supercontinent, Pangaea, which existed during the late Paleozoic.

Western Cordillera   The main mountain/tectonic region of the western part of the North American continent.

Wilson   In 1965 J. Tuzo Wilson proposed the existence of transform faults, a necessary element in plate tectonics.

Windward   The direction from which the wind is blowing.

Winnowing   A process whereby fine-grained particles are eroded from continental or marine surface sediments, leaving the coarser particles behind.

World Wide Standardized Seismograph Network (WWSSN)   A global seismic network built in the late 1950s and early 1960s to monitor global seismic activity; initially designed to monitor Soviet nuclear tests.

X

Xenolith   A block of country rock that has been incorporated into an intruding magma, identified as a distinctive lithology within the solidified igneous rock.

Z

Zeolite   A set of low-grade, hydrous, usually white or colorless metamorphic minerals.

Zone   A minor interval in any stratigraphic unit.

Zone of aeration   A zone in the subsurface, above the water table, that is not saturated with water.

Zone of saturation   A zone in the subsurface, below the water table, that is saturated with water.

Index

Contents

Introduction

The Lifespan Developmentalist

Competitors in the Boston Marathon run through several small towns before reaching "Heartbreak Hill" and the finish line. Some years ago I lived in one of these communities (Ashland) and, in company with three-year-old David, joined the local throng awaiting the appearance of the marathoners. The runners were greeted with encouragement and applause by the adult Ashlanders, while a number of children gleefully accompanied them for a short distance. David, however, had a different approach to the situation. To my surprise, he started running in the opposite direction. On catching up with him, I asked, "Why are you running this way?" His bright-eyed answer: "I want to see where they're all coming from!" And I had imagined myself the developmentalist in the family!

Understanding the course of human development requires attention both to where we are coming from and where we are going. Individuals at times state this principle for themselves when they admit, "I didn't know whether I was coming or going!" Theoreticians and researchers face exceptional challenges in trying to comprehend the patterns, meanings, and potentials of development from infancy through old age. This is quite a different task than, say, describing the exploratory behavior of a year-old child, the career anxieties of a thirty-year-old,

or the political attitudes of a seventy-year-old. The lifespan developmentalist hopes to understand the individual along his or her entire journey through time.

In one sense, the individual is "real" only at a particular point in time. In another sense, however, the individual is most fully actualized or "real" only when the total shape of his or her life can be discerned. Two decades after he ran "the wrong way" with the marathoners, David is a different and more complex person, and the same will probably hold true years from now. And yet the child and the young man will all have contributed to the person he continues to become. The lifespan developmentalist, then, takes on the dual responsibility of trying to do justice by the particular individual at a particular moment and the continuing, growing, and aging person throughout all the moments of his or her life.

As you will discover throughout the *Encyclopedia of Adult Development*, the challenge is even greater than what has already been implied. Each of us not only emerges as an unique individual, but we also have unique experiences with the world every step of the way. Today's researchers, theoreticians, and service providers are usually very much aware of the socio-environmental side of human development. Did you grow up in rural Wyoming? A Baltimore housing project? An affluent suburb of Cleveland? Were you accustomed to living with a houseful of brothers and sisters and boisterous interactions with more relatives than you could count? Or were you the only child of a single working parent who had few relatives to offer social support? Were you an anxious teen or an anxious parent when the Vietnam conflict was also generating a conflict of values and allegiances at home? These are but a few of the questions that illustrate the developmentalist's concern for understanding individual development within particular socio-environmental contexts at particular moments in history.

For our own particular shared moment in history, the *Encyclopedia of Adult Development* offers the opportunity to rethink the relationship between the individual's journey through life and the social, symbolic, and physical forces that provide the context for our existence. Surely, this journey and this relationship differ markedly from the pattern of lifespan development experienced in a past that was concerned more with infant and youth mortality than with medical care for the aged, had little doubt that a woman's place was in the home, and considered a man's success more dependent on his ability to lift a shovel or pitchfork than to process information at a computer station. Yet, surely there are universal events, challenges, and experiences that link us with all previous generations as well: establishing ourselves as independent and responsible adults; selecting a mate and creating a life together; creating new lives and sharing the anxieties as well as the pleasures of parenting; encountering loss, sorrow, threat, doubt, and the unknown.

About This Encyclopedia

With these thoughts in mind, I have called upon gifted researchers and service providers from a variety of fields who have reason to be interested in adult development: anthropology, communication, education, family studies, health sciences, history, nursing, physiology, psychiatry, psychology, sociology, and speech and hearing sciences. I have asked each contributor to share basic ideas and findings, but also concerns and controversies, within their topic areas. Some contributors found that they could call upon a set of well-documented conclusions that lent themselves to concise and straightforward presentation. Some contributors found that controversies dominate their topic at present, or that methodological concerns could not be ignored if the material was to be presented in an accurate and judicious manner. It was my own judgment that each contributor should be free to present his or her topic in the way that was best attuned to the current status of this field. This seemed preferable to forcing a "standard operating procedure." Therefore, some of the 106 articles chiefly report well-established findings, while others open relatively new areas of inquiry. Rest assured that the contributors are neither withholding firmly established conclusions from you nor claiming that all the answers are in when this happens to be far from the truth. The length of contributions was also allowed to vary in accord with the amount of information or explanation that was needed to deal usefully with the topic at hand at this point in time.

There may be as many ways to use this book as there are readers. As with any encyclopedia, the present volume is intended as a convenient "look-it-upper."

What is really accomplished by exercise in adult life? What do we know about driving behavior through the adult years? What are the main risks to life in adulthood? These are all obvious "look-it-ups." Like the more interesting encyclopedias in various fields, this book also rewards the curious scanner. "Hmmm—twins! I've always sort of wondered about that, and 'Native American Perspectives on the Lifespan'—that sounds interesting too."

Perhaps most satisfying is the opportunity for each reader to create his or her own set of mini-books by linking together several entries that share a theme or concern of particular interest. Several of these implicit mini-books have been built in. For example, one might sequence through "Age 21," "Age 40," "Age 65," and "Centenarians" to survey symbolic markers of adult development and aging. This cruise might then be supplemented by "Slowing of Behavior with Age," a phenomenon that offers an objectivistic counterpoint to the symbolic marking of the life course by identifying certain ages as especially meaningful. One might instead give priority to understanding how social attitudes and political-economic processes affect the way we perceive and treat people of different ages. Here one could choose to begin with "Development and Aging in Historical Perspective" and then move on to "Metaphors of the Life Course" and "Generational Equity," to name just two of the cognate entries.

To make it easier for readers to construct their own sequences or sets, each article is followed by a listing of "See also" references that pick up on one or more of the issues discussed in that entry. There is also a "Guide to Related Topics" encompassing the entire book. The contributors have provided fairly extensive references for further reading. Although hundreds of other references had to be cut, we have retained many citations because readers of a preceding work (*Encyclopedia of Death*) told us they appreciated this convenience and stimulation.

Overall, the *Encyclopedia of Adult Development* offers background information on earlier phases of development when this seems necessary, but our focus is upon the adult years. Similarly, we deal with some medical and pathological issues when these seem integral to understanding developmental processes and experiences, but one will need to look elsewhere for comprehensive coverage of all that can go wrong with our wiring and plumbing when exposed to time's not invariably tender mercies. It is also worth noting that this is not a treatise on aging and the aged. The contributors do have much to say about the later adult years, and some of their observations are new, vital, and challenging. We see aging and the aged, however, not as a specialized topic, but as part of the total adult lifespan within the social, symbolic, and physical environment. By and large, scholars have given less attention to the middle adult years than to early developmental processes and aging. Here we take a step toward closing the gap.

Preparing an encyclopedia is something of a marathon project, and I wish to express my gratitude to the many contributors who have been my companions, to the patient souls at Oryx Press, and to Bunny Kastenbaum, who sacrificed some precious free time to work her magic with the computer codes at the last moment.

Contributors

Hiroko Akiyama, Ph.D.
Institute for Social Research
University of Michigan
P. O. Box 1248
Ann Arbor, MI 48106-1248

Jess Alberts, Ph.D.
Department of Communication
Arizona State University
Tempe, AZ 85287-1205

Carolyn M. Aldwin, Ph.D.
Human Development & Family Study
Department of Applied Behavior Science
University of California
Davis, CA 95616

Joseph Amato, Ph.D.
Department of History
Southwest State University
Marshall, MN 56258

Lars Andersson, Ph.D.
Department of Stress Research
Karolinska Institute
Box 60 205
S-104 01 Stockholm, Sweden

Toni C. Antonucci, Ph.D.
Institute for Social Research
University of Michigan
426 Thompson Street
Ann Arbor, MI 48106

William E. Arnold, Ph.D.
Department of Communication
Arizona State University
Tempe, AZ 85287-1205

Robert C. Atchley, Ph.D.
Scripps Gerontology Center
Miami University
Oxford, OH 45056

Kurt W. Back, Ph.D.
Department of Sociology
Duke University
Durham, NC 27710

Jeanne E. Bader, Ph.D.
Home Economics Department
California State University, Long Beach
1250 Bellflower Boulevard
Long Beach, CA 90804-0501

Charles R. Bantz, Ph.D., Chair
Department of Communication
Arizona State University
Tempe, AZ 85287-1205

Marty Birkholt
Department of Communication
Arizona State University
Tempe, AZ 85287-1205

Thomas O. Blank, Ph.D.
Travelers Center on Aging, U-58
348 Mansfield Road
University of Connecticut
Storrs, CT 06269-2058

Stanley H. Brandes, Ph.D.
Department of Anthropology
University of California
Berkeley, CA 94720

James L. Case, Ph.D.
Department of Speech and Hearing Science
Arizona State University
Tempe, AZ 85287-0102

Allan B. Chinen, M.D.
1437 11th Avenue
San Francisco, CA 94122

David A. Chiriboga, Ph.D.
School of Allied Health
The University of Texas Medical Branch at
    Galveston
Galveston, TX 77550

Wotjek J. Chodzko-Zajko, Ph.D.
Applied Physiology Research Laboratory
School of Physical Education, Recreation,
    and Dance
Kent State University
Kent, OH 44242-0001

Thomas R. Cole, Ph.D.
Institute for the Medical Humanities
The University of Texas Medical Branch at
    Galveston
Galveston, Texas 77550

Frederick C. Corey, Ph.D.
Department of Communication
Arizona State University
Tempe, AZ 85287-1205

Harold Cox, Ph.D.
Sociology and Social Work
Indiana State University
Terre Haute, IN 47809

John Crawford, Ph.D.
Department of Communication
Arizona State University
Tempe, AZ 85287-1205

Maureen Edwards
Department of Health Education
University of Maryland
College Park, MD 20742

Ruth Ann Erdner, Ph.D.
Department of Sociology and Social Work
Memphis State University
Memphis, TN 38152

Barbara J. Felton, Ph.D.
Department of Psychology
New York University
6 Washington Place, Room 279
New York, NY 10003

Linda K. George, Ph.D.
Duke University Medical Center
Duke University
Durham, NC 27710

Kathy Goff, Ph.D.
Georgia Studies of Creative Behavior
183 Cherokee Avenue
Athens, GA 30606

M. Christina Gonzalez, Ph.D.
Department of Communication
Arizona State University
Tempe, AZ 85287-1205

David Gutmann, Ph.D.
Professor of Psychiatry and Education
Northwestern University
633 Clark Street
Evanston, IL 60201

Rebecca F. Guy, Ph.D., Chair
Department of Sociology and Social Work
The College of Arts and Sciences
Memphis State University
Memphis, TN 38152

Barbara K. Haight, RNC, Dr. PH.
College of Nursing
Medical University of South Carolina
171 Ashley Avenue
Charleston, SC 29425-2403

Joel Haycock, Ph.D.
Forensic Training and Research Center
University of Massachusetts Medical
    Center
55 Lake Avenue North
Worcester, MA 01655

Bert Hayslip, Jr., Ph.D.
Department of Psychology
College of Arts and Sciences
University of North Texas
P. O. Box 13587
Denton, TX 76203-3587

Michael L. Hecht, Ph.D.
Department of Communication
Arizona State University
Tempe, AZ 85287-1205

Jon Hendricks, Ph.D.
Department of Sociology
Oregon State University
Corvallis, OR 97331

Gregory A. Hinrichsen, Ph.D.
Hillside Hospital
Long Island Jewish Medical Center
Glen Oaks, NY 11004

James S. Jackson, Ph.D.
Institute for Social Research
University of Michigan
Ann Arbor, MI 48106

Beatrice K. Kastenbaum, R.N., M.S.N.
College of Nursing
Arizona State University
Tempe, AZ 85287

Robert Kastenbaum, Ph.D.
Department of Communication
Arizona State University
Tempe, AZ 85287-1205

Richard C. Keefe, Ph.D.
Department of Psychology
Scottsdale Community College
Scottsdale, AZ 85250

Douglas T. Kenrick, Ph.D.
Department of Psychology
Arizona State University
Tempe, AZ 85287

Helen Q. Kivnick, Ph.D.
21 Seymour Avenue SE
Minneapolis, MN 55414-3521

Dan Leviton, Ph.D.
Department of Health Education
University of Maryland
College Park, MD 20742

Dale A. Lund, Ph.D., Director
Gerontology Center
College of Nursing
University of Utah
Salt Lake City, UT 84112

Susan MacLaury
Department of Health Sciences
William Paterson College
Wayne, NJ 07470

Kyriakos S. Markides, Ph.D., Director
Department of Preventive Medicine and
    Community Health
1.128 Ewing Hall–Rt. J53
The University of Texas Medical Branch at
    Galveston
Galveston, TX 77550

Peter J. Marston, Ph.D.
Department of Speech Communication
California State University
Northridge, CA 91330

Judith Martin, Ph.D.
Department of Communication
Arizona State University
Tempe, AZ 85287-1205

John L. McIntosh, Ph.D.
Department of Psychology
Indiana University at South Bend
South Bend, IN 46615

William A. McKim, Ph.D.
Department of Psychology
Memorial University of Newfoundland
St. John's, Newfoundland
Canada

Brian L. Mishara
Psychology Department
University of Quebec at Montreal
Montreal, Quebec P2 H3C 3P8
Canada

Michael D. Mumford, Ph.D.
Center for Behavioral and Cognitive
    Studies
Department of Psychology
George Mason University
Fairfax, VA 22030-4444

Thomas K. Nakayama, Ph.D.
Department of Communication
Arizona State University
Tempe, AZ 85287-1205

William E. Oriol, Director
Communications Department
The National Council on the Aging, Inc.
409 Third Street SW
Washington, DC 20024

Erdman B. Palmore, Ph.D.
Professor Emeritus
Duke University Medical Center
Durham, NC 27710

Candida C. Peterson, Ph.D.
Department of Psychology
The University of Queensland
Brisbane, Queensland 4072 Australia

Sandra Petronio, Ph.D.
Department of Communication
Arizona State University
Tempe, AZ 85287-1205

Leonard W. Poon, Ph.D., Director
Gerontology Center
The University of Georgia
100 Candler Hall
Athens, GA 30602

Karen A. Roberto, Ph.D., Coordinator
Gerontology Program
Department of Human Services
University of Northern Colorado
Greeley, CO 80639

Ken J. Rotenberg, Ph.D.
Department of Psychology
Lakeland University
Thunderbird, Ontario P7B 5E1
Canada

Timothy A. Salthouse
Professor of Psychology
Georgia Institute of Technology
Atlanta, GA 30332-0170

Rick J. Scheidt, Ph.D.
Human Development and Family Studies
College of Human Ecology
Kansas State University
Manhattan, KS 66506

Paula P. Schnurr, Ph.D.
National Center for PTSD (116D)
Veterans Administration
Medical and Regional Office Center
White River Junction, VT 05001

Mildred M. Seltzer, Ph.D.
Director of Education and Training
Scripps Gerontology Center
Miami University
Oxford, OH 45056

Ilene C. Siegler, Ph.D.
Duke University Medical Center
Department of Psychiatry
Duke University
Durham, NC 27710

Bernard Spilka, Ph.D.
Department of Psychology
University of Denver
University Park
Denver, CO 80208-0204

Bernard D. Starr, Ph.D., Director
Advanced Certificate Program in
    Gerontological Studies
City University of New York
Brooklyn College
Brooklyn, NY 11210

Robert D. Strom, Ph.D.
Curriculum and Instruction
College of Education
Arizona State University
Tempe, AZ 85287-1205

Shirley K. Strom, Ph.D.
Office of Parent Development International
Arizona State University
Tempe, AZ 85287

L. Eugene Thomas, Ph.D.
Travelers Center on Aging
348 Mansfield Road
University of Connecticut
Storrs, CT 06269-2058

E. Paul Torrance, Ph.D., Director
Georgia Studies of Creative Behavior
183 Cherokee Avenue
Athens, GA 30606-4305

Guus Van Heck, Ph.D.
Department of Psychology
University of Tilburg
Tilburg, The Netherlands

Max Vercruyssen, Ph.D., Director
Aging and Ergonomics Laboratory
Center on Aging
University of Hawaii at Manoa
2545 The Mall, Bilger Annex B
Honolulu, HI 96822

Hannelore L. Wass, Ph.D.
Educational Psychology
1418 Norman Hall
University of Florida
Gainesville, FL 32611

Alan T. Welford, ScD.
187A High Street
Aldeburgh
Suffolk, England IP15 5AL

Mary G. Winkler
Institute for the Medical Humanities
The University of Texas Medical Branch at
    Galveston
Galveston, TX 77550

Steven K. Wisensale, Ph.D.
School of Family Studies, U-58
University of Connecticut
Storrs, CT 06268

Alphabetical List of Articles

Guide to Related Topics

Adulthood as a Time of Life

Adult Development and Aging, Models of
Age 21
Age 40
Age 65
Centenarians
Development and Aging in Historical
    Perspective
Maturity
Metaphors of the Life Course
Mid-Life Crisis

Communicating

Adult Children and Their Parents
Attachment Across the Lifespan
Communication
Creativity
Expressive Arts
Friendships through the Adult Years
Grandparent Communication Skills
Interethnic Relationships
Language Development
Listening
Privacy Issues
Reminiscence and Life Review
Sibling Relationships
Social Relationships, Convoys of
Travel: Stimulus to Adult Development
The Voice
Work Organization Membership and
    Behavior

Competency and Coping

Caregiving
Communication
Conflict as a Challenge in Adult
    Development

Control and Vulnerability
Driving: A Lifespan Challenge
Exercise
Grandparent Education to Enhance Family
    Strength
Habituation: A Key to Lifespan
    Development and Aging?
Humor
Intelligence
Learning and Memory in Everyday Life
Life Events
Listening
Loving and Losing
Mastery Types, Development, and Aging
Maturity
Mid-Life Crisis
Political Beliefs and Activities
Religion and Coping with Crisis
Retirement: An Emerging Challenge for
    Women
Retirement Preparation
Risk to Life through the Adult Years
Spiritual Development and Wisdom: A
    Vedantic Perspective
Stress
Suffering
Trust as a Challenge in Adult
    Development
Widowhood: The Coping Response
Work Capacity Across the Adult Years

Gender and Sex Influences

Adult Children and Their Parents
African-American Experiences through the
    Adult Years
Age and Mate Choice
Caregiving
Development and Aging in Historical
    Perspective

Mental Health and Illness

Alcohol Use and Abuse
Conflict as a Challenge in Adult
    Development
Control and Vulnerability
Depression
Drug Use and Abuse
Exercise
Expressive Arts
Grandparent Communication Skills
Humor
Loneliness
Mental Health in the Adult Years
Mental Health Resources for Adults
Mid-Life Crisis
Narcissism
Premature Aging
Sex in the Later Adult Years
Sexuality
Sleep and Dreams
Spiritual Development and Wisdom: A
    Vedantic Perspective
Suffering
Suicide

Paths through Life

Adult Development and Aging, Models of
African-American Experiences through the
    Adult Years
Creativity
Criminal Behavior
Development and Aging in Historical
    Perspective
Gender as a Shaping Force in Adult
    Development and Aging
Gender Differences in the Workplace
Homosexuality
Individuality
Interethnic Relationships
Japanese Perspectives on Adult
    Development
Loneliness
Metaphors of the Life Course
Mexican Americans: Life Course
    Transitions and Adjustment
Mid-Life Crisis
Military Service: Long-Term Effects on
    Adult Development

Narcissism
Premature Aging
Social Class and Adult Development
Spiritual Development and Wisdom: A
    Vedantic Perspective
Subcultural Influences on Lifelong
    Development
Twins through the Lifespan

Perceiving Self and World

Body Senses
Listening
Sleep and Dreams
Taste and Smell
Vision

Personal Change and Development

Age 40
Communication
Creativity
Development and Aging in Historical
    Perspective
Expressive Arts
Fairy Tales as Commentary on Adult
    Development and Aging
Grandparent Education to Enhance Family
    Strength
Habituation: A Key to Lifespan
    Development and Aging?
Health Education and Adult Development
Humor
Individuality
Language Development
Loving and Losing
Marital Development
Mastery Types, Development, and Aging
Maturity
Menopause
Metaphors of the Life Course
Parental Imperative
Reminiscence and Life Review
Sexuality
Spiritual Development and Wisdom: A
    Vedantic Perspective
Suffering
Travel: Stimulus to Adult Development

Work Organization Membership and
  Behavior

Social and Environmental Influences

African-American Experiences through the
  Adult Years
Cohort and Generational Effects
Contextualism
Gender as a Shaping Force in Adult
  Development and Aging
Gender Differences in the Workplace
Generational Equity
Happiness, Cohort Measures of
Housing as a Factor in Adult Life
Japanese Perspectives on Adult
  Development
Mexican Americans: Life Course
  Transitions and Adjustment
Military Service: Long-Term Effects on
  Adult Development
Native American Perspectives on the
  Lifespan
Place and Personality in Adult
  Development
Rural Living: What Influence on Adult
  Development?
Social Class and Adult Development
Social Relationships, Convoys of

Subcultural Influences on Lifelong
  Development
Travel: Stimulus to Adult Development

Theoretical Perspectives

Contextualism
Control and Vulnerability
Development and Aging
Disengagement Theory
Habituation: A Key to Lifespan
  Development and Aging?
Mastery Types, Development, and Aging
Maturity
Parental Imperative

Thought Processes

Communication
Control and Vulnerability
Creativity
Habituation: A Key to Lifespan
  Development and Aging?
Information Processing
Intelligence
Intelligence—Crystallized and Fluid
Language Development
Learning and Memory in Everyday Life
Memory
Reminiscence and Life Review
Slowing of Behavior with Age

ADULT CHILDREN AND THEIR PARENTS

Humans, unlike some other species, typically do not lose all contact with their mature offspring. The parent-child bond is often the longest-lasting close relationship in an individual's life, continuing even after the parent's death. As a leading developmental psychologist explains, "A person's response to the death of a parent usually demonstrates that the attachment bond has endured. Even after mourning has been resolved, internal models of the lost figure continue to be an influence" (Ainsworth, 1989, pp. 710-11).

To understand the development of the individual adult through the lifespan, it is important to have accurate knowledge of the overt dealings and inner feelings between a parent and child throughout the child's adulthood. Therefore, this article begins with a brief descriptive account of the contemporary parent-adult child relationship. Two major theoretical perspectives on the adult family—equity theory and Erikson's (1968) epigenetic developmental theory—are then outlined and evaluated in light of current research.

Changing Family Patterns

In Western cultures, recent social, economic, and childbearing changes have created "a quiet revolution in the demography of intergenerational family life" (Bengston, Rosenthal, & Burton, 1990). This "revolu-tion" has in turn generated new possibilities and challenges for the psychosocial relationship between adult children and their middle-aged or elderly parents over the lifespan.

One demographic innovation is described as the extended family's new "bean-pole" shape (Bengston, Rosenthal, & Burton, 1990). Families of four and five generations are replacing those of only two or three generations. Thus, Shanas estimated in 1980 that approximately 50 percent of adults in the United States aged 65 and over belonged to families with at least four living generations. Failing health in the eldest cohorts of these families is apt to place new burdens of care upon aged parents and elderly offspring alike. For example, one recent urban Australian survey revealed that "among the current generation of disabled old people in the community, ten percent of their children are themselves aged 60 or over" (Kendig, 1986, p. 180). The family's new "bean-pole" shape also means that contemporary young adults have fewer siblings and siblings-in-law to relate to than their parents did, possibly making their relationship with their parents a more salient and emotionally significant one throughout the lifespan.

Another change in family life noted by Hagestad (1981) is the new phenomenon of *social-age peership*, an increasing pursuit by both generations of similar lifestyles. This trend is a sharp departure from the past when each adult generation in a family usually had a distinctly different role to play. Today, by contrast, we find university students, workers, retirees, and newlyweds in all of a family's adult generations. Rises in the rates of divorce and remarriage have like-wise meant that some older parents (especially fathers) re-experience the bearing and rearing of young children simultaneously with their adult offspring's first venture into parenthood.

What the full impact of such changes will be upon the nature and quality of parent-offspring ties through the lifespan is still unclear. But one consequence of social-age peership may be a rapprochement of values and attitudes between adult children and their parents. This was the perhaps unexpected finding from Miller and Glass's (1989) assessment of attitudes toward supposed "generation gap" issues such as religion, politics, and sexual equality across three adult generations within the same U.S. families. Attitudes were compared in 1971 and again in 1985. Although the attitudes of elderly grandparents (mean age 78 in 1985) and their middle-aged offspring (mean age 57) diverged increasingly over time, the same middle-aged parents expressed attitudes almost identical to those of their young adult children (mean age 33) at both testing points. There was no suggestion of an increasing gap between the middle and young adult generations as might have been expected on the basis either of the offspring's departure from home or the parents' aging.

In fact, the timing of the fledgling offspring's departure from the parental "nest" is yet another area in which contemporary social changes have been occurring. During the economic prosperity of the 1950s, 1960s, and early 1970s, the nest emptied earlier and earlier. Families became more compact than in previous generations. The ages of marriage dropped steadily lower, and rates of first marriage reached all-time highs (Clemens & Axelson, 1985). Most wives in the mid to late 1970s could count on having all their children launched by about the time they themselves reached menopause (Fiske-Lowenthal, Thurner, & Chiriboga, 1977).

This pattern has begun to change in recent decades with dramatic increases in youth unemployment, early divorce, and unmarried parenthood. Today, fewer adult children leave home in the first place. Of those who do, more eventually return to the parental household. Clemens and Axelson (1985) estimated a 23 percent increase between 1968 and 1983 in the number of older couples in the U.S. with a child aged 18 to 24 in their homes. Those with one or two children over the age of 25 in residence increased by more than 60 percent over the same period.

Older parents are inclined to greet this new pattern with alarm. Eighty percent of the older full-nest couples Clemens and Axelson (1985) interviewed had planned on an empty nest by the time their youngest was 22. Furthermore, a full 60 percent of the older parents were adamantly opposed to their children continuing to live with them. Serious conflicts with at least one resident adult child were reported by 42 percent of this sample. The divisive issues were mainly "times of coming and going," "cleaning and maintenance of the house," "mealtimes," and "money" (Clemens & Axelson, 1985, pp. 261-62).

Researchers have suggested that having offspring at home is contributing to the declines that are typically observed in spouses' feelings of affection, intimacy, and commitment for one another during the mature years of marriage (e.g., Anderson et al., 1983; Arbyle, 1986). However, this adverse impact of resident children upon the parents' marriage evidently ceases once the offspring reach adulthood. In fact, Suitor and Pillemer (1987) found no differences when they compared levels of marital happiness between older couples with a completely empty nest with a matched set of older spouses who still had at least one adult child in residence. Furthermore, the study found no overall difference in the amount of marital conflict or discord between empty-nest and full-nest spouses. Thus, at least as far as their own marital happiness is concerned, it seems that older couples have little to fear from the continued presence of adult offspring in their homes.

Proximity and Contact

Systematic research has dispelled another myth about adult family life—the belief that old age isolates elderly parents from their adult offspring. In fact, multigenerational households have never been the norm in Western societies (Kendig, 1986). Despite not living together under one roof, aging parents and their married adult children have frequent contact. For example, one cross-cultural survey conducted in the middle part of this century showed that more than 80 percent of the aged parents in the U.S., Great Britain, and Denmark had contact with at least one adult child at least once a week (Riley & Foner, 1968).

More recent surveys have confirmed that these patterns are still true today (Kendig, 1986; Shanas, 1980). Furthermore, the high frequency of contact between elderly parents and their adult sons and daughters can be explained partly by the two generations living so near to one another. A majority of older parents are found to live within an easy 5- to 10-minute walk of at least one adult child, even in large nations such as Australia (Kendig, 1986) and the U.S. (Shanas, 1980), where children's occupational mobility can take them far from their parents' homes.

Although frequent contact between adult family generations is clearly desirable, this contact does not, in itself, guarantee the benefit that older parents usually desire most from their adult children—intimacy and emotional closeness (Kelley, 1981). In fact, the Sydney survey (Kendig, 1986) showed that despite living so near, less than 30 percent of older fathers (whether widowed, divorced, or married) could describe an adult son or daughter as a confidante. The pattern for older Australian married women was very similar. Indeed, confiding in a child (usually a daughter) was reported by less than 40 percent of elderly widows, even though they were the ones who chose an offspring as a confidante more often than any of the other elderly groups surveyed in this study. On the other hand, the fact that these older Australians were even less prone to confide in their friends, neighbors, older kin, or other social contacts suggests that the parent-child bond, while not guaranteeing intimacy, may nonetheless serve as a necessary precondition for it in the mind of many elderly parents.

Despite careful research, the missing link explaining how some elderly parents manage to grow emotionally close to an adult child while others retain an aloof and superficial relationship remains somewhat elusive. Indeed, even the commonsense notion that the more frequent their contact, the closer the two generations should feel toward one another has not been reliably confirmed. For example, one recent U.S. study found that the older parent's state of health and financial security were far more important predictors of strong affection and close feelings of intimacy with an adult child than how often the parent and this child saw one another or communicated by letter or telephone (Peterson, 1989). Perhaps the fear of becoming a burden on one's child, either through failing health or financial hardship, inhibits an older parent's willingness to confide, even when frequent contact might otherwise make this possible.

Equity Theory and Intergenerational Exchange

Another possible explanation for these inconsistencies between levels of intergenerational social contact and degree of emotional closeness can be derived from *equity theory* (Hatfield et al., 1979). According to equity theory, an objective factor such as frequency of contact has far less to do with family members' mutual feelings of intimacy or satisfaction than the subjective variable of how "fair" the balance is seen to be between the two partners' net investments and net profits from their relationship to one another. In an "equitable" relationship both parties feel they are receiving a "fair deal." Their respective levels of input into the relationship seem to be balanced by comparable relative outcomes from it. The exchanges involved are likely to include the giving of love, money, and assistance "minus" the negative inputs of hostility, anxiety, and so forth.

Two distinct types of "inequitable" intimate relationships are possible, each with similarly deleterious effects upon the partners' satisfaction and mutual affection. An "underbenefited" relationship is one in which the respondent feels exploited. For example, an elderly father might feel that although he contributes as much or more than his son does to their relationship, it is the son who derives the lion's share of benefits from it. The other type of inequitable relationship, "overbenefit," is one in which the respondent gains proportionately more than he or she should, relative to net contribution. For example, an aged mother might perceive that her daughter does all the giving and helping in a relationship that mainly benefits the mother.

Do parent-offspring relationships tend to become inequitable with the children's maturation and the parents' aging? Evidently not, according to the results of a recent study of Australian adult families (Peterson & Peterson, 1988). In fact, 64 percent of older fathers and 68 percent of older mothers rated their relationships with their adult sons and daughters as being strictly equitable. An even larger majority of older mothers (100 percent) and older fathers (75 percent) who still had frequent contact with an aged parent perceived strict equity in such relationships. The majority of adult sons (57 percent) and adult daughters (65 percent) also described their dealings with their parents as a strictly equitable balance. The results of this study were therefore consistent with the equity theory prediction (Hatfield et al., 1979) that intimates continually strive throughout their lives to maintain or restore equity in their relationships with intimate kin.

Furthermore, the pattern of results suggests that the parent-child equity arrangement tends to balance itself out again over time even when there have been disruptions by such early adult life events as career entry, marriage, childbearing, or college (Peterson & Peterson, 1988). The same long-term perspective may likewise help to maintain famil-

ial equity when parents encounter frail health or disability in extreme old age. Antonucci suggests that "in the United States deposits are made early in one's life in anticipation of future needs or withdrawals...This permits an escape from negative feelings of overbenefit or exploitation" (Antonucci, 1990, p. 213).

However, inequities can arise out of an incapacitated aged parent's unrequited dependency upon an adult child, with consequences that are emotionally distressing even to the overbenefited adult parent (Kendig, 1986). In fact, a comparison among elderly widows and widowers residing in an adult child's home showed that only 5 percent of those who were fit, active, healthy, and contributing to the household felt depressed as compared to 62 percent of those who were disabled and dependent upon their child's care. This latter result suggests that when physical capacity prevents an older parent from keeping up his or her own end of the equity balance, the consequence can be a drastic loss of morale.

Developmental Changes in the Parent-Adult Child Relationship

Equity theory suggests that the parent-child relationship should change very little over time in normal, healthy families (Hatfield et al., 1979), allowing for a few minor fluctuations or temporary upheavals.

By contrast, Erikson's (1968) lifespan developmental theory predicts a regular sequence of changes in the way that parents and their adult offspring will relate to each other as they traverse the milestones, life crises, and turning points of mature adulthood.

After the adolescent achieves a secure *identity*, the next developmental task for the young adult revolves around a crisis over *intimacy* (Erikson, 1968). Here, the emotional bond to parents needs to be replaced by a mature couple-bond with a spouse or lover. Eventually, a new, more empathic, and more egalitarian mode of relating to parents should replace the dependencies, hostilities, and conflicts that often arise during the process of emancipation. Some empirical evidence supports this idea. Barber (1989) found that a majority of parents whose adult offspring had left home believed the relationship had improved because the children now appreciated their parents more and were less argumentative, judgmental, and critical. Bengston and Black (1973) likewise found that young adults' perceived solidarity with parents increased after the young adults married.

Two other studies, however, found that this progression in the parent-adult child relationship was not so easily accomplished. Cox (1970) followed a group of American men and women for a decade after their university graduation. She found that even by age 30, a majority still seemed to be locked into mutual misperceptions, conflicts, and independence struggles. "A state of uneasy truce is characteristic of the relationship with parents of more than half of these young people, while a sixth of them are angry and condemning" (p. 221). Similarly, Gould (1972, p. 525) concluded that between the ages of 22 and 28, most offspring "see their parents as people...to whom they still have to prove their competence as adults."

Gender and life experience appear to account for some of the variation in adult children's ability to establish empathic, egalitarian friendships with their parents. White, Speisman, and Costos (1983) found clear evidence of maturation in modes of relating to their mothers for married women by age 24 and single women by age 26. Sons, however, displayed no clear growth over the same age period, and no clear ability to perceive their parents as separate individuals. A more recent study (Frank, Avery, & Laman, 1988) similarly found that most adults aged 22 to 32 were still locked into various types of immature relationships with both parents (dependent, conflicted, or pseudoautonomous). Nevertheless, some 40 percent of the women in this age group had managed to develop a more "connected" style of relating with their mothers where empathy, involvement, and mutual understanding were tempered by a clear independence of views. Only half as many sons (20 percent) had developed simi-

larly mature relationships with their fathers. Overall, the evidence suggests that more mature modes of relating do develop as adult children grow a little older (see also Frank, Avery, & Laman, 1988; White, Speisman, & Costos, 1983), but there is clearly a need for further studies, especially of offspring over 30.

Erikson's (1968) theory views the middle-aged adult's developmental task as the attainment of *generativity*—a mature capacity to be productive and to care for and guide younger generations in an altruistic way. Much of the difficulty young adults have in attaining mature relationships with their parents is attributed to the latter's failure to achieve generativity. Empirical research supports the view that the older parent's successful establishment of a warm, interconnected mode of relating to a son or daughter can assist the growth of generativity. A longitudinal study spanning four decades and involving 343 married men showed that the highest levels of generativity were attained by adoptive fathers who had compensated for the initial frustrations over infertility by investing deeply in warm, caring patterns of fathering (Snarey et al., 1987). Next in the ranks of success came biological fathers, followed by childless men whose response to infertility had often been a self-absorbed investment in such interests as automobiles or bodybuilding, which had less potential to stimulate mature personality growth.

Summary

Although further studies are clearly needed, the existing research literature on parent-offspring interaction during adulthood already offers a rich blend of sociological, social-psychological, developmental, and demographic-historical ideas. These studies suggest that typical fathers or mothers who embark on parenthood in the 1990s can count on staying in close and frequent contact with at least one of their children for up to six decades of a dynamically changing partnership. The data reviewed here indicate that, like their counterparts today, most parents and adult children in the next century will probably describe their relationships as reasonably satisfying. Nevertheless, wide variations almost certainly will emerge among adult families of the future, just as they do among contemporary parent-adult child dyads. ▼ CANDIDA C. PETERSON

See also: Attachment Across the Lifespan; Divorce; Fictive Kin; Gender as a Shaping Force in Adult Development and Aging; Generational Equity; Grandparent Communication Skills; Japanese Perspectives on Adult Development; Marital Development; Parental Imperative; Social Relationships, Convoys of.

References

Ainsworth, M.D.S. (1989). Attachments beyond infancy. *American Psychologist, 44:* 709-16.

Anderson, S.A., et al. (1983). Perceived marital equality and family life-cycle categories: A further analysis. *Journal of Marriage and the Family, 46:* 127-39.

Antonucci, T. (1990). Social supports and relationships. In R.H. Binstock & L.K. George (Eds.), *Handbook of Aging and the Social Sciences.* New York: Academic Press.

Arbyle, M. (1986). Social behavior problems in adolescence. In R.K. Silbereisen (Ed.), *Development as Action in Context.* Berlin: Springer-Verlag.

Barber, C.E. (1989). Transition to the empty nest. In S. Bahr & E.T. Peterson (Eds.), *Aging and the Family.* Lexington, MA: Lexington.

Bengston, V., & Black, K.D. (1973). Solidarity between parents and children. Paper presented at the National Council on Family Relations Theory Development Workshop (October).

Bengston, V., Rosenthal, C., & Burton, L. (1990). Intergenerational relations. In R.H. Binstock & L.K. George (Eds.), *Handbook of Aging and the Social Sciences.* New York: Academic Press.

Clemens, A.W., & Axelson, L.J. (1985). The not-so-empty nest: The return of the fledgling adult. *Family Relations, Journal of Applied Family and Child Studies, 34:* 259-64.

Cox, R.O. (1970). *Youth into Maturity: A Study of Men and Women in the First Ten Years After College.* New York: Mental Health Materials Center.

Erikson, E.H. (1968). *Identity, Youth, and Crisis.* New York: Norton.

Fiske-Lowenthal, M., Thurner, M., & Chiriboga, D. (1977). *Four Stages of Life.* San Francisco: Jossey-Bass.

Frank, S.J., Avery, C.B., & Laman, M.S. (1988). Young adults' perceptions of their relationships with their parents. *Developmental Psychology, 24:* 729-37.

Gould, R.L. (1972). The phases of adult life: A study in developmental psychology. *American Journal of Psychiatry, 129:* 521-31.

Hagestad, G. (1981). Problems and promises in the social psychology of intergenerational relations. In R. Fogel, E. Hatfield, S. Kiesler, & E. Shanas. (Eds.), *Aging: Stability and Change in the Family.* New York: Academic Press, pp. 11-46.

Hatfield, E., et al. (1979). Equity theory and intimate relationships. In R.L. Burgess & T.L. Huston (Eds.), *Social Exchange in Developing Relationships.* New York: Academic Press.

Kelley, H.H. (1981). Marriage relationships and aging. In R. Fogel, E. Hatfield, S. Kiesler, & E. Shanas (Eds.), *Aging: Stability and Change in the Family.* New York: Academic Press.

Kendig, H. (1986). *Ageing and Family.* Sydney: Allen & Unwin.

Miller, R., & Glass, J. (1989). Parent-child attitude similarity across the life course. *Journal of Marriage and the Family, 51:* 991-97.

Peterson, C.C., & Peterson, J.L. (1988). Older men's and women's relationships with adult kin: How equitable are they? *International Journal of Aging and Human Development, 27:* 221-31.

Peterson, E.T. (1989). Elderly parents and their offspring. In S. Bahr & E.T. Peterson (Eds.), *Aging and the Family.* Lexington, MA: Lexington.

Riley, M.W., & Foner, A. (1968). *Aging and Society.* (Volume 2). New York: Russell Sage Publications.

Shanas, E. (1980). Older people and their families: The new pioneers. *Journal of Marriage and the Family, 42:* 9-15.

Suitor, J., & Pillemer, K. (1987). The presence of adult children: A source of stress for elderly couples' marriages? *Journal of Marriage and the Family, 49:* 717-25.

White, K.M., Speisman, J.C., & Costos, D. (1983). Young Adults and Their Parents: Individuation to Maturity. In H.D. Grotevant & C.R. Cooper (Eds), *Adolescent Development in the Family.* San Francisco: Jossey-Bass, pp. 43-59.

ADULT DEVELOPMENT AND AGING, MODELS OF

Many people believe that description is based on neutral, objective observation and is prior to theory building. However, even direct observation is guided...by theoretical notions that distort the flow of behavior in some way. The observer records certain behaviors and ignores others. She divides the stream of behavior into units. She encodes the behavior into words that add connotations. She allows inference to creep into her observations. (Miller, 1983, p. 6)

Most scholars in the field of human development would agree with this statement. The choice is not between theory and no-theory. It is a choice between a clearly articulated theory that one can test, revise, improve, or reject, and fuzzy, half-formed ideas that influence observations without themselves being subject to observation and critical review. Theories of adult development serve much the same purposes as theories in other domains: guiding observations, suggesting connections among various phenomena, predicting outcomes, and providing a general overview or image of the subject matter. The availability of several competing theories or models tends to have a stimulating effect on the design and interpretation of scientific research. Questions are sharpened and improved methodologies are devised as advocates attempt to demonstrate the superiority of their particular models.

It would be naive, however, to assume that theoretical models are limited to the scientist's workshop and operate in isolation from all the other currents of social force. Bruno offered an alternative view of the universe and was first imprisoned by the church and then burned at the stake. Darwin was vilified for generating an evolutionary model of adaptation and survival from his naturalistic investigations. Freud was subjected to abuse for proposing a model of early development that included sexuality. Some theoretical models seem to support the beliefs and wishes of society, while others are eyed as dangerous and disagreeable.

To the task of determining the scientific status of a theoretical model, then, must be added the challenge of disentangling the model from society's prevailing hopes, wishes, beliefs, and fears. A theoretical model that is comforting to vested interests can prove very resistant to extinction by "mere facts," as illustrated by the perseverance of racist and sexist ideologies in the face of disconfirming evidence. Given that political, religious, and other powerful agendas interact even in the somewhat distant realms of the physical and natural sciences, it is only to be expected that intense social pressures may be exerted on competing models of human development.

It will be useful, then, to keep the social context in mind as attention is given to the overall quest for adequate theoretical models of adult development.

General Characteristics and Themes of Developmental Models

The course of life has often been likened to a journey through time. Most theoretical models focus on the first steps of the journey. Indeed, "development" traditionally has been regarded as virtually identical to "child development." Other models extend themselves to the adolescent years. Charlotte Buhler (1953, 1962; Buhler & Massarik, 1968) was among the few to encompass the entire life course within a single model. It is only in recent years, however, that a large number of researchers have become interested in the adult sector of the life course. In the search for a viable model to guide research throughout the entire life course, some investigators have attempted to extend the range of theories that concern themselves chiefly with early development. There have also been some stimulating but not entirely successful attempts to formulate new theories that deal specifically with adult experience. Disengagement theory (Cumming & Henry, 1961) was the first substantive and innovative theory to consider the middle and later adult years; "mid-life crisis" emerged as an influential alternative a few years later (Gould, 1978; Levinson, 1978; Sheehy, 1976).

The status of theoretical activity today might be likened to the challenge of reconstructing a bridge after one has belatedly discovered that it must cover a much larger span than originally envisioned. One might proceed by expanding on the short-span plans that were generated when the bridge was first conceived, i.e., by infant-child developmentalists. Instead, one might stand at the other shore and insist that it is wiser to construct the span with the destination side (old age) in mind. Still again, it might be argued that primary attention should be given to the large middle section (adulthood) in which travelers will spend most of their time. In retrospect, of course, it might be viewed as unfortunate that so little attention had been given to conceptualizing the entire span before the enterprise had come this far.

As matters now stand, researchers and theoreticians are working toward an overall conception of the life course from various starting points. The eventual winner might be some version of the "push theory": Adult behavior and experience is propelled ahead by forces and influences that were at work in early development. Alternatively, a "pull theory" might prevail: The potentials and possibilities of adult experience give direction and momentum to early development. Perhaps, however, a completely new type of theoretical model will emerge, one in which the "push" of the past and the "pull" of the future are integrated within a larger perspective of what it means to become fully human throughout the total life course.

There are several other ongoing conflicts in the attempt to establish an acceptable theoretical model for the life course in general, or adult development in particular. Among the most consequential areas of conflicts are the following:

1. *Is it actually "development" that is taking place through time?* The individual changes in many ways from infancy through old age. Change, however, is not necessarily identical to development. For example,

there is a consensus that the biological changes taking place at puberty can be regarded as expressions of a developmental process. By contrast, there is much less agreement about the changes that take place in cognitive functioning from childhood to young adulthood, or in attitudes and interests between early and later adulthood. Critics may contend that such changes can be explained more parsimoniously by the individual's increasing interaction and experience with the world, rather than by a rather vague concept of "development." The fact that the developmental approach can be called into question adds to the difficulties involved in conceptualizing the total life course. It is becoming fashionable to speak of "adult development" as though such a process has already been clearly identified and confirmed.

Nevertheless, there is still room for dissenting opinions and alternative models.

2. *Should the focus of developmental studies be on the individual or the family/society?* The most familiar and seminal models tend to emphasize the individual. For example, it is the cognitive and emotional development of the individual child that has been so extensively conceptualized and studied by Piaget (1926, 1954) and his colleagues—and Skinner's (1953) resolutely nondevelopmental approach also focuses on the individual. Piaget, Skinner, and other individual-oriented theorists are attentive to social and environmental influences, but the focus is usually placed on the individual's acquisition of thoughts and behaviors.

The alternative view has yet to impress itself on mainstream developmental theory and research, but may do so in the future. This is the view that it is artificial and inappropriate to consider individuals as though independent units that can be detached from their socio-environmental contexts. How much would have been learned about "beeness," for example, if scientists had studied only the lives and loves of individual bees, instead of looking at the total hive organization and culture? Would it not be more reasonable and productive, then, to study the interconnectedness of development in humans as well? There have already been numerous contributions that reveal how one individual's path through life has been created and shared with others. For example, the life stories told by elderly people (Mullen, 1992) would lose much of their proportion and meaning unless understood within their distinctive interpersonal contexts. The developmental theory implicit in the teachings of Confucius strongly emphasizes the cultivation of virtues within a social context (Ya, 1988). To study individual development without consideration of the cultural values placed on courtesy, moderation, and obedience would be to miss a crucial part of the picture—actually, the frame in which the portrait of the individual is to be displayed. A pioneering attempt to emphasize social rather than individual factors in adult development (Rose, 1965) attracted some attention, but did not lead to a substantial body of confirmatory research. Nevertheless, there is still advocacy for the family / social side of adult development (e. g., Connidis, 1988), and it is certainly possible that stimulating new theoretical models representing this viewpoint may be forthcoming before long.

3. *What methods should be used to test theoretical models?* As already noted, "neutral, objective observation" rarely if ever occurs. Research is conducted within either an implicit or explicit model that emphasizes some observations and tends to exclude others. Accordingly, there is an interaction between a particular theoretical model of change / development and the methods selected to test that model. For example, several influential personality theories have proposed that each individual continues to maintain certain "traits" or enduring characteristics throughout the life course (e. g.,

Allport, 1955; Murray, 1981). This approach encourages the use of self-report questionnaires in which individual traits are assessed through simple true / false responses. By contrast, Piaget's approach seems better served by designing experimental tasks that challenge the individual's ability to comprehend cognitive principles, and Erikson's (1959, 1969, 1979) version of psychoanalytic theory may be better served by individual case studies that are sensitive to the cultural context.

It is difficult to compare models of the life course when each model seems to have its own configuration of methodologies, and yet it may not be appropriate or productive to expect all models to agree upon one standard methodological approach. Basically, the conflict is between quantitative and qualitative studies on the one hand, and individual versus contextualized studies on the other. Should the journey from youth to age be regarded primarily as a challenge to precise measurement and sophisticated mathematical models (e.g., Buss, 1979) or as a complex story that needs to be read by a knowing and alert reader? (e.g., Wyatt-Brown, 1992). This question may never be resolved to everybody's satisfaction.

4. *Do models reveal what is or what should be?* There is an inherent ambiguity in theoretical models of development—or, at least, in the ways these models are interpreted by society. The word "model" itself embodies this ambiguity. A well-crafted toy automobile may be regarded as a model in the sense that it is a reasonably faithful representation of its full-sized and operational counterpart. This usage of the term differs appreciably from the Gilbert and Sullivan character who proclaims himself to be "the very model of a model major general." In such instances, "model" refers to an ideal or desired state of affairs (e. g., a "role model"). A "model home" that is available for inspection is likely to combine

both meanings: It serves as an example of the type of house available for purchase, but also as a lavishly decorated and furnished version.

There is potential for confusion, misunderstanding, and mixed signals when a scientific model is taken to be a model for "right" or moral behavior. Example: Disengagement theory proposed that with aging, interactions tend to decrease and become more distant between individuals and society. Does this mean that people *should* withdraw from social participation as they grow older? Example: An information-processing model is likely to include the core proposition that there is a slowing in complex cognitive operations with advancing age. Does this mean that people *should* slow down? It is not always this easy to identify where one usage of "model" ends and the other usage begins. Furthermore, political and economic pressures may accentuate the difficulty. For example, a government agency may be under the gun to reduce costs. One way to accomplish this reduction would be to restrict services and entitlements for elderly men and women. Disengagement theory could be interpreted as providing a rationale for reducing such services: "If people *do* disengage, then they *should* disengage; therefore, our agency should not interfere with this desirable state of affairs." It is a matter of self-protection that consumers of information about human development hone their sensitivity to possible confusions between "is" and "should."

Selected Models of Adult Development

Attention will be given now to several models of adult development. These models vary in their scope, data base, and influence. No two models give equal attention to the same problems. Taken together, however, this collection of theoretical models offers a variety of perspectives on many of the issues involved in the adult life course.

Psychoanalytic Models. Psychoanalytic theory originated with Freud (1900, 1933). Many concepts and observations associated with psychoanalytic theory soon "escaped" into society at large and exercised so much influence in various quarters that many who do not consider themselves sympathetic to psychoanalysis—and who have not even read the original sources—nevertheless make daily use of some of these views and assumptions. For example, it has become relatively commonplace to be aware that "Freudian slips" of the tongue and nonverbal behaviors (e. g., forgetting a key) may represent conflicted states of mind in which a strong but unacceptable impulse breaks through in a disguised form. Even should one reject psychoanalytic theory as the most dependable guide to adult development, it would be virtually impossible to eliminate or reverse the influence this model continues to exert.

There is a strong developmental component in psychoanalytic theory. Central to this model is the contention that adult personality is strongly influenced by one's experiences during infancy and early childhood. Every person is thought to proceed through a series of developmental stages. At each stage, the infant or child is dominated by a particular biosocial need, and parents respond to this need in various ways. In the ideal case, each developmental need is fulfilled in its turn, and the child grows into a confident and psychologically balanced adult who can face reality, derive pleasure from life, and truly love other people. This ideal scenario is quite rare, however. According to Freud's observations, it is far more common for the infant and child's impulses to be thwarted and distorted within families beset with their own conflicts. These families themselves suffer from fears, conflicts, and destructive attitudes that they have absorbed from society at large. In consequence, the typical pattern of early child-parent interaction is a kind of schooling for neurosis.

Freud's own observations and theoretical formulations emphasized ways in which processes and events in early development might lead to unhappiness and self-defeating behavior in adulthood. His writings on the middle and later adult years are less extensive, although he gave increasing attention to these realms as he himself moved into his seventh decade and beyond. Freud's writings can be used as the source for a theoretical model of adult development. For example, much of what Freud observed about early development and adult neurosis would support the hypothesis that people who appear rigid, distant, and incapable of loving relationships in their later adult years are not the victims of a mysterious process known as "aging," but, rather, victims of their own early-formed neurotic personalities that have not developed further over time.

Significantly more attention to adult development has been given by two other major theoreticians within the psychoanalytic tradition. Carl C. Jung, a Swiss psychiatrist, was one Freud's closest friends and disciples until he created his own version of psychoanalysis. Erik H. Erikson, described by his friends as "a wandering artist" (Coles, 1970, p. 118), became an influential theorist when he generalized from his observations of child development among Native-American peoples (Erikson, 1950). All three—Freud, Jung, and Erikson—often found insights and inspiration in literature, myth, and art, but each also had his distinctive approach to human development.

Jung's Model of Adult Development. Jung agrees that the early years of life are important and deserve careful attention. However, he differs from Freud in viewing development as a lifelong process. During the first half of life there is an expansion of the personality. The individual continues to gain in mastery over the external world. This growth includes the ability to form viable relationships (such as marital partnerships), win the approval of society, and become increasingly competent in work and other activities. This overall pattern is both similar and different according to gender. It is similar for each gender in that both men and women tend to emphasize their most defining characteristics, i.e., the men accentuate their masculinity, the women their feminin-

ity. This "similarity in emphasizing differences" leads to a one-sidedness in personality development. Up to mid-life (about age 35-40), men and women are locked into the expectations for their gender roles.

Development takes a different turn from mid-life onward. People become aware that they are not as enthusiastic about the course of their everyday lives as they once had been: Family, job, and other activities remain important, but there is a vague and growing sense that something is missing. Jung's (1933) description of this period of psychological uneasiness would later become the source for contemporary ideas of a "mid-life crisis" (e.g., Sheehy, 1976).

A new direction of development will emerge if the individual is able to respond to his or her inner promptings. These messages arise from facets of one's personality that had previously been denied free expression. Dreams provide one of the most insistent and informative modalities of self-speaking-to-self. Jung places great reliance on his own version of dream language to describe and explain how the mid-life adult becomes aware that he or she has potentials that have yet to be actualized. The person who continues to develop through the second half of the life course is the person who dares to face the new challenge of exploring his or her previously ignored potentials. This often means that women will start to cultivate the masculine side of their natures, and men the feminine. In practical terms, this shifting emphasis might be seen in women becoming more independent and aggressive in pursuing their own interests, and men becoming less aggressive and competitive, but more reflective and nurturant.

For both men and women, this process of continued individuation can result in people who are wiser and more complete than they were during the energetic strivings of their youth. Such a person can also face the prospect of death with greater equanimity because there is the sense of having become a complete person and having lived a full life.

Erikson's Model of Adult Development. Erikson's "Eight Stages of Life" have become widely familiar to students of education, psychology, sociology, and related fields. This stage theory builds upon many of Freud's observations, but offers a view that encompasses the entire course of life, rather than dwelling mostly on childhood. It also places more emphasis on sociocultural influences and somewhat less on biological factors in development. One reason for the popularity of Erikson's approach may be its attention to the infant and child as an active self that is trying to understand and cope with the world, as distinguished from a packet of biological impulses and instincts from which a rational self may eventually emerge.

Erikson's approach is often characterized as an *ego psychology*, meaning that the emphasis is upon an active rational self that is attempting to cope with inner and outer demands. The stages are known as *epigenetic*, meaning that personality continues to emerge (-genetic) from one phase to the next.

Each sector of the total life course is centered on a particular stage. All eight of these stages are already prefigured in the infant. One might think of them as seeds that will mature or as scripts that will be performed as their own time comes. Early-appearing stages do not disappear when a later stage becomes central. Instead, all previous stages contribute something to personality development that is encompassed and utilized by succeeding stages. The epigenetic stages constitute a series of challenges through which the individual may continue to become more emotionally mature, rational, and moral. There are no guarantees, however, and any success may be overtaken by subsequent failures.

A "psychosocial crisis" sets the problem for each stage. By "crisis" is meant a confrontation with emerging needs and demands that might turn out in either a positive or a negative way. Upon reaching a new point in one's developmental progress, one experiences new demands from self and others. If the immediate psychosocial crisis is mastered, one is then able to move more confidently toward the next challenge. A person who is adult in terms of age may still be struggling with earlier epigenetic crises rather than with the issues that are more appropriate for his or her age.

Each of the "Eight Stages of Life" is seen from two perspectives by Erikson: as a physiological status (similar to Freud's stages), and as a psychological conflict. Both the physiological status and the psychological conflict occur within a psychosocial context that either fosters or impedes successful resolution of the problem set by the particular stage.

Stage 1. Oral-Sensory: Basic Trust vs. Mistrust. The infant takes in the world through its mouth and its senses. It will have some experiences that lead to trusting others (e.g., "Somebody comes and makes me feel better when things go wrong") and some experiences that lead to mistrust ("You just can't count on getting everything you want just when you want it"). As in all subsequent stages, it is normal and useful to experience both sides of the situation. When early developmental interactions have gone well, the infant is more likely to move ahead with a basic sense of trust in the world and other people—but with also a degree of realistic doubt and caution.

Stage 2. Muscular-Anal: Autonomy vs. Shame & Doubt. The young child can do more than incorporate the world. He or she also is developing self-regulatory capacity and, accordingly, some ability to assert one's own will. At the same time, parents are expecting or demanding that the child become toilet-trained. When all goes (mostly) well, the child will come away with a sense of hope in his or her ability to meet demands while also asserting his or her own individuality.

Stage 3. Locomotor-Genital: Initiative vs. Guilt. The child is up and running now, actively exploring the possibilities of his or her world. Both boys and girls are carried by their curiosity and emerging impulses into new domains of experience. A particular issue here is one that Freud was first to identify: vague but exciting sexual urges that become provisionally attached to the parent (boys intend to marry Mother; girls intend to wed Father). Successful resolution of this stage involves reorienting the sexual urge into play, physical activity, and other nonthreatening realms. By dealing well with the dominant crisis of this stage, the child develops a stronger sense of his or her own ability to initiate and take responsibility for actions; failure leads to the burden of guilt, even though the child is not clear about "the charge" on which he or she has been found guilty.

Stage 4. Latency: Industry vs. Inferiority. Prior to adolescence there is a period during which children are exceptionally open to learning and social influence. It is an excellent time for children to learn cultural values such as industriousness, loyalty, and responsibility. However, the many new opportunities also bring many occasions for experiencing failure. One can learn from failure as well as success, so a sense of personal competence represents a positive resolution of the latency stage crisis.

Stage 5. Adolescence: Identity vs. Role Confusion. The biological changes are profound here, as puberty introduces the new sexualized self. One does not switch immediately from thinking of oneself as a child to thinking of oneself as an adult. Adolescence is a transitional period whose successful resolution involves developing a new sense of identity. Peers—and their approval—become very important in helping to make this transition. Successful resolution of this crisis requires the ability to make choices and commitments in a world that is now recognized as more complex and conflictive. Failure can result in the individual continuing to act like an adolescent well into the adult years, and avoiding his or her share of responsibility.

Stage 6. Young Adulthood: Intimacy vs. Isolation. With his or her new adult identity, the individual must now cope with the challenge of letting other people become close. There will be a temptation to hold oneself apart, cherishing and protecting the recently earned individual identity. Choosing this alternative leads to continued self-absorption and, eventually, a profound sense of loneliness and isolation. The epigenetic crisis of young adulthood, then, is to yield some of one's own identity and independence in or-

der to develop an intimate, sharing relationship with other people.

Stage 7. Adulthood: Generativity vs. Stagnation. Erikson's concept of this stage is one of his major contributions to current thinking about adult development. People do not just settle into adulthood: There is a major challenge inherent at this time of life. It is the adult's responsibility to guide and protect the future of humankind. This can be accomplished by being an effective parent for one's own children, but also in many other ways (e.g., as teacher or advocate for a safer world). People who do not rise to this challenge tend to indulge themselves as though they were their own child. This retreat into self-absorption results in a stagnation of development. Erikson (1959) notes that Western technological cultures seem to be less conducive to the development of community-minded generativity in the adult years than do Native-American and other tradition-oriented cultures. The intense emphasis on individual achievement in the United States, then, is seen as an additional challenge to adult development.

Stage 8. Maturity: Ego Integrity vs. Despair. During the later adult years one is increasingly faced the reality of limits, loss, and mortality. Paradoxically, perhaps, those who have mastered the various crises associated with growth and expansion of personality are in the best position to cope resourcefully with the prospect of loss and endings. Awareness of aging and death leads to a sense of despair when people feel that they have not lived a satisfactory or meaningful life. Those who can accept the kind of lives they have lived and the kind of people they have become are better able to face the prospect of death with equanimity. Accordingly, the aging adult is likely to reflect on the total shape of his or her life in attempting to master the final crisis and thereby achieve the final reward: wisdom.

Piaget's Genetic Epistemology. All of the psychoanalytic theories described above see human development as taking place through qualitatively different stages, each with their own challenges. The leading cognitive model of development also emphasizes a stage approach, but draws upon a much different methodology and data base and offers a much different set of stages. Piaget and his colleagues devised and conducted a great many studies of cognitive functioning in children and adults. He did not entirely neglect emotions, relationships, and contextual factors, but this model of adult development is constructed largely from Piaget's ideas about mental growth and the studies intended to demonstrate and advance this view.

Piaget's approach has often been characterized as *genetic epistemology*. This term refers to the individual's development (genesis) of ways of knowing (epistemology). The major underlying differences between the young child, the older child, and the adult are to be discovered in the different ways that each goes about "constructing reality" (Piaget, 1954). Each may also be said to live in a different reality. The "same" environment, the "same" messages, the "same" situation will be perceived and conceived quite differently from people who function at various cognitive stages or levels.

One of Piaget's guiding concepts is sometimes found difficult to grasp. It is commonplace to speak of the structure of a crystal, a building, or any other physical entity. Piaget asserts that there are mental structures as well. These structures are as "real" as structures examined in the physical realm, although for the experimental epistemologist it is mental performance that provides data for determining structure. The structure of cognition develops in an orderly manner from childhood through adolescence. It is possible that more subtle changes continue to occur through the adult years.

Piaget identified four general phases or periods of development. The first period (birth to two years) is exceptionally active and rapidly changing. There are six stages recognized within this period alone, and some of these stages have their own substages. Attention will be given here only to the major qualitative differences that Piaget believes characterizes the four basic periods of cognitive development.

Period I. Sensory-Motor Intelligence. Newborns are already attempting to organize the stimulation that reaches them through sight, sound, touch, temperature, smell, taste, and the internal senses. In a rudimentary but systematic way, they are *assimilating* experience, *accommodating* themselves to the environment, and *organizing* (or constructing) their own *schema*. Schemata can be regarded as the individual's own model of the world or, more specifically, his or her own "plan" for dealing with this world. A hungry baby who begins to relax at the sight of his mother approaching—before she starts to feed him—is demonstrating that he has developed an action structure or scheme that gives meaning to this event. By the end of this very productive period of development (at about age 2), the young child recognizes that he or she is a separate individual in a world that operates according to some helpfully reliable principles (e. g., the stuffed bear still exists—maintains *object permanence*—even when it is hidden under a pillow).

Period II. Preoperational Thought (ages 2 to 7). During this rather lengthy period, children become increasingly adept in using language and other symbols. This provides access to a larger world of meanings and a more flexible mental tool for dealing with these meanings. One of the lines of development at this time may be especially consequential for adult behavior: Children gradually develop the ability to see the world as others do; that is, they become liberated to some extent from *egocentrism* and come to realize that everybody else has a point of view, not just themselves. Another related line of development involves moral judgment. With continued mental development, children begin to comprehend rules and the value of playing by the rules. Later social development will build upon these new abilities to recognize the other person's perspective and behave in a rule-governed manner. The child at this time, however, is still situation-bound and unable to organize his or her experiences in a logically systematic way.

Period III. Concrete Operations (7 to 11 years). Children are now able to carry out a number of mental operations that are useful in everyday life. Their understanding of the world also approaches more closely the adult or scientific way of thinking. These operations remain "concrete," however, in the sense that they are applied only to objects that exist in the world or whose existence is represented in the mind. The child does not yet have the ability to think in high-level abstractions. Nevertheless, there are also important advances in this period that are related to subsequent adult development: For example, children now realize that others have intentions and purposes that must be taken into account in understanding their words and actions.

Period IV. Formal Operations (ages 11 to 15). Cognitive development finally reaches the upper levels of thought at this stage. Before adolescence is completed, most people have demonstrated the ability to think about possibilities and alternatives: not just how things are but how things might be. This new freedom includes the ability to think about thought: to criticize ideas, raise fresh questions, and put together one's own thoughts in unusual and creative ways. There is also a firmer sense of logical relationships. In short, by activating his or her ability to engage in formal operations, the adolescent is able to bring high-level processes to bear both on reality and possibility. Because thinking is now much more systematic as well, it can be said that the cognitive structure has been completed.

Attempts are now being made to extend Piaget's theory through the total life course. Piaget himself was open to the possibility that cognitive development might continue in a refined and subtle manner after adolescence. There is also the possibility, however, that not all people remain at the intellectual heights that they have reached in adolescence. Along with these and other specific questions about cognitive development throughout the adult years, there are questions regarding the appeal of this model to life-span developmentalists. Relatively little effort has been made to apply the Piagetian model to the full spectrum of adult experience (including interpersonal relationships,

self-concept, coping ability, political behavior, etc.). It may be that this approach is felt to be rather demanding in its complexity without providing easy "handles" for application to emotional and social aspects of development.

Self-Actualization Models. The self-actualization alternative usually is rooted in existing theories of personality, but branches off in a different direction. Perhaps the two most influential self-actualization approaches to adult development are those associated with Carl Rogers (1951, 1961) and Abraham Maslow (1968, 1970). Each of these psychologists drew extensively from psychoanalytic theory, and neither offered their models in much detail. Nevertheless, many educators and others concerned with personality development and mental health have found stimulation and encouragement in these theories. Rogers' theory found direct expression in his highly influential "client-centered" technique of counseling and psychotherapy. Maslow's theory has had perhaps its greatest influence through the concept of a "peak experience" as well as the indirect effect of suggesting a more optimistic approach to human potentiality in education and a variety of other realms.

Both theoretical models of self-actualization hold that all humans have the capacity to become positive, loving, and creative individuals. The quest for becoming a deeper, richer, and more complete person is often blocked, frustrated, and distorted by painful and disappointing experiences with the world. Nevertheless, the flame of self-actualization exists even within people who have fallen into stagnant and neurotic patterns of life. Rogers and Maslow do not deny that people face many conflicts from infancy onward, but they suggest that the positive side of human nature has been somewhat neglected by traditional psychoanalytic approaches. A loving and productive adult is not necessarily a person who has somehow overcome or compensated for his or her early developmental crises; rather, this person has actualized loving and productive potentials that existed from the beginning.

Roger's model gives priority to the role of the self, rather than to competing pressures and impulses (psychoanalysis), or a purely intellectual understanding of the world (genetic epistemology). Each individual—child or adult—has a unique way of experiencing life. Those who would interact in a moral and effective manner will respect the other person's distinctive self. Helpful people, recognizing that the other person has a right to his or her feelings, will refrain from making value judgements about them. Instead of rushing to offer advice, issue commands, or engage in other interventions, helpful people are skilled in the art of active listening. Frustrated and troubled people are able to discover their own best developmental path through interactions with a caring and nonauthoritarian companion. Rogers' concept of "unconditional positive regard" refers to this gift that one person can give another: accepting that person for who he or she is, rather than requiring that person to meet expectations and demands that are not favorable to his or her own self-development. The child whose energies were devoted to trying to please parents in order to win their love (conditional positive regard) may become a dissatisfied and anxious adult because his or her own individual potentialities have not had a chance to flourish. Unconditional positive regard, emphasizing respect for the individual's unique self, can help the adult to renew his or her personal development.

Maslow's model introduces the concept of two types of motives. *Deficiency motives* are those that are aimed at the reduction of basic needs such as thirst, hunger, safety, and approval from others. All people have such needs all through their lives. It is normal for much of human behavior to be directed toward the fulfillment of these needs. But people also have another set of needs that Maslow characterizes as *growth motives*. These motives differ in that they represent activities or states of mind that people want to experience, rather than to decrease. For example, hungry people do not want to feel hungrier; they desire relief from their pangs. By contrast, dedicated violinists do not seek reduc-

tion or elimination of the need to play this instrument: They want to continue to make music, but to do so at a higher and higher level. All people have growth motives as well, but these take many individual forms.

It is difficult to attend to one's growth motives if basic needs have not been met. Much human distress can be attributed to the failure to actualize one's higher or growth-oriented motives. In turn, this situation has often resulted from frustration or compromises involved in attempting to satisfy deficiency motives. Although their terminology differs, Maslow and Rogers offer almost identical views here: It is "natural" for people to continue to develop throughout the entire life course, but this process can be brought to a standstill at any point and become a source of distress.

Maslow goes somewhat further than Rogers into the interior of the self with his concept of *peak experience:* This is an almost indescribable private experience about which such terms as "intense," "mystical," and "awesome" tend to be used. Although peak experiences are fleeting and uncommon, they may have a powerful effect on the individual, leaving the impression that one has been granted a glimpse of a higher self in a more clearly perceived reality.

Some aspects of Rogers' theory have led to extensive research, i. e., on the process of counseling and psychotherapy. The broader and more philosophical facets of both theories have attracted relatively little research. Researchers and theoreticians of the "hard-nosed" school sometimes dismiss these theoretical models as lacking in sufficient detail and precision to guide definitive studies. Others are attracted to the optimistic tone of self-actualization theories and find them useful in their lives and work.

Other Approaches

Each of the theoretical models reviewed here takes a strong developmental approach; many also feature the idea of "stages" through which individuals must pass in order to complete the developmental progression. As